Advanced Quantitative Microbiology for Foods and Biosystems

Models for Predicting Growth and Inactivation

CRC Series in
CONTEMPORARY FOOD SCIENCE

Fergus M. Clydesdale, Series Editor
University of Massachusetts, Amherst

Published Titles:

Advanced Quantitative Microbiology for Foods and Biosystems:
Models for Predicting Growth and Inactivation
Micha Peleg

Antioxidant Status, Diet, Nutrition, and Health
Andreas M. Papas

Aseptic Processing and Packaging of Foods: Food Industry Perspectives
Jarius David, V. R. Carlson, and Ralph Graves

Automation for Food Engineering: Food Quality Quantization and Process Control
Yanbo Huang, A. Dale Whittaker, and Ronald E. Lacey

Bread Staling
Pavinee Chinachoti and Yael Vodovotz

The Food Chemistry Laboratory: A Manual for Experimental Foods,
Dietetics, and Food Scientists, Second Edition
Connie Weaver and James Reuben Daniel

Food Consumers and the Food Industry
Gordon W. Fuller

Food Emulsions: Principles, Practice, and Techniques
David Julian McClements

Food Microboilogy Laboratory
Lynn McLandsborough

Food Properties Handbook
Shafiur Rahman

Food Shelf Life Stability
N.A. Michael Eskin and David S. Robinson

Getting the Most Out of Your Consultant: A Guide to Selection
Through Implementation
Gordon W. Fuller

Handbook of Food Spoilage Yeasts
Tibor Deak and Larry R. Beauchat

Interdisciplinary Food Safety Research
Neal M. Hooker and Elsa A. Murano

Introduction to Food Biotechnology
Perry Johnson-Green

Modeling Microbial Responses in Food
Robin C. McKellar and Xuewen Lu

Advanced Quantitative Microbiology for Foods and Biosystems

Models for Predicting Growth and Inactivation

Micha Peleg

CRC Press
Taylor & Francis Group
Boca Raton London New York

CRC Press is an imprint of the
Taylor & Francis Group, an **informa** business
A TAYLOR & FRANCIS BOOK

CRC Press
Taylor & Francis Group
6000 Broken Sound Parkway NW, Suite 300
Boca Raton, FL 33487-2742

First issued in paperback 2019

ISBN-13: 978-0-8493-3645-4 (hbk)
ISBN-13: 978-0-367-39095-2 (pbk)

Library of Congress Card Number 2005046679

Library of Congress Cataloging-in-Publication Data

Peleg, Micha.
    Advanced quantitative microbiology for foods and biosystems : models for predicting growth and inactivation / by Micha Peleg.
        p. cm. -- (CRC series in contemporary food science)
    Includes bibliographical references and index.
    ISBN 0-8493-3645-7 (alk. paper)
    1. Food--Microbiology--Mathematical models. 2. Biological systems--Mathematical models. 3. Microbial growth--Mathematical models. I. Title. II. Series.

QR115.P45 2006
664.001'579--dc22   2005046679

Visit the Taylor & Francis Web site at
http://www.taylorandfrancis.com

and the CRC Press Web site at
http://www.crcpress.com

Dedication

In memory of my late parents and brother, and to my family, friends, teachers, and students, from all of whom I have received so much.

A theory is believed by no one except the person who created it. Experimental results are believed by everyone except the person who got them.

Harlow Shapely

Predictive, or perhaps more accurately, quantitative microbiology has been an active field of research in recent years. Numerous papers have been written on the subject, as well as many review articles and book chapters. A large amount of tabulated quantitative data and simulated growth and survival curves can now be downloaded from Websites, notably those posted by the USDA–ERRC (Eastern Regional Research Center) in the United States and IFR (Institute of Food Research) in the United Kingdom. Also posted on the Web are long lists of references that have dozens and sometimes hundreds of entries. Most recently, Robin C. McKellar and Xuewen Lu have edited *Modeling Microbial Responses in Foods* (CRC Press, 2003) — an update to the classic *Predictive Microbiology* by Tom. A. McMeekin, June N. Olley, Thomas Ross, and David A. Ratkowsky (John Wiley & Sons, 1983, 1993). Together with numerous other publications, they provide existing comprehensive coverage of the mathematical properties of the various existing quantitative models of microbial growth and inactivation, their origins and development during the years, and their application to specific organisms of interest in food and water safety or in disease control or eradication.

A large number of the publications in the food literature addresses the statistical aspects of a model derivation from experimental data, complementing the statistics textbooks that deal with sampling, data analysis, regression, distributions, quality control charts, and the like. Therefore, the purpose this book, by an author who is neither a practicing food or water microbiologist nor a statistician, is certainly not to add another compilation of inactivation and growth models and data or to provide an updated references list. The book is also not intended to discuss the strictly statistical aspects of mathematical models derivation and curve fitting. Discussion of these, as already stated, are readily available to the reader in many convenient forms.

The reason for writing this book is the feeling that research in the field of quantitative microbiology, especially of foods but also of other biosystems, needs new directions. In most scientific disciplines reaching maturity in which a massive body of literature exists, certain thought patterns become so ingrained that the foundations of the prevalent concepts, theories, and models are rarely questioned. A main objective of this book is to do just this — that is, to re-examine and challenge some of the dogmatic concepts that have dominated the field of quantitative microbiology for many years.

Another objective is to offer an alternative approach to modeling certain aspects of microbial growth and inactivation. The discussion will primarily focus on the *mathematical forms* of the proposed alternative models and on the rationale of their introduction as substitutes to those currently in use. Only when it is absolutely necessary will reference to biological aspects of the modeled phenomena be made. The mechanisms of microbial cell division and death and of spore formation, germination, and inactivation have been studied in great detail by professional microbiologists and other scientists. They should not concern us here, except when they have a quantitative manifestation and/or affect the shape of a growth or survival curve.

It is well known that different experimental procedures to grow, isolate, and count microorganisms can yield somewhat different results. Still, the published microbial count records to be analyzed and interpreted in this volume will be always considered as correctly determined and faithful representatives of the systems in question. The roles of sampling and uncertainties, for example, even when pertinent to the data interpretation, will only be assessed in terms of their possible effect on the mathematical model's structure and the magnitude of its parameters. The reliability of the reported experimental data is an issue that has been intentionally left out.

This book is primarily a summary of microbial modeling work done at the Department of Food Science of the University of Massachusetts Amherst in the last 10 years. Many individuals participated in the concept and model development and we have been helped in various ways by experts from outside the department. Several food companies and other institutions helped us considerably by allowing us to share records and, in one case, to create new data. Their contributions, without which this book could have never been written, are gratefully acknowledged.

We have been fortunate to be provided not only with challenging data but also with crucial mathematical ideas and technical assistance in programming. In writing this volume, no attempt has been made to offer an updated comprehensive list of pertinent works published by others and an assessment of their merits. Long lists of related publications can now be found easily in books and reviews, as well as in various sites on the Internet. Unfortunately, very few of these sources of information contain critical assessment of the cited publications even though, at least in some cases, they have obvious shortcomings. Whenever we deal with the publications of others, the emphasis will be primarily on the mathematical properties of the models that they present or propose. Only rarely will their data quality be addressed. Also, and as already stated, we have not tried to document the historical roots of published models and thus credit to the original authors might not have been given. For these omissions, I take full responsibility and apologize in advance to everyone whose work might not have received the proper acknowledgment that it deserves.

Many if not all of the concepts presented and discussed in this book's chapters will probably be controversial and even objectionable to some. This is quite understandable. In fact, although most of the ideas presented have been welcomed in mathematically oriented biological, food, and engineering publications, some were initially turned down by leading food and general microbiology journals for reasons that are still hard to understand. That the rejection came from expert referees demonstrates that a critical re-evaluation of the field's foundations is necessary and timely.

Comments made in several reviews have raised doubts about the openness of the field to any criticism of its long-held beliefs. The same can be said about the current attitude of certain governmental programs that fund food safety research. The verdicts of their review panels and administrators explicitly stated that revision of the currently held concepts of microbial inactivation, although acknowledged to be deficient, is not a welcome proposition, let alone a research priority. This attitude may be changing now and it is possible that this change is partly due to issues raised in the publications on which the first part of this book is based.

A growing circle of microbiologists and scientists in industry, academia, and government share our concerns about the quantitative models and calculation methods now in use in the food industry. Some have actively encouraged us to search for new models and to develop our nontraditional approach. This book will present growth, inactivation, and fluctuation models based on a departure from many of the established concepts in the field. Because the models that we propose are now available to professionals and students together in a single volume rather than scattered in many journals, they might invite criticism as well as trigger a debate on whether some of the theories that currently dominate the field of quantitative microbiology should be replaced.

It is to be hoped that the debate will result in the abandonment of some old ideas and open the way to novel and more effective solutions to the outstanding problems of predicting microbial growth and inactivation. Even if initiating such a debate is all that this book will ever achieve, the effort invested in writing it would be worthwhile. Still, I hope that some readers will discover the utility of the proposed approach and find that at least some of the models described here can be useful to their work. I also hope that the models originally developed for food and water will find applications in other fields, notably in environmental, pharmaceutical, and perhaps even clinical medical microbiology.

Acknowledgments

This book would have never seen the light and the studies on which it is based never come to fruition without the contributions and assistance of many individuals and several institutions. The help came in many forms: suggested key ideas, solutions to mathematical problems, programming, sharing or producing essential experimental data, computer simulations and graphing, and patient typing and retyping of the many drafts of the book's chapters and of the original journal articles summarized in them. During the years in which the research was done, we received much encouragement from friends and colleagues in other institutions. Although they did not participate directly in any of the projects, their moral support was invaluable to us, especially at times when some of our ideas have received an irrationally hostile response in certain quarters. At the same time, we have always appreciated the sympathetic and constructive comments of several anonymous referees and lament, of course, that we cannot thank them in person.

The first part of the book (Chapter 1 through Chapter 8) summarizes the results of research that, by and large, has never been funded, except for modest internal support from the Massachusetts Agricultural Experiment Station at Amherst. We have benefited greatly from a project funded by Nabisco (now Kraft); although it did not support us directly, it allowed us to participate in some of the experiments and gave us access to crucial data that we could not obtain from other sources. We have also been allowed to take a modest part in projects sponsored by Unilever, which also gave us access to helpful data and boosted our morale at the time.

Much of the second part of the book (Chapter 9 through Chapter 12) reports the results of work funded by the USDA–NRICGP (National Research Initiative Competitive Grants Program). (The project was approved when the program had been under a previous management, before a concept development could be labeled "passive research.") We gratefully acknowledge this old program's support.

The list of individuals and institutions to whom I am indebted is long and I wonder if it will ever be complete. It includes principal collaborators Joseph Horowitz, Claude Penchina, programmer Mark Normand, and Maria Corradini, who produced many of the figures that appear in this book. It also includes Martin Cole, Amos Nussinovitch, Osvaldo Campanella, Ora Hadas, and former graduate students Robert Engel and Karen Mattick, who have made significant direct contributions to our research. Among those who have showed their support by inviting us to present our ideas and by other means are Peter ter Steeg, David Legan, Cindy Stewart, Betsy Reilley–Mathews, Helmar Schubert, Walter Spiess, Gustavo Barbosa–Canovas, Jorge Welti, Miguel Aguillera, Ricardo Simpson, Isreal Saguy, Michael Davidson, Tiny van Boekel, Peter McClure, Donald Schaffner, Christopher Doona, Pilar Cano, Janet Luna, Giovanna Ferrari, Gustavo Gutiérez, and Joy Gaze, the students of UDLA (Univerisad de Las Américas), the organizers of the IFT (Institute of Food Technologists) summit on microbial modeling of a USDA sponsored workshop on non-thermal food preservation, and those of the IFTPS (Institute for Thermal Processing Specialists) meeting on thermal processing.

I also express my gratitude to those in industry who provided us with very important data but have preferred that their sources would not become public. I thank Beverly Kokoski and Frances Kostek for patiently typing and editing the manuscript and for accepting the endless revisions and corrections with a smile. I also want to thank Judith Simon and Jill Jurgensen of Taylor & Francis for their help in editing the book and bringing it to press. Last but not least, I want to thank my department head, Fergus Clydesdale, for his continued moral and material support, especially at the most difficult times of the research.

I have been fortunate and privileged to have the cooperation of so many.

About the Author

Micha Peleg has been a professor of food engineering at the University of Massachusetts at Amherst since 1975. He holds a B.Sc. in chemical engineering and M.Sc. and D.Sc. in food engineering and biotechnology from the Technion-Israel Institute of Technology. He teaches unit operations and food processing. Dr. Peleg's current research interests are in the rheology of semiliquid and foamy foods, the mechanical properties of particulated brittle food materials, powder technology, and mathematical modeling of microbial growth and inactivation. He is an editorial board member of several food journals and has been a reviewer for many scientific journals in a variety of fields. Dr. Peleg has more than 300 technical publications, and is listed by ISI (Information Sciences Institute) as a highly cited researcher. He has been elected a member of the International Academy of Food Science and Technology and a Fellow of the World Innovation Foundation.

A theory is a good theory if it satisfies two requirements: it must accurately describe a large class of observations on the basis of a model that contains only a few arbitrary elements, and it must make definite predictions about the results of future observations.

Stephen Hawking

Contents

Introduction

Science may be described as the art of systematic oversimplification.

Karl Popper

According to a popular cliché sometimes seriously cited on both sides of the Atlantic, "all models are wrong but some are more wrong than others." The first part of this statement stems from a misconception of what a mathematical model should be. The essence of a model is that it capture the relevant or important features of the phenomenon or process at hand and deliberately neglect all peripheral aspects. If a model were detailed to the utmost, it would be as complicated as reality itself and thus totally useless. This is known as the "map paradox," alluding to the fact that if a map were an exact match to the physical landscape, its utility as an orientation and navigation tool would disappear and its user remain lost.

The issue can also be viewed from another angle. If a model is so elaborate that identifying its mathematical structure and determining its parameters requires more effort than the actual experiment does, why bother with a model at all? Yet, some mathematical models can be useful even though the magnitude of their parameters cannot be feasibly or accurately determined. Even in an abstract form, certain models can be used to describe and sometimes to predict a system's response or evolution qualitatively. By establishing trends or theoretical limits, for example, such abstract models can help in the study of microbial systems and in the design of processes in which microbial growth and/or inactivation is of concern.

The term "mathematical model" has various interpretations and what constitutes a valid model has been and will remain a debatable issue. Most people would agree, though, that in the context of food, water, and medical microbiology, at least, quantitative models can belong to the following five classifications.

Empirical Models

These are ad hoc mathematical expressions employed to fit a set or sets of experimental data; they can be used for interpolation. No physical significance is assigned to their parameters, although their magnitude can sometimes reflect the relative importance of underlying mechanisms or phenomena. A power law equation, a polynom, and a series of exponential terms are typical examples of such empirical models. There is nothing inherently wrong in using them, as long as they are identified as such. However, one must be careful not to use them for extrapolation and they should be treated as valid only in the range of experimental conditions under which they were determined. If this range of conditions had been extended, the magnitude of the model's parameters might be different, and probably the model's mathematical structure would need to be modified as well.

Fundamental Models

These are models derived from a set of assumptions supposedly anchored in basic principles and the physicochemical fundamentals of the described phenomenon. The most familiar example is the prevalent concept that microbial inactivation follows a first-order kinetics and the models that it has produced. Another is that the rate constant's temperature dependence follows the Arrhenius equation or that its reciprocal, the "D value," has a log linear relationship with temperature. (The last two models are mutually exclusive.)

The first-order kinetics model, which will be again mentioned in Chapter 1, has produced an equation that implies an analogy between microbial destruction and radioactive decay, i.e., that the probability of microbial cell death or a spore's inactivation depends on temperature (and other conditions, like pressure pH, or water activity, for example) but not on the exposure time. This is a very strong assumption and one would expect to find a mechanistic explanation of why such an analogy should exist. Surprisingly, such an explanation is very difficult to find in the literature in which the first-order kinetics is the starting point rather than a consequence of an underlying mechanism.

The Arrhenius equation, which will also be discussed later, implies that all microbial destruction processes, regardless of the organism kind, state, temperature, and the medium in which it is treated, have a single characteristic "energy of activation" as if the inactivation process has been a simple chemical reaction. One should therefore expect that this energy of activation could also be determined by an independent assay to confirm this important assumption. However, again, in most if not all cases in which the energy of activation has been reported or incorporated into a microbial kinetic model, the "evidence" of its existence is only the ostensible, sometimes forced linearity of the log rate vs. $1/T$ plot (see following); by itself, this does not prove that such an energy of activation even exists.

Reductionism is a powerful concept in science and some food safety specialists seriously believe that, eventually, the mortality pattern of microbial cells and inactivation of bacterial spores could be derived directly from the kinetics of individual processes at molecular and cellular levels. No doubt, understanding what happens at the molecular, organelle, and cellular levels would be enlightening and extremely useful. However, *how* such knowledge can be translated into a survival curve at a given pH, for example, is still an open question that may remain unsolved for years to come.

This is not only because of the complexities of the biophysical mechanisms that underlie microbial growth and inactivation, which obviously can be enormous, but also because certain patterns are a manifestation of *rules that operate at the assembly level only.* Here are two familiar examples: The validity of the famous gas law equation $PV = nRT$ (at low pressures) is *independent* of the gas molecules species and a phase transition may follow the same pattern independently of the nature of the physical changes that actually take place in the examined system. The latter has been demonstrated by Kaufmann (1992), for example, who expressed the universality of the phase transition phenomenon in terms of the connectivity of nodes that can be devoid of any physical counterpart. (To be sure, there is a continuing search in physics for a "Theory of Everything" — string theory is a case in point. But at the current state of knowledge, there are still rules that cannot be derived directly from laws that govern the system at a more fundamental level.) Some fundamental models developed in other fields, notably physical chemistry, may be applicable to microbial growth and inactivation under certain circumstances. Before they are actually used, however, their underlying assumptions must be confirmed and the meaning of their parameters validated by *independent tests* — a theme that will appear repeatedly in this book's pages.

Population Dynamic Models

These models are based on rate or balance equations constrained by preservation laws. Familiar examples are the growth rate models based on the logistic equation and the sporal activation–inactivation models, which will be discussed in more detail in Chapter 1 that deals with microbial survival during heat treatments. The model developed by Taub et al (2003), where cells growth, division, and mortality are handled simultaneously, is another good example. The second and third models are based on the observation that the overall rate of change in a growing or inactivated microbial population continuously changes together with the population's *composition*. As a result, such models provide a realistic account of microbial growth or decline patterns at least qualitatively.

Mathematically, some of the dynamic population balance models can be quite complicated, especially when they require the simultaneous solution of two or more differential or difference equations. Also, several of these models' parameters are usually unknown and thus must be assumed or estimated from the data. The same can be said about the model's mathematical structure. Because the exact growth pattern, or "inactivation order," of the contemplated subpopulation is usually also unknown *a priori*, it too must be assumed before the model can be assigned a mathematical expression. Consequently, the strength of the resulting model and its ability to predict, rather than merely to describe growth or survival patterns, largely depends on the correctness of the assumptions from which the model was derived.

Probabilistic Models

The basis of these models is the admission that we do not really know what actually happens at the cellular and molecular levels but we can monitor the overall manifestation of the processes at the population level. For example, without elaborate complementary microscopic analysis and biochemical tests, we cannot readily tell how many heat-treated dead *Salmonella* cells were killed by disruption of their DNA, inactivation of certain essential enzymes in their cytoplasm, and/or changes in their cell walls' physical properties. However, we can relatively easily count the number of survivors after any chosen time even though we do not know exactly what caused the death of those individual cells that did succumb to the treatment. Although this idea has received a scornful reception in certain quarters of food microbiology, it is quite accepted and prevalent in other disciplines, such as public health, actuary, and epidemiology.

Phenomenological Models

These models are constructed for the sole purpose of studying the evolution of microbial systems *qualitatively*. They are used to investigate general trends and patterns, but not to describe the behavior of any particular system in detail, although this can be accomplished sometimes. The magnitude of such models' parameters need not be unique and one can introduce as many terms as required so that the resulting model captures the important features of the process at hand. The main application of such models is in computer simulations. These can emulate observed patterns and suggest tests that will confirm or refute an existing or proposed theory. They can also be used to test qualitatively the assumptions on which a model of a certain type is based.

Obviously, the preceding classification is neither exhaustive nor exclusive. However, it seems that it can be used to characterize most of the main kinds of microbial growth and inactivation models, at least when it comes to food and water, and perhaps in other biosystems.

Prediction, Curve Fitting, and Significant Digits

One of the most unfortunate features of commercial statistical software is the flagrant abuse of the verb 'to predict.' Literally, according to various dictionaries and in the scientific sense, 'to predict' means to foretell, i.e., *to state in advance that something will happen in the future, not in retrospect.* Perhaps, the most celebrated scientific example of a prediction is the demonstration of the works of Einstein's theory of relativity that has received worldwide attention. According to the theory of relativity, mass as large as that of the sun curves the space–time continuum around it to such an extent that light rays passing near it will be bent to a measurable extent. Thus, during a solar eclipse viewed from Earth, stars that happen to be behind the Sun would appear shifted from their usual positions in the sky. This, as well as the magnitude of the shift, had been stated *before* the now famous solar eclipse took place in 1919 and the observation made, thus providing a direct test of the theory.*

In contrast, *numerical values calculated by regression are not predictions in the scientific sense.* They are only *fitted values.* It is a fundamental error to consider them predicted values, the common use of the term in many other fields notwithstanding. (It may come as a surprise to many readers that fitted values need not be unique even when the same mathematical model is used to calculate them using the very same set of experimental data. If one minimizes not the *mean square error,* as the users of standard regression procedures do, but rather the *mean absolute error,* the fitted line or curve and, consequently, the supposedly predicted values could be different to an extent that depends on the pattern of the experimental data's scatter.)

---

* The true story is a little more complicated and, in fact, the theory was conclusively confirmed only during a solar eclipse in 1922. See J.D. Fernie, 2005. Judging Einstein. *American Scientist* 93, 404–407.

Moreover, a good fit by statistical criteria, even if the residual's distribution is random around the fitted line or curve, does not mean that the tested mathematical expression is a viable model of the phenomenon at hand. It is a logical fallacy to treat a fit as if it were confirmation of a model or a theory. That the observed data might be *consistent* with a model or theory should not be confused with a proof of the model's validity. For a model or theory to be truly valid, even if only under a specified set of conditions, it has to be *both* logically consistent and predictive in the correct scientific sense of the term, i.e., it should be able to predict the results of experiments *not used in its formulation*.

All this has been well known but frequently overlooked. For example, if the familiar Arrhenius equation were a correct model of the temperature dependence of microbial mortality rates, it could be used to predict the survival ratio reached in a nonisothermal heating process. Until this is demonstrated, it should be treated as just an empirical secondary model (see following) of the kind $k(T) = a_1\exp{-(a_2/T)}$, where $k(T)$ is the temperature dependent "rate" and a_1 and a_2 are adjustable (regression) parameters. No physical significance should be assigned to these parameters before it is confirmed by an *independent physical test*.

As will be shown later, the Arrhenius model is neither unique nor a particularly useful secondary model, its popularity in the food research community not withstanding. (See also McMeekin et al., 1993, where the model's problems are viewed from another angle.) Obviously, the Arrhenius model and other traditional alternatives, like the log linear temperature dependence of the rate constant's reciprocal or the WLF equation, will always fail if applied to a process that involves enzymatic activity at temperatures that may cause the enzyme's denaturation.

A less frequent methodological error in the food microbiology literature is to consider a single deviant experimental point as evidence of a local extremum and several such points as evidence of the existence of several peaks or of oscillations. The confirmation of a local peak, or minimum, should come from several measurements around its contemplated location, not from a single data point.

In the food microbiology literature and in several Websites, there seems to be a confusion between the confidence levels of the regression, which is primarily determined by the scatter in the fitted measurements, and the range of values one would expect to find if the whole experiment were replicated, with cultures of different origin, fresh growth media, nutrients from different sources, etc.

Quantitative microbiology is a notoriously inexact field of science, as everyone knows. Numerous factors can affect microbial growth and inactivation. Many of them might not be fully known and some cannot be always tightly controlled even under the best laboratory practices. With quite a few exceptions, like the presence of fecal bacteria or certain pathogens, the microbial load of raw foods (especially meats or milk), affluent water, and other biosystems is counted in powers of ten. The same can be said about laboratory inocula used in growth and inactivation studies. Thus, the frequently noisy inactivation or growth data are almost always plotted on semilogarithmic coordinates. This, as well as the notorious irreproducibility of microbial counts, raises the issue of statistically *vis-à-vis* practically significant digits. Too often, survival or growth parameters are reported with a large number of insignificant digits, as judged by statistical criteria. But many would have no meaning even if they could pass the statistical test. Consider that microbial mortality is a process that follows a first-order kinetics (which most probably it does not — see Chapter 1) and that the D value has a log linear temperature dependence, as the standard textbooks on subject tell us (usually without an actual example). Even then, one must still doubt that a D value can be accurate to a tenth of second, and the z value to a few hundredths of a degree Celsius, as has been reported in numerous publications.

Interpolation and Extrapolation

Most experimental data sets, especially if they have a scatter, can be fitted successfully, from a statistical point of view, with more than one mathematical model. It can be shown that as long as these models are intended for interpolation, they can be used interchangeably (see Chapter 1, Chapter 2, Chapter 4, and Chapter 8). However, they can imply, qualitatively, a very different outcome if used for extrapolation. Therefore, unless a proven theory or, better still, direct evidence indicates that the model's validity extends to beyond the experimental range, the practice of extrapolating survival and growth curves can be very risky and thus should be discouraged.

The problem is further aggravated when the extrapolation is done over several orders of magnitude, as in the determination of the notorious thermal death time, which has rarely, if ever, been determined directly in any real low-acid food undergoing heat sterilization. In fact, even a slight true curvature in the original semilogarithmic survival data can result in a significant discrepancy between theory and practice. As will be shown in Chapter 1, depending on the concavity direction, the discrepancy might have safety or quality implications.

From a logical viewpoint, even if a certain trend has been unambiguously detected in a set of experimental results obtained within a range of conditions, there is no reason to assume that it will continue unchanged beyond this range. For example, a continuation of a survival or dose–response curve that is concave in one direction can have an opposite concavity if the exposure time is long enough, thus producing as sigmoid pattern (see Chapter 1 and Chapter 7). A similar situation may occur in microbial growth (Chapter 8) where what one might consider a population's "stationary stage" might well be the beginning of its decline.

Another point sometimes overlooked in a model's construction is its long-term implications. These ought to be examined *before* the model's fit is even tried. A typical example is the polynomial model used as such or as part of the popular response surface methodology. Consider that the inactivation rate measure of a certain organism can be described by the secondary model $p = a_0 + a_1T - a_2T^2$, where p is an inactivation rate parameter, T is the temperature and the a's are constants. Because the increase of the hypothetical parameter with temperature is not exactly linear, the "correction" for its nonlinear temperature dependence is done by adding the term $-a_2T^2$, where a_2 is much smaller than a_0 and a_1.

As long as this model is used for interpolation, it might be quite effective. Yet, because the term with the highest power, $-a_2T^2$, has a negative sign, this model implies that beyond a certain high temperature, the organism will grow very rapidly rather than be destroyed. The same problem can arise in more complicated models, which contain combined terms like, temperature–pH, temperature–salt, or acid concentration and the like. One needs to remember that the response surface methodology was originally developed to identify optimal combinations of conditions or factors in an economic manner, i.e., to locate an extremum with a minimal number of experiments. The response surface methodology has never been intended to and does not produce models that are predictive in the correct scientific sense, as certain papers' titles and abstracts imply. As already stated, a model's predictive ability must be confirmed by *independent tests* and a good fit by itself, especially when the model's number of terms is allowed to vary, does not provide such confirmation.

Logarithmic and Other Transformations

In many processes of interest to food microbiologists and technologists, the microbial population's size frequently changes by several orders of magnitude. Consequently, expressing microbial counts in log units makes perfect sense. This is not only because the microbial counts and variability dictate it, but also because a significant *practical difference* can be found between a survival ratio of 10^{-3} and 10^{-6}, for example. Although these two small ratios will appear as approximately zero if plotted on a linear coordinate with a scale from zero to one, the 10^{-6} survival ratio might be considered as guaranteeing a safe product and the 10^{-3} ratio might indicate an unacceptable safety risk. Thus, for many microbial populations, primary models formulated on the basis of logarithmic counts are most useful, their rationale unquestionable, and their interpretation straightforward.

This is not necessarily true in the case of secondary models (see Chapter 1 and Chapter 2), in which the dependent variable is not the microbial count but a primary model's parameter. The temperature dependence of a rate constant is a familiar example. Because many growth or inactivation rate parameters have been derived from already logarithmically transformed experimental data, it is unlikely that their magnitude in the pertinent temperature or pressure range will have a span of several orders of magnitudes. Thus, an additional logarithmic transformation rarely, if ever, serves any useful purpose. Moreover, because the logarithmic transformation gives equal relative weight to low and high values, errors can be amplified in this way (see the discussion of the Arrhenius and WLF models in Chapter 1).

The same applies to the presentation of temperature not in °C but, as its reciprocal in K^{-1}, as required by the popular Arrhenius equation. When using this model, the difference between 5 and 40°, for example, which for many microorganisms of food and water safety concern represents the difference between precarious survival and intensive growth, becomes a difference between 0.0036 and $0.0032K^{-1}$ after the (unnecessary) mathematical conversion of the temperature scale. The same applies to the Arrhenius model's application to microbial inactivation at high temperatures. For at least certain bacterial spores that are the target of thermal processes, a difference between 80 and 125°C means the difference between survival (and in some cases even activation — see Chapter 1 and Chapter 3) and destruction on a time scale of a few minutes. Yet, this huge temperature range, as far as the fate of bacterial spores is concerned, is compressed into the very modest 0.0028 to $0.0025K^{-1}$ interval as a result of the temperature scale transformation.

A scale should be transformed only when absolutely necessary, not because this is a common practice in other fields. In some cases, of course, a transformation, logarithmic or other, can be quite useful. A good example is the test of whether a given process follows the rules of classical diffusion theory, in which the distance reached is proportional to the square root of time. In such a case, the linearity of the distance vs. the square root of time relationship could serve as a test of the agreement between theory and experiment.

Similar techniques are used in a variety of scientific disciplines. A notable example is the Q–Q plot in statistics, which is used to test the conformity of a given set of data to a contemplated distribution function (see Chapter 9 and Chapter 10). A clear nonlinearity of such a plot can safely be used to *dismiss* a candidate model. However, a "slight deviation" from linearity can be of various kinds and therefore more difficult to interpret in terms of the shape of the distribution in question. This is just another demonstration of why the *appearance* of linearity should be always treated with caution.

Similar situations frequently arise in the interpretation of Arrhenius plots or data plotted on semilogarithmic coordinates. In most cases, arriving at a conclusion based on them requires additional statistical testing that the regression coefficient alone does not provide. Notice that a high regression coefficient or low mean square error, by itself, may be misleading in such situations. This is because a slight curvature in a logarithmic plot can result in a large discrepancy between the fitted and actual values.

Models that have nested exponentials, $\exp[\exp(x)]$ or logarithms, $\log[\log(x)]$, like those that appear in certain versions of the Gompertz model (see Chapter 8), almost always have a good fit to growth curves of a certain type. However, the rationale of describing a phenomenon with such transformation of the independent variable is unclear in most cases. Some of the parameters of such functions can rarely be interpreted even in terms of the curve's morphological characteristics, let alone of an underlying mechanism or pattern.

Analogies

Sometimes, very different physical phenomena have a striking similarity in form and therefore can be described by the same kinds of mathematical models. Heat and mass transfer equations are perhaps the most illustrative example. The similarity emerges because both are derived from what is basically the same conservation principle. Some similarities, however, do not stem from any common fundamentals and are therefore only superficial or incidental. In such cases, adapting a model derived for one system for use in another requires considerable caution.

The two most illustrative examples are the already mentioned implicit analogy between microbial death and radioactive decay, on which the famous first-order kinetic model is based (Chapter 1), and the idea that all microbial inactivation processes have a single energy of activation, as if a cell's death or a spore inactivation is equivalent to a reaction between two simple chemical species. Another, not as widely held, is that the kinetics of sporal destruction is the outcome of glass transition and thus can be described by the WLF equation as an inactivation model. The WLF equation was originally developed for describing the temperature effect on the viscosity of heat-plasticized synthetic polymers. If it is literally accepted as an inactivation kinetic model, a survival ratio and the spore's material viscosity are not only intimately related but also governed by the same rules.

Although certain kinds of models may apply to simple chemical reactions and certain biological processes (primarily due to these models' mathematical flexibility), this does not mean that the corresponding phenomena share any similarity or analogous mechanisms. Suffice it to say that microorganisms and spores are not molecules but elaborate structures. Sometimes, certain bacteria can even employ defense tactics when exposed to a hostile environment. Thus, any reference to a "mole," in a term such as kcal/mole or kJ/mole, makes little sense when used in the context of microbial growth or destruction.

To demonstrate the point, let us assume that the volume of a bacterial endospore is on the order of 0.25 μm^3 ($1 \times 0.5 \times 0.5$ μm) and that its density is on the order of a typical biological material —for example, 1300 kg·m^{-3}. Because Avogadro's number is about 6×10^{23}, a mole of spores will have a mass on the order of $6 \times 10^{23} \times 0.25 \times 10^{-18} \times 1.3 \times 10^3 \approx 2 \times 10^8$ kg, i.e., on the order of 200,000 metric tons! (It is not surprising that very few microbiology laboratories keep even a micromole, let alone a millimole, of spores for calibration purposes…)

This brief introduction clearly illustrates that several practices and held concepts in quantitative microbiology deserve a second look and a thorough revision. To an outsider, the literature in the field seems to be a mixture of sound and intriguing ideas, on the one hand, and theories that should have been abandoned long ago, on the other. What follows in this book is not intended to sort the grain from the chaff — this should be left to the reader — but rather to present an alternative view on several issues with which I have been involved during the past years. In certain instances, the applicability of the proposed models has yet to be demonstrated. However, they are all testable models, i.e., their validity can be confirmed or refuted by experiment, at least in principle. None of the presented models is claimed to have universal applicability and, in many cases, their description is accompanied by an explanation of how alternative models can be chosen and tested.

Some of the models discussed in the book will be totally unfamiliar to most practicing microbiologists and in many cases will be (correctly) considered totally incongruent with those that appear in textbooks and are taught in universities and colleges around the world. Thus, wherever possible, demonstrations of the new models' applicability to old problems will be provided, allowing the microorganisms to speak on the models' behalf. The demonstrations will show that the new models are not merely mathematical creations, but that they can be used to predict growth and inactivation of real microbial populations, at least under certain conditions.

Regrettably, for reasons beyond our control, the number of experimental demonstrations is limited to a selected group of organisms or products. In all the cases discussed in Chapter 7 through Chapter 12, we had to rely on others' data by choice. When it came to issues involving inactivation (Chapter 1 through Chapter 6), we simply had no other alternative. Only in a single case (the inactivation of *Salmonella*) did we have a certain degree of control over the nonisothermal experiments, thanks to the cooperation of the Kraft Corporation (then Nabisco). We also got very useful unpublished experimental data from various sources, including other food companies.

All of our attempts to obtain funding for developing the inactivation models and including them in user friendly freeware on the Web (see list of Websites in the "Freeware" section of this book) have been denied by the current food safety program of the USDA–NRICGP. Its panels and present administrators concluded that improving the method to calculate sterility "is not an urgent issue in food safety" and classified the proposed study as a project that should "not be supported even if funds were available for three years in a row."

However, the final judgment on the utility of all the concepts and methods described in this book, that of the inactivation models included, should come not from reviewers' and panelists' opinions, but rather from the microorganisms themselves. Hopefully, the microorganisms will be given the opportunity to pass an approving verdict on the resulting models in future research, as they have done consistently until now.

General Approach

Two main themes are covered in this book's chapters.

Survival and Growth Patterns of *Large* Microbial Populations

The emphasis here is on 'large,' that is, the number of spores and cells in question is sufficiently high so that continuous models can be used to describe them. Another key assumption is that the habitat is sufficiently uniform at the pertinent level of scrutiny so that issues concerning local microenvironments need not be addressed. This, of course, is only permissible if one uses phenomenological models, that is, models applicable *only* at the population level. Although such models can still be truly predictive, as will be demonstrated, they might not be useful for isolated small populations.

Thus, the outcome of the presence of five spores (for example, at the surface of an air bubble in a food can) or of a few organisms attached to a tiny solid particle suspended in water is unlikely to be predicted from a survival model developed on the basis of experiments in which spores or cells have been suspended or inoculated in a culture medium at levels of millions per gram or milliliter. The statistics of small numbers may well be an area that should be included in quantitative microbiology (see Chapter 13). It has obvious practical implications, especially when injured cells are involved. However, the topic is outside the scope of this particular book and thus will not be discussed in detail.

Another starting point of the alternative growth and survival models is that, at the present stage of knowledge, the exact quantitative aspects of a microbial population's fate, especially under changing conditions, cannot be directly derived from events at the molecular and cellular levels. Moreover, it will be assumed that the only available source of information for the models' derivation is the experimentally observed patterns of growth and decay. For what follows, it is not required that the observed survival or growth patterns should conform to any preconceived theory. It is always the organism at hand that will dictate the model. All the cells or spores have to do is to obey physical laws. As already mentioned, we have no reason to assume that a single universal growth or inactivation model would describe all the observed patterns of microbial multiplication and death. We can expect, though, that certain ubiquitous growth and inactivation modes can be described and even predicted, by the same *kinds* of general mathematical models.

Situations in Which Intervals between Successive Counts Are Large Relative to the Time Scale of Microbial Growth or Decline

Routine microbial counts taken several hours apart in meat and poultry or water-treatment plants are typical examples. The result is a microbial record in the form of counts plotted vs. time or date, which may appear to have an irregular random pattern of oscillations. Apart from sampling issues, which will not be considered here, the ostensibly sharp fluctuations are, in most cases, evidence that significant events in the evolution of the monitored population have been undetected because of the time lapse between successive observations. Again, and contrary to the belief of some, such situations cannot be effectively dealt with by employing conventional kinetic or population dynamics modes or even models based on chaos theories. This is especially the case when relevant information like a corresponding change in temperature and pH, for example, is missing.

Still, purely probabilistic models can be used, not only to characterize the counts' fluctuation patterns, but also, under certain circumstances, to predict the frequencies of outbursts that might become a health hazard. Here, too, the only reliable source of information is the counts record. Certain population dynamics models developed in population biology, epidemiology, and other fields might be helpful in explaining certain patterns but they can rarely, if ever, be used to predict *aperiodic* outbursts of varying magnitude and duration.

The alternative approach we offer to both classes of problems (that is, to those associated with continuous growth and inactivation and those with discontinuous records) is based on the notion that the absence of complete knowledge is not necessarily an insurmountable barrier to a quantitative model development. The departure from the traditional approach, as already mentioned, is that our starting point is the acknowledgment that we do not really know exactly how a particular microbial population responds to environmental changes.

Therefore, we cannot take for granted the applicability of any previously suggested dynamic model, especially if it has been derived on the basis of a presumed analogy to a nonliving system. The underlying principle is that we must always leave it to the microorganisms to decide whether a set of assumptions is correct and whether the mathematical model based on them is appropriate. This can be done convincingly in only one way: by testing the predictive power of the model and not simply its fit.

All this does not mean that phenomenological modeling and studies of molecular and biophysical processes at the cellular or sporal level should be considered as separate pursuits. Although none of the models prescribed in this book's chapters has been derived from the kinetics of processes and events at the microscopic level, they can still be used to *quantify these processes' manifestation at the population level*. Thus, if a model's parameters can be expressed in meaningful terms, they can become a useful tool in the study of microbial populations as such and of the macroscopic consequences of microscopic events. This, needless to say, is true for *all* phenomenological models, traditional and nontraditional alike.

A Note to the Reader

Certain equations and occasionally even graphs will appear more than once in the text with a different equation or figure number in different chapters. This has been done deliberately so that most chapters or groups of chapters could be read as covering independent topics. This will eliminate the need to search previous chapters in order to find an equation or an explanatory schematic figure. The same can be said about certain repeated statements. They too are intended to eliminate the need to go to another chapter for explanation of the text. I apologize to readers who will find these redundancies unnecessary or annoying. The mention of commercial software, books, or any other commercial products in this book does not imply endorsement to the exclusion of other products of their kind by the author or the University of Massachusetts. All the opinions expressed in this volume are those of the author. They do not necessarily reflect the University of Massachusetts' position on the respective issues or that of any other institution or company involved in the research that has produced this treatise.

The quotations that start each chapter have been taken from various sources, but mostly from the Internet. Some have been attributed to more than one person and I am not in a position to affirm their authenticity. Yet, I hope that the reader will enjoy the quotations despite the uncertainty in some cases as to whom the idea had occurred first.

1

Isothermal Microbial Heat Inactivation

> The great tragedy of science — the slaying of a beautiful theory by an ugly fact.
>
> **Thomas H. Huxley**

Primary Models — the Traditional Approach

The First-Order Kinetics and the *D* Value

According to almost every textbook on general and food microbiology (e.g., Davis et al., 1990; Brock et al., 1994; Prescot et al., 1996; Jay, 1996; Holdsworth, 1997), microbial inactivation is a process that follows a first-order kinetics. The same fact is repeated in numerous Websites, whether they are tutorial or an official document, like the FDA–IFT 2000 comprehensive panel summary that lists thermal and nonthermal inactivation kinetic parameters of a large number of microorganisms and bacterial spores in a variety of media. The starting point of this theory is that, upon exposure to a uniform lethal temperature, the number of the affected microbial cells or spores decreases exponentially with time. Or, stated differently, the rate of inactivation is proportional to the number of cells that are still alive or the number of bacterial spores that are still viable. In what follows, we will use the term 'organism' or 'microorganism' inclusively — that is, it will refer to microbial cells and spores, especially bacterial.

Let N_0 be the initial number of organisms in the heat-treated population and $N(t)$ their number after time t. According to the first-order kinetics model, under isothermal conditions (T = const), the inactivation rate $dN(t)/dt$ is:

$$\frac{dN(t)}{dt} = -k'(T)N(t) \tag{1.1}$$

where $k'(T)$ is a temperature-dependent exponential rate constant.
Integrating Equation 1.1 yields:

$$\frac{N(t)}{N_0} = \exp[-k'(T)t] \tag{1.2}$$

or

$$\log_e\left[\frac{N(t)}{N_0}\right] = -k'(T)t \tag{1.3}$$

Because microbial counts are usually expressed in powers of 10, Equation 1.3 is commonly presented in the form:

$$\log_{10}\left[\frac{N(t)}{N_0}\right] = -k(T)t \tag{1.4}$$

where $k = k'/\log_e 10$.

By definition, $N(t)/N_0$ is the momentary *survival ratio*, $S(t)$, and thus Equation 1.4 can also be written as:

$$\log_{10}S(t) = -k(T)t \tag{1.5}$$

a convention to be followed in this and successive chapters.

The reciprocal of the rate constant $k(T)$ is known as the *D value*. It is the time, usually in minutes, needed to reduce the size of a given microbial population by a factor of 10. This *D* value has been traditionally treated as a measure of the organism's heat resistance in the particular medium at which the inactivation has been monitored. One can find lists of the *D* values of a variety of microorganisms, in different media and at different temperatures, published in books, numerous research articles, and several Websites, including the one mentioned in the opening sentences of this chapter.

One of the implications of a first-order kinetics is that a plot of $\log_{10}S(t)$ vs. *t* (Equation 1.5) must be a straight line. Whether this is true in practice is a debatable issue; growing evidence suggests that log linear survival curves, i.e., linear $\log_{10}S(t)$ vs. *t* plots, are the exception rather than the rule (van Boekel, 2002). Now, let us examine the *meaning* of the first-order kinetics model. If all the exposed organisms, cells and spores alike, had been completely identical, then one would expect that in a medium of uniform lethal temperature they would all die or be inactivated at exactly the same time. In reality, even in a capillary — the traditional device used to contain the cells or spores during a microbial survival curve determination — the temperature is not perfectly uniform. Thus, one would expect that not all the organisms will die simultaneously, but rather within a very narrow time span determined by the medium's homogeneity.

However, such a mortality pattern has rarely if ever been observed and there is a legitimate reason to question the hypothesis that microbial populations are always composed of identical members as far as heat resistance is concerned. If the population's members have a spectrum of heat resistances, then any mechanistic explanation of microbial inactivation and death will need to take this factor into account. The inevitable question in this case will be of *how* the nonuniformity of the organisms' response could be identified and quantified. Regardless of the issue of the cells' or spores' genetic and physiological homogeneity, however, the first-order kinetic theory maintains that the *probability of a mortality or an inactivation event is time independent.* Thus, according to this theory, isothermal microbial inactivation is analogous to radioactive decay, in the sense that the exposure time has no effect whatsoever on the probability of death of an individual cell or of a spore's inactivation, the equivalent of an atom's nucleus disintegration or the emission of a particle.

To many, the implied analogy between microbial inactivation and radioactive decay might come as a surprise, but this is exactly what the first-order kinetics means. Obviously, there must be several different biophysical processes that can cause the death of a microbial cell or the inactivation of a bacterial spore directly or through initiation of a cascade of events that will have the same result. Therefore, it would be reasonable to assume that two or more of such processes will proceed simultaneously and interactively within the same cell or spore while it is exposed to a lethal temperature. However, it will be very difficult to come up with a ubiquitous biophysical mechanism that will coordinate all such processes to produce a *universal* first-order inactivation kinetics, i.e., regardless of the microorganism kind, its growth stage, the medium in which it is treated, and the temperature to which it is exposed.

This is, of course, a purely formalistic argument, but it cannot be dismissed offhand. Criticism of the first-order kinetic theory based on mechanistic considerations can be found in many publications in the literature of microbiology (e.g., Casolari, 1988). However, because the focus of this chapter is on the mathematical properties of inactivation models, the biophysics of cell mortality and spore inactivation should not concern us here and the issue will not be further discussed. All that can be added at this point is that if a fundamental biophysical principle can support the *universality* of the first-order kinetics, it has yet to be demonstrated.

A more likely explanation of the first-order kinetics model introduction to microbiology is that it was formulated at times when computation was done with tables of logarithms, mechanical desk calculators, or graphical methods. Therefore, there was a premium on linearization methods, not only in microbiology but also in many other disciplines, especially in engineering. However, once the theory became established, it turned into a dogma that has not been frequently challenged. This statement is not as outrageous as it might sound. The correctness of the first-order kinetics theory has been taken for granted to such an extent that the majority of the textbooks in which it is introduced have only a schematic drawing of a linear $\log_{10}S(t)$ vs. t plot. In some cases, the plot is supplemented by computer-generated "points," but very rarely is the figure accompanied by experimental survival curves that would demonstrate the relationship's linearity.*

As has already been stated and as examples will show, most microbial survival curves do not follow the first-order kinetics model and there is no reason that they should. Still, quite a few microbiologists insist that the D value is a useful measure of microbial heat resistance even when the actual survival data from which it is derived are not really log linear. The argument, which does carry some weight, is that it is still important to know whether the time for a decimal reduction at a given temperature is on the order of 1 minute, for example, or 5. Yet, one could also state, for example, that under specified conditions, a 4 decades reduction, say, had been observed after 10 minutes. It would be a statement of fact and therefore cannot be challenged. This is because such a statement does not imply that the reduction had followed the particular pathway dictated by the first-order kinetics hypothesis which most probably it had not, or any other pathway for that matter.

The "Thermal Death Time"

According to the first-order kinetics model, absolute sterility can never be achieved because, in exponential inactivation, the number of survivors can only approach zero asymptotically but never reach it. Stated differently, because $\log_{10}S(t) = -kt$, the logarithmic survival ratio must decrease indefinitely with time. Consequently, one can only achieve what is called 'commercial sterility.' It is defined as the outcome of a thermal process where the survival ratio of the targeted spores, if they had been present,

---

* I attended a meeting in which a professor of microbiology made the formal statement that if an experimental semilogarithmic survival curve turns out to be curvilinear, the person who has recorded it should be retrained. If this attitude has prevailed in other laboratories, one can only wonder how many of the D values reported in the literature and posted on Websites have been obtained from truly log linear survival curves.

would have been reduced by so many of orders of magnitudes that the heat-treated product (primarily a low-acid food) could be considered microbially safe.

The traditional criterion for the commercial sterility of heat-processed low-acid foods has been 12 orders of magnitude reduction in a hypothetical population of C. *botulinum* spores, i.e., $\log_{10}S(t) = -12$. The corresponding time has been called the "thermal death time," and it has been assumed, correctly, that once this level of reduction has been achieved, the probability of actual survival is so low that it can be ignored. However, the problem with the concept, known from its inception, is that counting survivors after their population has been reduced by 12 orders of magnitude is not an easy task. In fact, for technical reasons, most experimental survival curves in foods only cover six to eight decades' reductions. Thus, the thermal death time has been estimated by *extrapolating* the "linearized" isothermal semilogarithmic survival curve to $\log_{10}S(t) = -12$.

In most scientific disciplines, extrapolation over four to six orders of magnitude would be unthinkable. However, because of the rigor that the 12 orders of magnitude reduction requirement imposes on any given thermal sterilization process, even a considerable deviation from the correct value would still result in a microbially safe product. The record of the food industry testifies that this is indeed the case, especially because a substantial safety factor is added to a process based on the 12D reduction requirement.

Because of numerous reports that neither the isothermal survival curves of microbial cells and bacterial spores nor the temperature dependence of the D values (whatever they mean in such cases) are necessarily log linear, one must conclude that the safety of canned foods cannot be attributed to the accuracy of the survival models, but to an arbitrary and largely unknown degree of overprocessing. Obviously, when the isothermal survival curves have a downward concavity (Figure 1.1), the extrapolated thermal death time will be an overestimate of the true time needed to reach the 12 decades' reduction. If the isothermal semilogarithmic survival curves have an upward concavity (i.e., they show a considerable "tailing"), however, then the extrapolated thermal death time will be an underestimate (see figure), which might have safety implications in certain cases, at least theoretically.

Biphasic and Multiexponential Decay Models and Their Limitations

The isothermal semilogarithmic survival curves of numerous organisms and spores have a noticeable upper concavity. Because they are clearly not linear, even the proponents of the first-order kinetic model admit that the log linear model in its original form cannot be used to describe them. Therefore, it has been suggested that the curvature of all concave upward

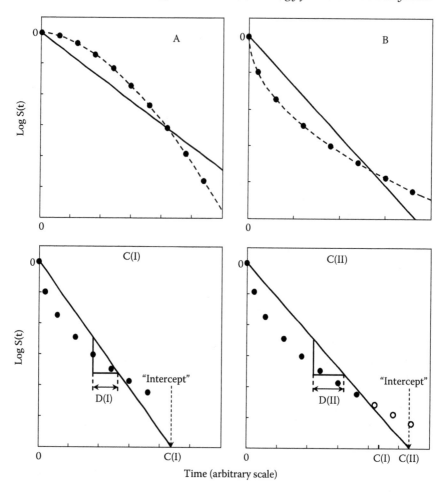

FIGURE 1.1
Passing a straight line through curved data in order to calculate a D value. Notice that extrapolation can lead (theoretically) to over- or underprocessing, depending on the experimental concavity direction and that the magnitude of the calculated D value will depend on the number of points take for the regression.

semilogarithmic survival curves is evidence that the affected population is a mixture of two or more subpopulations whose mortality follows the first-order kinetics, each with its own rate constant k (Stumbo, 1973; Geeraerd et al., 2000). Following this reasoning, the survival curve of certain organisms has been described by the double exponential model:

$$S(t) = a\exp(-k_1 t) + (1 - a)\exp(-k_2 t) \qquad (1.6)$$

This is known as the "biphasic model," where a, k_1, and k_2 are constants.

If three adjustable parameters are insufficient for fitting the experimental survival curve, more terms can be added, in which case the model becomes:

$$S(t) = a_1\exp(-k_1t) + a_2\exp(-k_2t) + a_3\exp(-k_3t) + \ldots \tag{1.7}$$

where the k's are the presumed subpopulations' first-order kinetics rate constants and the a's their respective relative weight fractions in the mixed population.

Notice that, always, $a_1 + a_2 + a_3 + \ldots = 1$. Therefore, if the model has two exponential terms (Equation 1.6), it will have three adjustable parameters; if it has three terms (Equation 1.7), it will have five adjustable parameters and so on. If the number of exponential terms can be freely chosen, this model will fit well almost any monotonic decay curve having upward concavity of the kind found in microbial inactivation, as well as in mechanical stress relaxation, electrostatic charge dissipation, and numerous other such phenomena.

One should always remember, however, that the fit of Equation 1.6 or Equation 1.7 stems from its mathematical flexibility and not from the physical existence of subpopulations. If a microbial population is truly a mixture, then to validate the model, *the subpopulations must be isolated and their rate constant determined independently*. Or, alternatively, if the survival curve is considered as being the outcome of two distinct inactivation kinetic modes (Equation 1.6), then one should be able to identify the corresponding mechanisms by especially designed assays before they can be used as a post hoc explanation of the survival curve's shape.

The weakness of Equation 1.6 and Equation 1.7 as kinetic models is evident in the following simple fact. If the experiment from which their parameters are determined had been shorter or longer, the magnitude of their parameters — namely, the a's and k's — would have been different. However, it would be very difficult to understand how the experiment duration can affect the mixed populations' composition, as would be dictated by the relative magnitudes of the a's in Equation 1.7 or the ratio $a/(1 - a)$ if Equation 1.6 is used as a survival model. Similarly, it would be very difficult to explain how the heat resistance of the subpopulations as manifested by the absolute magnitudes of the k's in both models can be influenced by extending or shortening the test. Moreover, it can be easily demonstrated (Figure 1.2) that almost every set of decay data that can be fitted with Equation 1.7 as a model can be also fitted, with the same or better degree of fit, by the alternative model (Equation 1.8) where $a_1 + a_2 + a_3 + \ldots = 1$ and the c's are constants:

$$S(t) = -\left(\frac{a_1 t}{c_1 + t} + \frac{a_2 t}{c_2 + t} + \cdots \right) \tag{1.8}$$

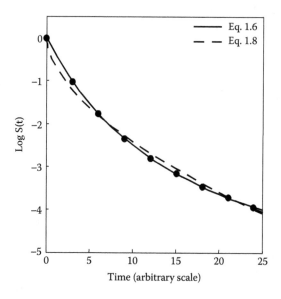

FIGURE 1.2
A simulated demonstration of the fit of Equation 1.8, dashed line, to data generated with
Equation 1.6, dots and solid line. Notice that if Equation 1.6 could be considered a mechanistic
inactivation model, it would imply the existence of a mixture of *different composition* than
the one that would be if Equation 1.8 were used instead. The presumed mixture's composition would also change if the experiment was prolonged or shortened.

In many cases, a model based on Equation 1.8 will achieve the same
degree of fit with fewer terms than one based on Equation 1.7. However,
a model based on Equation 1.8 will imply a totally different composition
of the subpopulations, characterized by the new set of a's, with very
different spectrum of heat sensitivities now characterized by the cs instead
of the k's. Notice that a prerequisite of Equation 1.6 and Equation 1.7 is
that the isothermal semilogarithmic survival curve must have an upward
concavity. van Boekel (2002) showed, however, that about a third of the
hundred or so survival curves that he examined happened to have a
downward concavity; in this case, Equation 1.6 or Equation 1.7 will fail
not only as a kinetic model but also as an empirical curve-fitting model.
    As already stated and as will be repeatedly stressed in this book, the
information contained in a survival (as well as a growth) curve's shape
is limited. Thus, certain features of the microbial population can only be
revealed by complementary independent tests, not through curve fitting.
The assumption that a given population whose survival curve can be
described by Equation 1.6 or Equation 1.7 is a mixture of subpopulations
might be correct. That the subpopulations' inactivation follows the first-
order kinetic might be correct too (although very unlikely, as has already

been explained), but the assumption that the equation's coefficients obtained by regression are sufficient to identify the mixture's composition and the heat sensitivities of the subpopulations is simply false.*

The Logistic Models

These models provide another example of the need to examine analogies critically before they are invoked in a mathematical model development. The Baranyi–Roberts model (Baranyi and Roberts, 1994) and the logistic equation from which it has been derived are discussed in great detail in Chapter 8. The model had been originally proposed for sigmoid growth curves, on the correct premise that a microbial population introduced into a new habitat quite often experiences an adjustment period before the onset of massive (exponential) growth. This period of acclimatization in the new growth medium, known as the "lag time," has been documented in many studies of microbial growth. Its existence has obvious implications in food spoilage, fermentation, the life history of diseases, and contamination of water sources. The time scale of the delay is in many cases on the order of at least hours, but lags of several days, weeks, and, in some cases, even months are also not unheard of.

One of the most salient features of the Baranyi–Roberts growth model is that the growth curve has an identifiable lag produced by the organism's physiological state at the time of its transfer to the new habitat and that the maximum specific growth rate is the curve's slope at its inflection point. What should concern us here is that what is essentially a microbial growth model has been applied by certain researches to describe microbial *inactivation* with what has been in essence a sign change only. In other words, the population's decreasing size (negative growth) has replaced the increasing numbers of the growing population as the dependent variable, leaving the rest of the model's structure and terminology basically intact. According to this model, all survival curves must have a finite residual survival ratio, there is always a "specific maximum inactivation rate" at the inflection point, and there is a definite lag phase determined by the distribution of lags in the inactivation of the individual cells.

---

* An erroneous interpretation of concave upward decay curves is not unique to food microbiology. It is most common in food rheology — probably the origin of these models in food research — where the "successive residual method" was used to determine this model's coefficients. To this date, the discrete Maxwell model, the equivalent of Equation 1.7, is used to calculate the spectrum of relaxation times of solid foods. However, a spectrum calculated in this way, supposedly a material property, can vary dramatically depending on the experiment's length and the rate at which the specimen has been strained. Consequently, the reality of such a spectrum of relaxation times, like the spectrum of the inactivation rate constants produced by Equation 1.7, is in serious doubt, all claims to the contrary not withstanding.

Unfortunately, quite often none of this model's qualitative predictions can be observed experimentally. Even on the rare occasions when an isothermal survival curve can be described by the Baranyi–Roberts inactivation, its shape can have an alternative and much simpler explanation. For example, all monotonic concave upward curves, or even linear semilogarithmic survival curves for that matter, have no lag time at all. If the survival ratio's drop is preceded by a discernible short "shoulder," this would be most likely an artifact of the come-up time (see Peleg, 2003a), i.e., a reflection that a lethal temperature has not been instantaneously reached. Thus, before the targeted temperature was reached, no inactivation had taken place and thus the apparent lag. Steep concave downward survival curves also do not have a lag time and, in most cases, they indicate a continuous monotonic destruction, with no sign of a residual survival level and/or of an inflection point — the pillar of the Baranyi–Roberts inactivation (and growth) model.

As will be shown in Chapter 8, any attempt to identify a definite segment in a continuous curve, like the lag time here, is a risky enterprise and the result will depend on the mathematical model used. Again, unless confirmed by *independent* experimental evidence, the Baranyi–Roberts inactivation model should be considered no more fundamental than any other empirical model and no physical significance should be assigned to its parameters if solely determined by regression. All the preceding applies to similar logistic' models — that is, to growth models adapted for inactivation like those described by Geeraerd et al. (2000). Although they can describe sigmoid survival curves, they are not unique (see Sigmoid and Other Kinds of Semilogarithmic Survival Curves). And unless their coefficients can be determined independently, they do not provide any deeper mechanistic insight than any empirical model with the same or smaller number of adjustable parameters.

Concluding Remarks to This Section

Despite widespread belief, first-order kinetics is *not* and cannot be a universal model of microbial inactivation, regardless of whether the mortality or destruction is caused by heat or by any other lethal agent. This model should have been abandoned years ago because ample evidence has shown that, upon scrutiny, it fails almost every criterion of validity. True, some experimental semilogarithmic isothermal survival curves *do appear linear*, especially when the number of data points is small and their scatter hides their true curvature. However, even if some survival curves can be judged as being linear by statistical criteria, this does not validate the basic assumptions on which the model is based.

No doubt the first-order kinetic is the simplest inactivation model. But 'simple' and 'fundamental' are not synonymous terms and should not be used interchangeably. Survival curves are encountered in other fields, notably epidemiology and actuary. In neither discipline is mortality treated as following a first-order kinetics or any other "order" for this matter. One can only imagine the response of a public health biostatistician or an insurance company executive if told that people who have a certain cancer have a D value of 2.35 years, for example. As will be demonstrated in the following chapters, the concept of the first-order kinetics and the D and z values that it has produced (see "The 'z' Value and the Arrhenius Equation") can be safely discarded. More realistic concepts and adequate mathematical methods can deal with the class of nonlinear semilogarithmic survival curves, of which the log linear survival curves are just a small and perhaps insignificant subgroup.

There is also sufficient computing power today to deal with nonlinear survival patterns and readily available commercial software to do the required calculations. John Maynard Keynes said, "The difficulty is not in the new idea, but in escaping the old ones." The same applies to other rate models. One can always write a differential rate equation or set of equations based on dynamic population balance and other considerations that will *describe* the response of a given organism when exposed to a lethal agent. However, unless the parameters of the model can be determined *independently* so that it can predict the organism's response, it will fall short of expectation and its validity will remain uncertain.

The Survival Curve as a Cumulative Form of the Heat Distribution Resistances

A survival curve, by definition, is the cumulative form of a temporal distribution of mortality or destruction events. This is how it is treated in most disciplines, including actuary, epidemiology, and certain fields of material science and engineering. Although, this is not the accepted view in food microbiology, (see the first section), the idea is not new to the field and has been introduced repeatedly during the years. Recent examples are Casolari (1988), Anderson et al. (1996), Augustin et al. (1998), and Peleg and Cole (1998).

The kinetics–distribution duality emerges from the fact that the local slope of the survival curve has rate units and therefore it can be viewed as the *momentary inactivation rate* (see following). When survival data are presented as a survival ratio, $S(t)$, vs. time relationship, the plot shows what portion of the treated population's members is still alive or, in the case of spores, viable after any given time, t. A plot of the survival curve's slope vs. time, i.e., $dS(t)/dt$ vs. t, would be the frequency or density plot of the particular temporal distribution, which can be presented in the form of a histogram or as a continuous curve. Either way, it can be used to determine the portion of the treated population that will die or be inactivated in any given time interval, e.g., in the first minute, between 2 and 3 min, etc. However, this form of presentation is rather rare in the literature of microbiology and, with few exceptions (Peleg and Cole, 1998), it is customary to report microbial survival data in the cumulative form, with the survival ratio represented by its 10 base logarithm.

Now consider a microbial inactivation process on a time scale that is too short for adaptation and selection. In the absence of injury and subsequent delayed recovery, the survival curve of an individual cell or a single spore, i, is characterized by the time of its death or inactivation, respectively, marked as t_{ci}, as depicted in Figure 1.3. The plot shows that prior to the time t_{ci}, a particular cell or spore, i, is viable and beyond that time, it is dead or inactivated or, mathematically:

$$\text{If } t \le t_{ci} \quad S_i = 1 \text{ and } \log S_i(t) = 0$$

$$(1.9)$$

$$\text{and If } t > t_{ci} \quad S_i = 0 \text{ and } \log S_i(t) = -\infty$$

The figure also shows that if one cell or spore is more resistant than another, its t_c will be longer, meaning that it will remain alive or viable for a longer time. Although what follows will primarily focus on heat inactivation, the reasoning and terminology can be extended to microbial inactivation by other means. This discrete relationship described by Equation 1.9 can be approximated by the Fermi distribution function. (Fermi originally developed this to describe the atomic density as a function of the distance from the atom's center. It was assumed to be uniform and very high inside the nucleus, but dropping to almost zero outside it [Seaborg and Loveland, 1990].) For our purpose, the model, known as Fermi's distribution function, can be written in the form:

$$S_i(t) = \frac{1}{1 + \exp\left(\dfrac{t - t_{ci}}{a}\right)} \tag{1.10}$$

where a is an arbitrary, very small number relative to $t - t_{ci}$.

It is easy to see that when $t < t_{ci}$, $\exp[(t - t_{ci})/a] \ll 1$ and thus $S_i(t) \approx 1$ and $\log S_i(t) \approx 0$. However, when $t > t_{ci}$, $(t - t_{ci})/a$ becomes very large and $\exp[(t - t_{ci})/a]$ much larger still, i.e., $\exp[(t - t_{ci})/a] \gg 1$. Thus, from a time

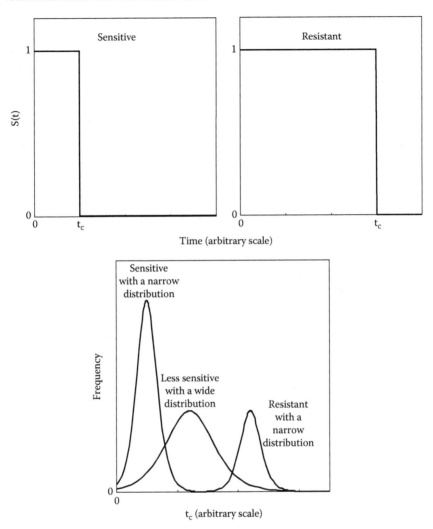

FIGURE 1.3
Schematic view of the survival curves of an individual heat-sensitive and heat-resistant microbial cell or spore.

only very slightly exceeding t_{ci}, $S_i(t) \approx 0$ and log $S_i(t) \approx -\infty$. In fact, Figure 1.3 was plotted using Equation 1.10 as a model, with $a = 0.001$, and it demonstrates that a survival curve thus created is indistinguishable from a true step function for all practical purposes. (The advantage of Fermi's model as an approximation to a step function is that Equation 1.10 is a continuous function and differentiable everywhere.)

If all the members of the microbial population had been totally identical, then when exposed to a perfectly uniform lethal temperature or any other lethal agent for that matter, they would be expected to die or be inactivated at exactly the same time, t_{ci}, as has already been said (provided that the agent's intensity is uniform). In other words, if all the cells or spores in a given inoculum had exactly the same heat resistance, the population's survival curve should be identical to that of an individual cell or spore considerations. In such a hypothetical case, the survival curve of the populations, too, could be described by Equation 1.9 or Equation 1.10, except that $S(t)$ would replace $S_i(t)$.

Because of heat transfer, it is impossible to increase and decrease temperature instantaneously (except adiabatically), even in a capillary. Therefore, the sudden inactivation shown schematically in Figure 1.3 cannot be achieved with any real microbial population, only approximated at best. One should expect, though, that under very fast heating and cooling conditions, the actual survival curves of a group of identical organisms will still closely resemble the step function shown in the figure. The resemblance should be unmistakable — that is, the survival ratio will drop sharply in the neighborhood of the characteristic inactivation time t_{ci}, which would be a measure of the heat resistance or sensitivity of the cells or spores at hand. As before, heat resistance will be then be characterized by a long t_{ci} and heat sensitivity by a short one.

When there is a number of countable individual cells or spores, each with its particular t_{ci}, the discrete survival curve of their assembly when exposed to a uniform lethal temperature will look like the schematic plot shown in Figure 1.4. Expressed mathematically:

$$S(t) = \Sigma \, S_i(t) \, \Delta\phi_i \qquad (1.11)$$

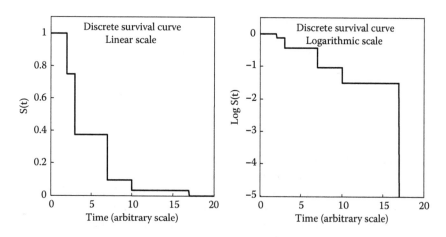

FIGURE 1.4
Schematic view of the survival curve of a microbial population with a discrete spectrum of heat resistances.

where $\Delta\phi_i$ is the fraction of the assembly's members sharing the same heat resistance, i.e., having the same inactivation time, t_{ci}, and $\Sigma\Delta\phi_i = 1$.

If the assembly is a microbial population sufficiently large so that the distribution of its members' t_{ci}s can be characterized by a continuous function, the survival curve could be described by:

$$S(t) = \int_0^t \frac{d\phi}{1+\exp\left[\dfrac{t-t_c(\phi)}{a}\right]} \qquad (1.12)$$

where $t_c(\phi)$ is the inactivation time that corresponds to the particular value of ϕ.

In practice, it is more convenient to use a known parametric distribution function as a survival model instead of Equation 1.11. Or alternatively, one can use an empirical mathematical term to express the $S(t)$ vs. time relationship, depending on the experimental survival data at hand. However, regardless of whether the chosen mathematical model is discrete or continuous, based on a known distribution function, or purely empirical, the shape of the survival curve will be always determined by the characteristics of the temporal distribution of the mortality or inactivation events. As will be shown later, the steepness of a monotonic survival curve, its concavity direction, and whether it shows a "flat shoulder" or not are all direct manifestations of the distribution's type and of the relative magnitudes of the distribution's parameters (i.e., its mean, mode, variance, and skewness).

When a microbial population is truly a mixture of two real subpopulations, each with a unimodal (bell-shaped) distribution of inactivation times with a different mean, mode, variance, and coefficient of skewness, its survival curve can have a discernible inflection point, depending on the distance between the modes and the magnitudes of the two distributions' variances. (Although this is outside the scope of this book, it is worth mentioning that a mixed distribution can appear indistinguishable from an asymmetric and sometimes even a symmetric unimodal distribution. Whether a mixed distribution will appear uni-, bi-, or multimodal will depend on the relative "weight" of the components and magnitudes of their respective modes and standard deviations [Everitt and Hand, 1981].)

Although the observation of a prominent concavity direction inversion is probably a sign that the population might be a mixture, at least as far as resistance to the lethal agent is concerned, the opposite is not true. Theoretically at least, a mixed microbial population can have a monotonic survival curve, depending on the characteristics of the subpopulations' inactivation time distributions. With a few extreme exceptions, however, whether a given microbial population is really a mixture or not cannot be decided on the basis of the survival curve's shape alone. If the issue is of practical significance, it should be resolved by independent tests in which the existence of the presumed subpopulations could be proven by other means.

All the preceding discussion is based on the ubiquitous observation that individual cells and spores are inactivated at different times, even under close to true isothermal conditions. There are several theories as to why this happens and different explanations of the links between the various modes by which heat can destroy a living cell or inactivate a bacterial spore. Still, because the survival curve of an organism in a specified temperature and medium is largely reproducible, one can surmise that the underlying distribution of inactivation times is a reflection of the organism's spectrum of heat resistances or sensitivities under the given conditions.

Thus, any change in the treatment's temperature or in the medium's properties (notably its pH) that affects the survival curve's shape can be interpreted *and quantified* in terms of corresponding changes in the underlying spectrum of the organism's resistances. As demonstrated in Figure 1.5, sensitizing a microorganism would typically result in a shift of its inactivation time's spectrum's mode to the left and narrowing of its span (Peleg and Cole, 1998, Peleg, 2000b). Potentially, sensitizing could also be manifested in a reversal of inactivation time's distribution's skewness direction.

In the explanation offered by the first-order kinetic theory, any change in the process's lethality must be reflected, universally, in a change of a single survival parameter — namely, the exponential rate constant. However, when it comes to real microbial populations, one can expect that *all three* distributions' characteristics — namely, the resistance spectrum's mean or mode, variance, and coefficient of skewness — might be affected and, at least theoretically, need not vary in unison. (It is very unlikely that parameters based on the higher moments of the inactivation times' distribution, like kurtosis, would be ever needed to characterize an organism's heat resistance. Still, the theoretical possibility exists and it should be kept in mind.) Translation of a microbial survival curve obtained *under nonisothermal conditions* into an underlying distribution parameter is totally unnecessary and if done may yield misleading results, as will be shown in the next chapter.

Also, the distribution of the times at which mortality or inactivation events occur determines an observed survival pattern — *not* a distribution of underlying exponential inactivation rates, as some assume. This is because the time it takes to inactivate a given number of cells or spores has clear physical reality and can be measured directly. In contrast, the supposedly "fundamental" but unobservable first-order kinetics inactivation rate is an abstract construction. If someone decided to determine such

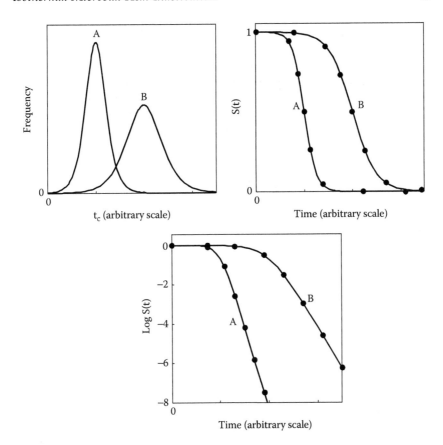

FIGURE 1.5
Schematic view of the survival curves of a microbial population with a continuous spectrum
of heat resistances. A = sensitive or sensitized, B = resistant.

a rate or rates, he or she would always need to obtain counts taken at
different times first. In other words, the fundamental source of information about inactivation kinetics will always be counts and times or vice
versa, both measurable quantities. The fundamental source cannot be
rates, which must be calculated from counts vs. time data, because the
rates themselves cannot be measured directly.

The Weibull Distribution

A microbial inactivation process can be viewed as a failure phenomenon:
the failure of an organism or spore to resist the harsh conditions beyond
a certain time (van Boekel, 2002). Failure phenomena, where the fraction of
survivors progressively decreases, are encountered in a variety of unrelated

fields (Abernethy, 1996). Particulate disintegration during grinding and the serviceable life of machine parts are two familiar examples. Despite the very different physical nature of the processes and their time scales, many can be characterized by the same mathematical model, commonly known as the Weibull distribution function. It is a variant of the exponential distribution and thus has an old ancestry. It was first formulated for size reduction in 1933 in a form now known as the Rosin–Rammler distribution (Schubert et al., 1984). In our context, this could be written as:

$$\frac{dS(t)}{dt} = b'nt^{n-1}\exp(-b't^{n})$$

(1.13)

or

$$S(t) = \exp(-b't^{n})$$

(1.14)

and

$$\log_{e}S(t) = -b't^{n}$$

(1.15)

or

$$\log_{10}S(t) = -bt^{n}$$

(1.16)

where $b = b'/\log_{e}10$, the form that we will use in all subsequent discussions of the model's use.

The more familiar version of the distribution, as its name indicates, was proposed by Weibull in 1936 (Abernethy, 1996). It is now the accepted form by the majority of statistical textbooks and commercial mathematical and statistical software, but not by all. In the context of microbial survival curves, the distribution can be written as:

$$S(t) = \exp\left[-\left(\frac{t}{\beta}\right)^{\alpha}\right]$$

(1.17)

or

$$\log_{e}S(t) = -\left(\frac{t}{\beta}\right)^{\alpha}$$

(1.18)

where α is known as the shape factor and β as the location factor.

Comparison of Equation 1.4 to Equation 1.15 and Equation 1.17 to Equation 1.18 shows that $n = \alpha$ and $b' = 1/\beta^{\alpha}$ and, thus, that the Rosin–Rammler and Weibull distributions are two ways of writing the same distribution function. Yet, if α and β are determined by regression using the transformed $\log_e S(t)$ or $\log_{10} S(t)$ vs. t rather than the original $S(t)$ vs. t data, then whenever the experimental survival data have a scatter, the location and shape factors thus calculated will only be an approximation of the distribution's real parameters. This is because the relative weight given to the different parts of the distribution and, consequently, to the deviations is distorted by the logarithmic transformation. When expressed in the form of Equation 1.16, the isothermal survival model can also be considered as an empirical power law model. And if viewed in this manner, the user is relieved from a strict commitment to the Weibull distribution function and may even consider alternative mathematical expressions that might not be as conveniently transformed into a familiar distribution parametric.

According to Equation 1.16, a concave upward semilogarithmic survival curve will be represented by $n < 1$ and a concave downward curve by $n > 1$ (Figure 1.6). A log linear survival curve or what has been traditionally considered a first-order kinetics is just a special case of the model where $n = 1$ (see figure). In such a case, the Weibullian b parameter will be the same as k, the (base 10) logarithmic rate constant in Equation 1.4 and Equation 1.5. Thus, all the methods to calculate sterility based on the Weibullian or power law model, where $n \neq 1$, will also be applicable to the linear case, i.e., where $n = 1$, but not vice versa.

The constant b in Equation 1.16 can be considered as a nonlinear rate parameter, which primarily reflects the overall steepness of the isothermal survival curve when the power n is fixed. Whenever applicable and properly determined, the Weibullian model's parameters b and n, or α and β, can be translated into the underlying distribution characteristics, i.e., its mode, mean, variance, and coefficient of skewness, using the formulas (Patel et al., 1976):

$$\text{The mode, } t_m = \left[\frac{(n-1)}{nb} \right]^{\frac{1}{n}} \tag{1.19}$$

$$\text{The mean, } \mu = \frac{\Gamma\left(\dfrac{n+1}{n} \right)}{b} \tag{1.20}$$

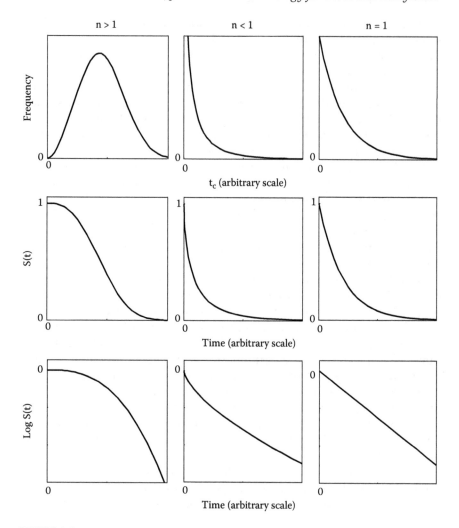

FIGURE 1.6
Weibullian survival curves with different shape factors. Notice that the log linear inactivation model is just a special case of the Weibullian model where $n = 1$.

$$\text{The variance, } \sigma^2 = \frac{\Gamma\!\left(\dfrac{n+2}{n}\right) - \left[\left(\dfrac{n+1}{n}\right)\right]^2}{b} \tag{1.21}$$

$$\text{The coefficient of skewness, } v: \; v = \frac{\mu_3}{\mu_2^{3/2}} \tag{1.22}$$

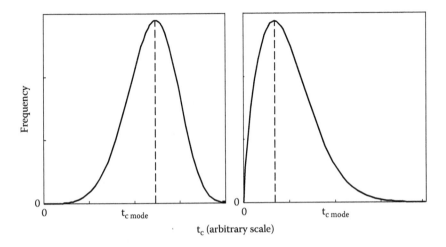

FIGURE 1.7
A simulated demonstration of how the skewness of the Weibullian heat resistance distribution can change its direction when the temperature increases or decreases.

where Γ is the gamma function, $\mu_3 = \Gamma(1 + 2/n)/b$, and $\mu_2 = \Gamma(1 + 2/n)/b^{2/n}$.

The great advantage of the Weibullian or power law model is its mathematical simplicity and great flexibility. It can describe, as already mentioned, linear survival curves and monotonic curves having upward or downward concavity — that is, distributions — with and without a mode ($n > 1$ and $n < 1$, respectively). In the first case ($n > 1$), the skewness can be to the right ($v > 0$) or the left ($v < 0$). Thus, the same model can be used to describe the survival curves of organisms that, at different temperatures, change their concavity direction (see Figure 1.4).

In terms of the underlying spectrum of resistances, the distribution can have not only a different mode and standard deviation, but also an inversion of its skewness direction (Figure 1.7). The Weibullian model can also describe survival curves with a notable shoulder (see following). Not surprisingly, a growing number of recent publications have reported the good fit of the Weibullian model to isothermal microbial survival curves. Notable among them is van Boekel (2002), where more than 100 such survival curves are reported.

Examples of the Weibullian–power law model's fit to the survival curves of *Salmonella* cells and *C. botulinum* spores are shown in Figure 1.8. The model, as will be shown later, is just as useful to describe microbial survival curves when the lethal agent is nonthermal — a chemical antimicrobial, for example. This is because the reasoning for the Weibullian model introduction is essentially the same as that for any lethal agent whose intensity remains uniform and constant in the treated medium

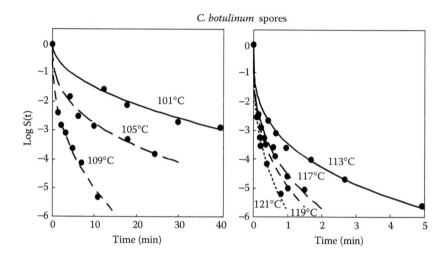

FIGURE 1.8

Demonstration of the fit of the Weibullian survival model (Equation 1.16) to isothermal inactivation data of *C. botulinum* spores. (From Campanella, O.H. and Peleg, M., 2001, *J. Sci. Food Agric.*, 81, 1069–1076. With permission.)

throughout the process duration (Peleg and Cole, 1998). For more on this point, see Chapter 4 and Chapter 5.

Although the fit of the Weibullian or power law model to a large number of survival curves is not at all surprising; there is no compelling reason to assume that it should always fit. In fact, other inactivation patterns are not uncommon. Examples are curve inhibiting, an extreme degree of tailing, an indication of the existence of survivors even after a long treatment, and sigmoid survival curves (see following). However, even if the Weibullian model is found perfectly suitable for a given organism in a certain medium, this does not establish the model's uniqueness. Probably, other models would have a comparable and perhaps even better fit to the same data, although they might not be as convenient mathematically and the meaning of their parameters might be less intuitively clear. For more on this point, see "Residual Survival (Strong 'Tailing')" and "Can an Absolute Thermal Death Time Exist?"

Interpretation of the Concavity Direction

Although many concave upward and concave downward survival curves can be described by the Weibullian or power law model (Equation 1.16) that has the same mathematical construction, their concavity direction is a manifestation of qualitatively different inactivation patterns. Upward concavity ($n < 1$), or tailing, indicates that sensitive members of the population perish rapidly. As the destruction process proceeds, it leaves behind progressively sturdier survivors. In other words, *tailing* means that it takes longer and longer to reduce the same fraction of the remaining survivors as the process continues.

In contrast, downward concavity ($n > 1$) means that it takes a progressively shorter time to destroy the same fraction of survivors as the survival ratio decreases. This is probably an indication that the continued exposure results in accumulated damage, which lowers the heat resistance of the surviving cells or spores.

In principle at least, the two modes of inactivation can coexist, in which case the observed degree of curvature in either direction will be determined by the relative weight of their effects. Theoretically, if these two modes are exactly or approximately balanced, the survival curve might appear log linear, which will be interpreted, erroneously in such a case, as evidence of a first-order inactivation kinetics. It is also not inconceivable that the mechanisms that result in weeding of the weak and damage accumulation in the survivors need not operate on the same time scale. Thus, if the first dominates the early inactivation and the second its later stages, or vice versa, the result would be a sigmoid semilogarithmic survival curve, i.e., one showing a concavity inversion (see "Sigmoid and Other Kinds of Semilogarithmic Survival Curves").

The Fermi (Logistic) Distribution Function

Certain survival curves exhibit a "shoulder" — that is, a period of up to several minutes in which no measurable inactivation takes place following the heat application (Figure 1.9). This observation, like the lag time in growth curves (see Chapter 8), has been attributed to a delayed response, in this case to the heat's damaging effects. Because the temperature of any finite volume cannot be raised instantaneously, as already mentioned, a certain lag in the inactivation must be *always* observed, although it might be too short to be noticed. What follows, therefore, only pertains to scenarios in which the observed delay is too long to be explained as an experimental artifact caused by the come-up time. Adhering to the traditional first-order kinetics concept, such curves have been described by the model (Toledo, 1999):

$$\text{If } t \le t_c, \quad \log_e S(t) = 0 \tag{1.23}$$

and

$$\text{If } t > t_c, \quad \log_e S(t) = -k(t - t_c)$$

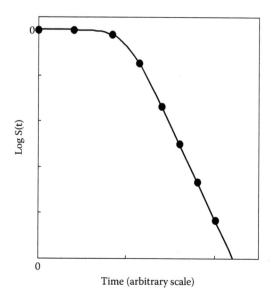

FIGURE 1.9
Schematic view of a bacterial survival curve exhibiting a flat shoulder. Notice that the data points were generated with the Fermi (logistic) distribution function, Equation 1.25, and fitted with Equation 1.23 as a model.

where t_c marks the length of the shoulder.

The explicit assumption here is that the inactivation follows the traditional first-order kinetics, but its onset is at the end of the lag time rather than at time zero. (More elaborate models have been recently developed by the group headed by Baranyi on the assumption that there is a distribution of lag times of inactivation among the heat-treated population analogous to the distribution of the lag times observed in cell division during bacterial growth.) However, an apparent lag followed by log linear decay also will be observed for any organism whose spectrum of heat resistances can be described by the already introduced Fermi distribution function, the mirror image of the logistic distribution function (Peleg, 2000a). Viewed in this way, the lag time has a much simpler explanation. To demonstrate the point, consider that the cumulative form of the organism's distribution inactivation times follows the model:

$$S(t) = \frac{1}{1 + \exp[k(t - t_c)]}$$

(1.24)

or

$$\log_e S(t) = -\log_e\{1 + \exp[k(t - t_c)]\} \qquad (1.25)$$

where the k and t_c are constants.

Notice that when $t \ll t_c$, $\exp[k(t - t_c)] \ll 1$ and thus $S(t) \approx 1$ and $\log_e S(t) \approx 0$. When $t \gg t_c$, $\exp[K(t - t_c)] \gg 1$ and thus $S(t) \approx \exp[-k(t - t_c)]$ and $\log_e S(t) = -k(t - t_c)$. This is exactly what Equation 1.25 describes, except that the transition between the two regions is smooth rather than abrupt as Equation 1.23 implies. In other words, what is called a shoulder is an integral part of the survival process and not a regime having independent reality.

Unlike the Weibull distribution, the Fermi (logistic) distribution is always symmetric around the mode, as can be seen in Figure 1.9. Therefore, its mode and mean are identical by definition, i.e., $\mu = t_m$ and its skewness coefficient (v) is zero. The distribution's variance (σ^2) can be calculated from:

$$\sigma^2 = \frac{\pi^2}{3k^2} \qquad (1.26)$$

The cumulative form of the Fermi distribution has the characteristic sigmoid shape, shown in the figure. The frequency (density) form of the distribution has a symmetric bell shape, reminiscent of that of the normal (Gaussian) distribution, but the two are not identical (Figure 1.10).

Notice that, according to the Fermi or logistic distribution model, the survival curve will always have a prominent flat shoulder whenever the distribution's variance (or standard deviation) is much smaller than the mean, as shown in Figure 1.10. However, this is also true for *any unimodal distribution*, the Weibull distribution with $n > 1$ included. The only difference is that, in such cases, the curve's continuation beyond the apparent lag will not be straight on semilogarithmic coordinates, but rather curved (see figure).

If the concept that an isothermal microbial survival curve's shape is a reflection of an underlying continuous spectrum of heat resistances or sensitivities is accepted, then one should not seek a special mechanistic explanation to what causes the shoulder as if it had independent existence and governed by special separate rules. A more relevant question would be *why the distribution as a whole* has the observed features, i.e., why it has the particular mode, variance, and degree of symmetry. Put differently, the question of why a given microbial population is inactivated at the observed rate after the shoulder is just as pertinent as the question of why a shoulder appears. This is because the shoulder duration and the post-shoulder inactivation rate are just two aspects, or manifestations, of the

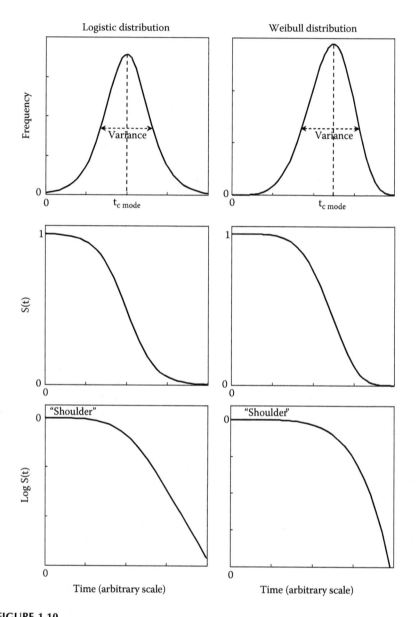

FIGURE 1.10
Simulated survival curves of hypothetical bacteria having a Fermian (logistic) and Weibullian spectrum of heat resistances. Notice that when the resistances distribution variance is much smaller than its mode (or mean), a flat shoulder will emerge regardless of the distribution type. (From Corradini, M.G. and Peleg, M., 2006, The nonlinear kinetics of microbial inactivation and growth, in Brul, S., Zwietering, M., and van Grewen, S. (Eds.). *Modelling Microorganisms in Food.* Woodhead Publishing, Cambridge, U.K. With permission.)

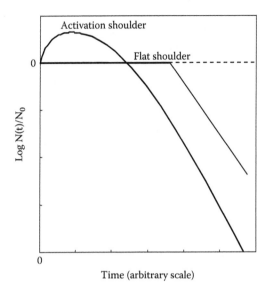

FIGURE 1.11
Schematic view of an isothermal survival curve exhibiting an activation shoulder.

very same destructive mechanisms that operate at the cellular or sporal level. Therefore, the different parts of the survival curves ought to be addressed together and not separately as if they had autonomous existence, unless there is a compelling reason to do otherwise.

The Activation Shoulder

The isothermal inactivation of certain bacterial spores, like those of *B. stearothermophilus*, exhibits what has been called an "activation shoulder" (Lewis and Heppel, 2000). The survival curve in such a case initially *rises* and then, after reaching a peak, starts its descent, as shown in Figure 1.11. A survival ratio larger than one, or $\log_{10}S(t) > 0$, would usually indicate growth. However, this is certainly not the case here simply because the time, on the order of minutes or even less, is too short for the spores to germinate and multiply. Therefore, the rise in the observed number has been attributed to *dormant spores' activation* by their exposure to the high temperature.

If these dormant spores had not been activated, they would not germinate during the recovery stage and would remain uncounted. This explains why, after a short heat treatment even at an otherwise lethal temperature, the number of spores appears to have grown. In fact, what appears to be the result of activation can be partly due to improved dispersion of attached spores. Thus, a lump of several spores that would have produced a single colony, if it had existed and remained intact, can produce several colonies after its breakup, which would also increase the overall count. (Dispersion can be achieved more effectively by sonication, but it is unclear whether this has always been done prior to the heating experiments.)

Because activation and improved dispersion can be responsible for the observed rise in the spores' counts, at least in principle, the term "recoverable spores" would be probably more appropriate than "activated spores" (Corradini and Peleg, 2003b). Here, too, the shape of a survival curve having shoulder alone does not contain enough information to decide the portion of the spores that have been really activated and not just better dispersed. To do that will require an independent test. For example, the same number of counted spores being recorded after sonification would be a strong indication that activation was indeed the major cause of the observed increase in their numbers after the short exposure to heat.

Assuming that any increase in the observed survival curves is due to activation only, several investigators (Shull et al., 1963; Rodriguez et al., 1992; Sapru et al., 1993) have proposed a survival model based on population dynamics. According to these models, which differ in their number of terms and degree of elaboration, the rate at which the number of spores changes is determined by the rate at which dormant spores are activated and by the inactivation rate(s) of the newly activated spores and those already in an active state. Invariably, the inactivation of the spores of either type had been assumed to follow a first-order kinetics with the same or different rate constant. Provided that improved recovery is not an issue, these models capture what actually happens qualitatively. Thus, their development and presentation should be considered an important step in the right direction.

Yet, whether the parameters of the proposed population dynamic model(s) can be calculated reliably from the survival curves without additional information is a debatable issue. Because there is no convincing evidence that sporal inactivation must follow a first-order kinetics (see the first section in this chapter) and the activation kinetics has yet to be independently confirmed, the main value of the previously mentioned models is that they provide a conceptual framework to the development of new and truly predictive models of spores' simultaneous activation and inactivation.

Survival curves with an activation shoulder can be described by empirical models that do not require the assumption of any particular kinetics order. For example, a model can be constructed by combining a general

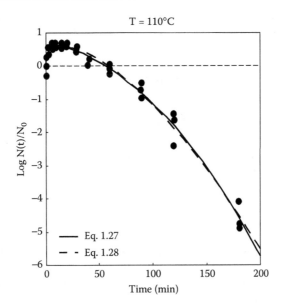

FIGURE 1.12

An experimental isothermal survival curve of *B. stearothermophilus* spores exhibiting an activation shoulder, fitted with Equation 1.27 (solid line) and Equation 1.28 (dashed line) as models. Notice that neither model implies that the inactivation of the viable and activated spores is a process that follows the first-order kinetics. The original data are from Sapru, V. et al., 1993, *J. Food Sci.*, 58, 223–228. (From Corradini, M.G. and Peleg, M., 2006, The nonlinear kinetics of microbial inactivation and growth, in Brul, S., Zwietering, M., and van Grewen, S. (Eds.). *Modelling Microorganisms in Food.* Woodhead Publishing, Cambridge, U.K. With permission.)

growth term of the $t/(k_1 + k_2 t)$ kind with a log logistic decay term of the kind $1 - \log_e[1 + \exp(kt)]^m$:

$$\log_{10} S(t) = \frac{t\{1 - \log_e[1 + \exp(kt)]^m\}}{k_1 + k_2 t} \qquad (1.27)$$

where the three k's and the m are adjustable parameters.

The fit of this model to activation/inactivation data of *B. stearothermophilus* reported by Sapru et al. (1993) is shown in Figure 1.12. Although the fit is better than that of the population dynamic model based on the first-order kinetics assumption, despite having the same number of adjustable parameters (four), it is by no means unique. In fact, the very same data also can be described by the "double Weibullian" model that has been proposed by van Boekel (2003):

$$\log_{10} S(t) = b_1 t^{n1} - b_2 t^{n2} \qquad (1.28)$$

where the b's and n's are characteristic constants. It, too, has four adjustable parameters and a similar or even slightly better fit, as shown in Figure 1.12. (Notice that the first term, bt^n, is positive and the second, b_2t^{n2}, negative. Thus, as long as $b_1t^{n1} > b_2t^{n2}$, the logarithmic survival ratio, $\log_{10}S(t)$, will be a positive number.)

Estimation of the Number of Recoverable Spores

The presence of dormant and possibility of aggregated spores as well, raises the question of whether the total size of the sporal population can be estimated even if not accurately determined. Because the spores of both types are initially unaccounted for and the applied heat has the dual and contradictory effect of making them countable while reducing their numbers at the same time, it is difficult to say how many of them there were at the start. Because we cannot distinguish between the activated and dispersed spores on the basis of the survival curve's shape alone, microscopic methods might be very helpful here. However, although it would be fairly easy to identify clumps in a micrograph, telling whether a given spore, alone or in an aggregate, is active or dormant might be another matter. This problem will remain regardless of whether the aggregates are intact or have been broken down by sonification.

Early modelers of the activation shoulder phenomenon estimated the initial inoculum size extrapolating the linear part of the semilogarithmic survival curve to time zero (Figure 1.13). The apparent rationale was that, because inactivation must follow the first-order kinetics, the initial part of the curves is a sort of a deviation from the "true" survival curve, which must be a straight line when plotted on semilogarithmic coordinates. The alternative view that we offer is based on the following: we do not really know how many spores become countable as a result of activation or improved dispersion and thus we will treat them as one group. Because, without additional information we cannot tell what their initial state was, we will consider the experimental counts as representing the momentary number of the total *recoverable spores*. Now, theoretically, there can be two main situations, depending on the relative rates of the spores' activation or dispersion and their inactivation:

- The spores had become countable before significant inactivation took place.
- Considerable inactivation had already taken place before all the spores could be counted.

In the scenario in the first bullet, the survival curve would be expected to show a plateau region or a wide, flat peak (Figure 1.14). The corresponding count could then be considered as an estimate of the total

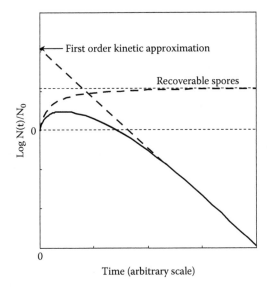

FIGURE 1.13
Schematic view of the traditional way to estimate the number of recoverable spores (viable and activated) from an isothermal survival curve exhibiting an activation shoulder, based on the first-order inactivation kinetics. (From Corradini, M.G. and Peleg, M., 2003, *Food Res. Int.*, 36, 1007–1013. With permission, courtesy of Elsevier Ltd.)

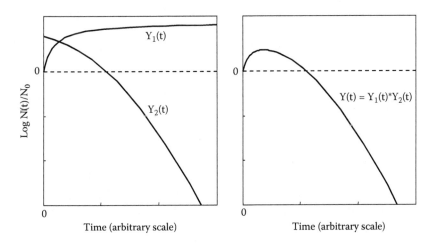

FIGURE 1.14
Schematic view of a proposed way to interpret the shapes of isothermal survival curves exhibiting an activation shoulder. (From Corradini, M.G. and Peleg, M., 2003, *Food Res. Int.*, 36, 1007–1013. With permission, courtesy of Elsevier Ltd.)

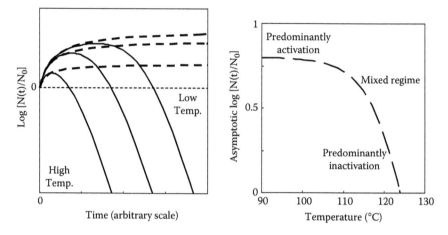

FIGURE 1.15

Schematic view of a proposed way to estimate the number of recoverable spores from isothermal survival curves exhibiting an activation shoulder. (From Corradini, M.G. and Peleg, M., 2003, *Food Res. Int.*, 36, 1007–1013. With permission, courtesy of Elsevier Ltd.)

number of recoverable spores in the treated population. In the second case, a plateau will not be observed because the decline precedes the complete transition of the spores from an uncountable to a countable form.

Suppose now that the test is repeated and the survival curves are recorded at *progressively decreasing temperatures* (Figure 1.15). One can expect that, if at lower temperatures the spores' inactivation is delayed or becomes less intense, there will be time for more spores to become countable, as shown in the figure. In principle at least, the temperature can reach a level low enough at which activation and dispersion might continue, but effective inactivation practically ceases. In such a hypothetical situation, the survival curve will have an asymptote, which will represent the total number of recoverable spores. Because we are unsure whether all this is true, we can only estimate this asymptote level by using the following procedure:

1. We fit the prepeak data with a growth model that has an asymptote, e.g., $\log[(N(t)/N_0)] = k_1 t/(k_2 + t)$ or $\log[(N(t)/N_0)] = a[1 - \exp(bt)]$, in which case the asymptotic level will be k_1 or a, respectively.

2. We plot the estimated asymptotic value of $\log[(N(t)/N_0)]$ vs. temperature and expect to obtain a relationship of the kind shown in Figure 1.15 (right).

3. The flat part of the curve corresponds to the lower temperature range and may provide a rough estimate of the number of recoverable spores.

TABLE 1.1

Recoverable Spore Estimates of *B. stearothermophilus* Spores

	Asymptotic ratio, $\log_{10}(N_{asymt}/N_0)$	
$T°C$	From $\log_{10}[(N(t)/N_0)] = k_1/(k_2 + t)$	From $\log_{10}[(N(t)/N_0)] = a[1 - \exp(-bt)]$
105	0.87	0.71
110	0.73	0.64
115	0.87	0.66
120	0.78	0.68

Source: Calculated by two models from Corradini and Peleg (2003b) using data from Sapru, V. et al., 1993, *J. Food Sci.*, 58, 223–228.

This estimate (see Figure 1.15) will be always much *lower* than the one reached by extrapolating the survival curves to time zero. The procedure can be repeated with different models and their results compared. This is demonstrated in the Table 1.1, based on data published by Sapru et al. (1993) that are shown in Figure 1.16.

The table indicates that the *B. stearothermophilus* spores become almost fully recoverable at all four reported temperatures. The asymptotic ratio of recoverable to initially counted spores was on the order of 5:7 according to the first model and on the order of 4:5 according to the second model. In light of the crudeness of the described estimation method and the considerable scatter in the original activation and inactivation data, the two estimates seem to be in reasonable agreement.

Although the procedure is based on a yet unproven set of assumptions, the hypothesis is testable. If the concept is correct, then the ratio between recoverable and initial number of *B. stearothermophilus* spores, heated in the same medium and prepared in the manner described by Sapru et al. (1993), would not be much higher than that indicated by the table regardless of whether the temperature would be below 105°C or higher than 120°C. At temperatures above 120°C, the ratio can decrease substantially to below the 4:7 level, but never rise to a level well above it. It would be left to future research to determine whether the estimates calculated in this manner are indeed close to the actual number of recoverable spores when these are counted by a direct method.

Sigmoid and Other Kinds of Semilogarithmic Survival Curves

Sigmoid Curves

In principle at least, heating can cause rapid elimination of the weaker members of the population, followed by sensitizing the sturdier survivors through damage accumulation. If this happens, then one would expect that the corresponding survival curve will have upward concavity initially, which will be gradually reversed and become downward concavity

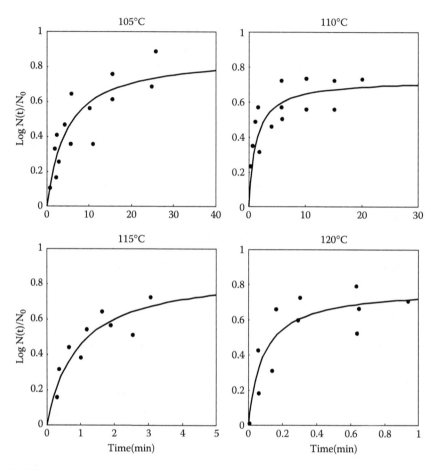

FIGURE 1.16
Demonstration of the procedure described schematically in Figure 1.15 to activation/inacti-
vation data of *B. stearothermophilus* spores. The results are reported in Table 1.1. The original
data are from Sapru, V. et al., 1993, *J. Food Sci.*, 58, 223–228. (From Corradini, M.G. and Peleg,
M., 2003, *Food Res. Int.*, 36, 107–1013. With permission, courtesy of Elsevier Ltd.)

as the heating continues. This type of sigmoid survival curve is shown
schematically in Figure 1.17 (left) and we will call it type A. An alternative
scenario is that accumulated damage lowers the heat resistance of the
weaker members of the population and, once these are eliminated, progressively sturdier survivors remain. This will be reflected in a semilogarithmic survival curve that is initially concave downward (and might
even exhibit a shoulder) and becomes concave upward as the heat process
goes on (Figure 1.17, right). We will refer to this kind of sigmoid curve as
type B.

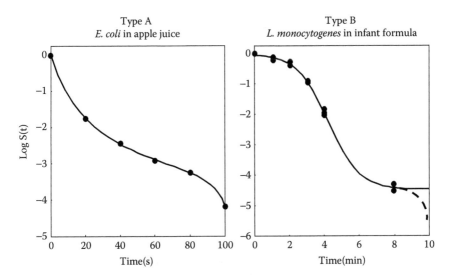

FIGURE 1.17
Examples of experimental isothermal sigmoid survival curves of types A and B fitted with empirical models. The original experimental data are from Vassoni Penna and Morales, 2002, and Ingham and Vljas, 1998, respectively. (From Peleg, M., 2003, *Crit. Rev. Food Sci.*, 43, 645–658. With permission, courtesy of CRC Press.)

One can argue, though, that if the heating had continued for a much longer time, accumulated damage in the sturdy survivors will eventually cause their demise, too; in this case, the survival curve's concavity direction will change once more and becomes downward again. The time for this to happen, however, can be too long for any practical consequences and thus, not surprisingly, reports on its actual occurrence are hard to find. Experimental sigmoid survival curves of type A or B are by far less common than those of the Weibullian kind. Nevertheless, both kinds have been reported in the literature. Examples of such curves are shown in Figure 1.18. They can be described by a variety of empirical models; some have a similar degree of fit as judged by statistical criteria (Peleg, 2003a). Examples are:

For sigmoid curves of type A:

$$\log_{10} S(t) = -\frac{a_1(T)t}{[1+a_2(T)t][a_3(T)-t]} \tag{1.29}$$

or

$$\log_{10} S(t) = -\frac{a_1(T)t}{a_2(T)+t} - \frac{a_3(T)t}{a_4-t} \tag{1.30}$$

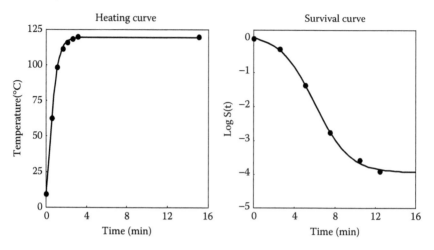

FIGURE 1.18
Example of a sigmoid curve produced by the come-up time. The original data are of *B. sporothermodurans* spores isolated from UHT milk published by Huemer, I.A. et al., 1998, *Int. Dairy J.*, 8, 851–855. (From Peleg, M., 2003, *Crit. Rev. Food Sci.*, 43, 645–658. With permission, courtesy of CRC Press.)

and for sigmoid curves of type B:

$$\log_{10} S(t) = -\frac{a_1(T)t^m}{a_2(T)+t^m} \tag{1.31}$$

or

$$\log_{10} S(t) = -a_1 t^{m_1}[1-\exp(a_2 t^{m_2})] \quad (m_1 < 1) \tag{1.32}$$

where the a's and m's are constants.

The proposal of certain investigators to describe sigmoid survival curves of type B by an inverted version of the Baranyi's and Robert's growth model is mentioned in "The Logistic Models." (The model and its merits as a growth model are discussed in Chapter 8.) Like the empirical models expressed in Equation 1.31 and Equation 1.32, the Baranyi–Roberts model in its original general form has four adjustable parameters. The difference is that it is written as a rate equation and thus, in principle, could be used for isothermal as well as nonisothermal condition. Although those who proposed the model acknowledge the differences in the kinetics of growth and inactivation they still attempt to show existence certain similarities. To me, though, the very rationale of using a growth model remains unclear and the inclusion of a term like $1 - N(t)/N_{max}$ in the model's rate equation is totally out of place. Similarly, the reason for using and the meaning of a term like "maximum specific rate" in the context of microbial inactivation must be better explained before the model is given the status of a mechanistic model.

Sigmoid survival curves can also be viewed as being the cumulative forms of *bimodal* distributions and these in turn as describing mixed populations. However, remember that the mathematical description of a bimodal distribution (representing two populations) may require *five* parameters: two modes (or means), two standard deviations, and one to account for the two populations' relative weight. These cannot be determined unambiguously from scattered experimental survival data using regression techniques. Validation of the mixed distribution's hypothesis will almost certainly require an experimental effort that very few microbiologists will be willing to invest, especially because of the relative rarity of sigmoid inactivation curves. Consequently, in light of the deficiency of the current mechanistic models, the description of sigmoid survival curves by purely empirical models will probably continue to be the best option for the time being, as will be demonstrated in Chapter 3.

Residual Survival (Strong "Tailing")

The Weibullian model, even with $n < 1$, i.e., when the semilogarithmic survival curve has upward concavity (tailing), entails that the number of survivors will decrease indefinitely, even at relatively low inactivation temperatures. However, at least theoretically, it is possible that under such circumstances, a certain number of survivors will remain even after a long exposure. Alternatively, the inactivation rate after a while can become so low that it will be practically undetectable. Such a scenario, if and when it exists, would require a survival model that allows for a finite asymptotic residual. A flexible empirical model that satisfies this requirement is:

$$\log_{10} S(t) = -\frac{at}{b+t} \qquad (1.33)$$

where a is the asymptotic residual survival ratio, i.e., when $t \to \infty$, $\log_{10}S(t) \to -a$, and b the time needed to reach half the asymptotic reduction level (i.e., when $t = b$, $\log_{10}S(t) = -b/2$). In some cases, this model has a fit comparable to that of the Weibullian–power law model (see Figure 1.19). Yet, as shown by Peleg and Penchina (2000), when the two models are used for extrapolation — a practice that is not recommended regardless of the model chosen — they will provide very different estimates of the

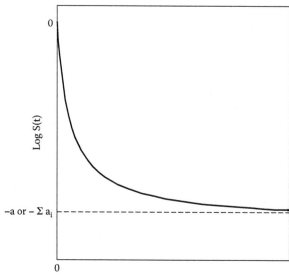

FIGURE 1.19
Schematic view of an isothermal survival curve generated with Equation 1.33 as a model. Notice that this model implies not only extreme tailing, but also a finite number of survivors even after a very long exposure.

survival ratios that will grow with time. All the preceding also pertains to more elaborate versions of the model like:

$$\log_{10} S(t) = -\Sigma \frac{a_i t}{b_i + t} \tag{1.34}$$

in which case the asymptotic logarithmic survival ratio will be $-\Sigma a_i$ (Corradini and Peleg, 2006).

Can an Absolute Thermal Death Time Exist?

Certain concave *downward* semilogarithmic survival curves that can be described by the Weibullian–power law model with $n > 1$ can also be described by the empirical model (Peleg, 2000a):

$$\log_{10} S(t) = -\frac{at}{b - t} \tag{1.35}$$

where a and b are constants.

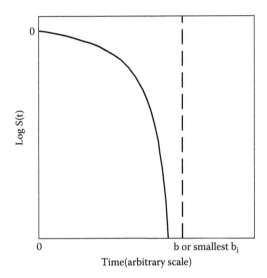

FIGURE 1.20
Schematic view of a hypothetical isothermal survival curve created with Equation 1.35 as a model that indicates the potential existence of an absolute thermal death time. (Adapted from Peleg, M., 2000, *Food Res. Int.*, 33, 531–538; and Corradini, M.G. and Peleg, M., 2006, in Brul, S., Zwietering, M., and van Grewen, S. (Eds.). *Modelling Microorganisms in Food*. Woodhead Publishing, Cambridge, U.K. With permission.)

This model (Figure 1.20) implies that as $t \to b$, $\log_{10}S(t) \to -\infty$, i.e., that no survivors will ever be found at times longer than b. This is, of course, a purely formalistic argument and it is yet to be proven that it holds for even a single microorganism or spore type. However, the hypothesis that an *absolute* thermal death time exists is not as outrageous as it might sound. If it really exists, then it could replace the traditional one defined by the arbitrary $12D$ requirement, at least for some spores. Recall that a semilogarithmic survival curve's having a downward concavity can be interpreted as evidence that damage accumulation progressively reduces the ability of the exposed cells or spores to maintain their vitality. If so, then it will not be unreasonable to expect that, at a certain point, none of the survivors would be able to resist the onslaught and absolute sterility will be achieved.

Proving the absence of survivors is much more difficult than proving their presence because of the inherent asymmetry between establishing and refuting existence. ("The absence of proof is not a proof of absence.") Thus, although finding survivors after a time exceeding b will invalidate the model, not finding any survivors can always be attributed to the failure of the experimental method to detect them. Still, the existence of an absolute thermal death time is a *testable hypothesis*, at least in principle (Peleg, 2001a). Persistent inability to find survivors even after long incubation of a large number of samples may be indicative that the thermal process has produced at least practical sterility. All the preceding points also pertain to more elaborate versions of the model:

$$\log_{10} S(t) = -\sum \frac{a_i t}{b_i - t} \qquad (1.36)$$

except that the time to reach practical sterility will be determined by the shortest b_i (Corradini and Peleg, 2006).

The same rationale applies to certain models that depict sigmoid semilogarithmic survival curves of type A (see Figure 1.17). According to Equation 1.29 or Equation 1.30, which can describe such curves, when $t \rightarrow a_3$, or $t \rightarrow a_4$, respectively, $\log_{10} S(t) \rightarrow -\infty$. Therefore, the parameters a_3 or a_4 of these models, like b in Equation 1.35, can serve as an estimate of the absolute thermal death time, provided that enough data points are in the fast inactivation region to estimate their values. However, even if data points in the region where the survival curve takes a steep dive are sufficient, a careful researcher will still use several alternative models to confirm the estimate.

An example is the two-parameter model $\log_{10} S(t) = a_1 \log[1 - t/a_2]$, where a_1 and a_2 are constants. According to this model, when $t \rightarrow a_2$, $\log_{10} S(t) = -\infty$, i.e., a_2 is the alternative marker of the absolute thermal death time. For an organism showing type A sigmoidal survival curve, the two alternative models (Equation 1.29 and Equation 1.30) can be used simultaneously. However, and again, that a concave downward survival might yield an absolute thermal death time is merely a hypothesis at this time. In principle at least, one can imagine the opposite scenario, in which a finite number of (undetectable) survivors will always remain, as required by the logistic type of models that have the term $1 - N(t)/N_{max}$.

Without experimental evidence, even the potential existence of an absolute thermal death time should be considered with caution, as well as its estimated value if solely determined by a mathematical model of the kind described. However, once the existence of an absolute thermal death time is confirmed one day, perhaps by a biochemical method or molecular assay, then the preceding data analysis procedure could assist in the choice of a *safety factor*, i.e., by how much the thermal process should be extended to make sure that absolute sterility had been indeed accomplished.

Secondary Models

A primary model describes the isothermal survival curves — i.e., the experimental inactivation data. The secondary model describes the temperature dependence of the primary model's parameters. (When the primary model's parameters depend on other factors that affect the isothermal survival pattern, like pH, salt concentration, etc., the model should be constructed to account for their influence too.) The same applies to nonthermal microbial inactivation, in which case the temperature is replaced by pressure, chemical agent's concentration, etc.

The "z" Value and the Arrhenius Equation

In almost every textbook of general and food microbiology, where thermal inactivation is discussed, one can find a figure depicting, schematically in the vast majority of cases, a log linear plot of the D value vs. temperature, i.e., governed by the relationship:

$$\log_{10}\left(\frac{D_1}{D_2}\right) = -\frac{T_1 - T_2}{z} \tag{1.37}$$

where z is the slope's reciprocal in temperature units.

The z value according to this model is the temperature difference needed to reduce (or increase) the D value by a factor of 10. Equation 1.37 is usually presented in a form that contains a reference temperature, T_{ref}, and a corresponding D value, D_{ref}, i.e.,

$$\log_{10}\left(\frac{D}{D_{ref}}\right) = -\frac{T - T_{ref}}{z} \tag{1.38}$$

which is frequently used to calculate the efficacy of heat processing of foods and pharmaceuticals. In canning of low-acid foods, the reference temperature has been traditionally 121.1°C (250°F). The efficacy of thermal sterilization processes has been determined in terms of the equivalent number of minutes at this temperature that will produce the desired mortality or inactivation level (see Chapter 2).

The utility of this log linear model immediately comes into question whenever the original isothermal semilogarithmic survival curves are curvilinear, in which case the D values are ill defined and their meaning unclear. However, suppose that there were microbial cells and spores whose isothermal inactivation indeed follows the first-order kinetics. In such a case, the survival curves must be log linear at all relevant temperatures and the temperature dependence of the resulting D values ought to be log linear too. Otherwise, the z value would lose its meaning and its magnitude would vary with the number of data points used in its calculation. Once the z value becomes suspect, the validity of any sterility calculation method based on it would become highly questionable too.

For more on this model's limitations and those of its traditional alternative, see "Other Empirical Models" and Chapter 2.

The second most popular secondary model of thermal inactivation — almost on par with the log linear model that has produced the z value is the already mentioned Arrhenius equation:

$$\log_e\left(\frac{k_1}{k_2}\right)=\frac{E}{R}\left(\frac{1}{T_1}-\frac{1}{T_2}\right) \tag{1.39}$$

where
k_1 and k_2 are the first-order kinetic rate constants at the absolute temperatures T_1 and T_2 (in °K), respectively
E is the energy of activation
R is the universal gas constant in the appropriate units of temperature and energy

Some of the theoretical deficiencies of the Arrhenius model when applied to microbial inactivation and other complex biological processes have already been discussed in the introduction to this book. Suffice it to say here that the validity of the Arrhenius model, like that of the log linear model that has produced the z value, becomes highly questionable, on of basic principles, whenever the isothermal survival curves from which the model is derived are not log linear. In *all* such cases, neither the D value nor the rate constant, k, can be determined unambiguously and thus the meaning of the secondary model's equation becomes unclear too.

However, even if the first-order inactivation kinetic could be established and the log linear model or Arrhenius equations considered empirical models applicable in a limited temperature range only, either would still be a poor choice as a secondary model. As already mentioned, the main reason is that because D and k are determined from a semilogarithmic relationship; the second logarithmic conversion that produces the terms log D or log k in the respective models is superfluous. It is very unlikely that any exponential rate parameter will vary by several orders of magnitude within the practical range of processing temperatures and thus there is no real reason for the double logarithmic transformation. Moreover, passing a straight line through data plotted on logarithmic coordinates carries the risk that an error at the low temperatures or even a slight deviation from linearity will result in significant error at the high temperatures, where most of the inactivation actually takes place. This is demonstrated in Figure 1.21, in a somewhat exaggerated manner. Needless to say, the predictive ability of kinetic models based on such data is highly questionable.

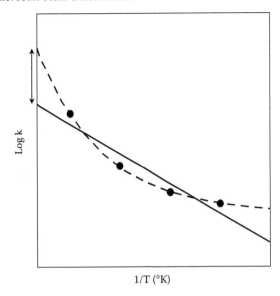

FIGURE 1.21
A schematic example of experimental Arrhenius plots treated as linear relationships. Notice that the error is particularly significant at the high temperatures (low $1/T$ values) because of the logarithmic scale of the ordinate.

That the Arrhenius model has survived for so long in microbiology (and other fields such as biochemistry and food chemistry) despite its obvious theoretical shortcomings and practical deficiencies is really surprising. The same can be said on the assumed log linearity of the D vs. T relationship, which is more or less limited to microbiology, although it has been extended to enzyme inactivation. Thus, one can find tabulated D and z values of enzymes together with those of microorganisms in the literature. (Both are usually unaccompanied by experimental data that would convincingly demonstrate that the temperature effect can be faithfully described by the log linear model.) It is also peculiar that many publications and a leading Website list D and z values and those of the energy of activation and k at a reference temperature side by side, although the two models that have produced them are mutually exclusive.

The relatively short-lived WLF equation that had been proposed as an alternative secondary model in thermobacteriology obviously shares the same problems that its two predecessors have. Adapted from polymer science, the model is also based on the assumption that isothermal survival curves are universally log linear and that the inactivation rate constant k is uniquely defined. Like the log linear and Arrhenius models, the WLF equation also requires double logarithmic transformation of the exponential rate constant, which would be unnecessary even if k could be determined unambiguously.

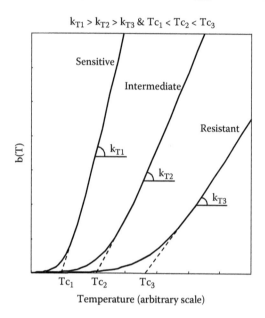

$$k_{T1} > k_{T2} > k_{T3} \ \& \ Tc_1 < Tc_2 < Tc_3$$

FIGURE 1.22

Schematic view of the Weibullian rate parameter, $b(T)$, temperature dependence of a resistant, intermediate, and sensitive organism when described by the log logistic model Equation 1.40. (Courtesy of Dr. Maria G. Corradini.)

The Log Logistic Model

Consider that an organism's isothermal survival curves all follow the Weibullian–power law model (Equation 1.6). Regardless of whether the power n is temperature dependent, $n(T)$, or fixed, in which case it can be bigger than, smaller than, or equal to one, the steepness of the overall survival curve or its scale will be primarily (but not exclusively) determined by the parameter $b(T)$. Thus, if $b(T) = 0$, no activation will take place regardless of the magnitude of n or $n(T)$. Measurable inactivation only starts when the temperature reaches a lethal level and a valid secondary model must account for this fact.

A convenient model that marks the onset of intensive inactivation is the log logistic model, which we have already encountered in this chapter, albeit in relation to a certain primary model. In the present context (Figure 1.22), the log logistic model assumes the form (Peleg et al., 2002):

$$b(T) = \log_e\{1 + \exp[k(T - T_c)]\} \tag{1.40}$$

where T_c marks the temperature level at which the inactivation starts to accelerate and k the rate (with respect to temperature) at which $b(T)$ climbs as the temperature rises to a level well above T_c.

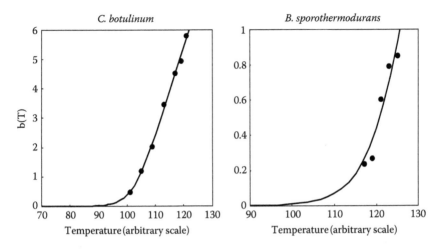

FIGURE 1.23
Demonstration of the fit of the log logistic secondary model (Equation 1.40) to data of *C. botulinum* and *B. sporothermodurans* spores. The original experimental isothermal data from which the shown curves were derived were reported by Anderson, W.A. et al., 1996, *J. Appl. Bacteriol.*, 80, 283–290; and Periago, P.M. et al., 2004, *Int. J. Food Microbiol.*, 95, 205–218, respectively.

Again, it is easy to see that when $T << T_c$, exp $[k(T - T_c)] << 1$ and $b(T) \approx 0$. However, when $T >> T_c$, exp $(T - T_c) >> 1$ and thus $b(T) \approx \log_e$ exp $[k(T - T_c)] = k(T - T_c)$. In other words, according to this model, $b(T)$ is practically zero at nonlethal temperatures, i.e., when $(T << T_c)$, and increases linearly with temperature at high lethal temperatures after passing a transitional temperature region around T_c, i.e., when $(T >> T_c)$. The fit of the model to experimental data is demonstrated in Figure 1.23. This model would be just as appropriate if the isothermal inactivation had truly followed a first-order kinetics, in which case $k(T)$ would replace $b(T)$ in the model's equation.

Unlike in the traditional secondary models, and because $b(T)$ is not expressed as a logarithmic transform, the temperatures of intensive lethality receive an appropriate weight relative to that of the low temperatures, where hardly any or no inactivation occurs. (Notice that the operation $\log_e\{1 + \exp[]\}$ is not a logarithmic transformation in the same sense as $\log_e[]$.) Also, because the model is solely derived from the *observed* temperature dependence of $b(T)$ and is treated as purely empirical, the unnecessary and unproven assumptions on which the traditional models are based are simply being avoided. The use of Equation 1.40 as a secondary model also eliminates the traditional models' troublesome theoretical implications.

Unlike the Arrhenius equation, the log logistic model does not require that every microbial thermal inactivation process should have a single fixed "energy of activation." Unlike the log linear, Arrhenius, and WLF models, the log logistic model is consistent with the fact that a qualitative difference exists between the effect of heat on microorganisms at high and low temperatures.

In case the continuation of $b(T)$ vs. T beyond T_c is nonlinear, the model can be modified by an added power, m, rendering it (Peleg et al., 2002):

$$b(T) = \log_e\{1 + \exp[k(T - T_c)]\}^m \tag{1.41}$$

If needed, the added power, m, would be most likely greater than one, $m > 1$, i.e., the continuation of $b(T)$ vs. T relationship would have upward concavity. This is because in the lethal regime, the inactivation rate parameter, $b(T)$, is expected to rise *at least linearly* with the temperature, if the other Weibullian parameter, $n(T)$, remains fixed. Nevertheless, a secondary model with three adjustable parameters is much less attractive than one having only two. Fortunately, in all the microbial data that we have examined, Equation 1.40, which is the same as Equation 1.41 with $m = 1$, had an adequate fit — shown in Figure 1.23 and in Chapter 2. There can be experimental data sets in which the log logistic model and the Arrhenius or WLF equation will have a very similar degree of fit as judged by statistical criteria (Peleg et al., 2002; Corradini et al., 2005). Yet, as has been emphasized in this chapter, the three models differ considerably when it comes to the validity of their theoretical foundations and the analogies that they do or do not imply.

A Discrete b(T) vs. T

The considerations leading to the formulation of the log logistic model (Peleg et al., 2003) can be also implemented in the formulation of a discrete model (Corradini et al., 2005). Using the syntax of Mathematica®, the program used to generate most of the figures shown in this book, it will have the form:

$$b(T) = \text{if } [T \le T_c, 0, k(T - T_c)] \tag{1.42}$$

or, explicitly, if $T \le T_c$, $b(T) = 0$; otherwise, $b(T) = k(T - T_c)$.

In Mathematica®, $b(T)$ thus defined is treated as a normal function that can be used as a model for nonlinear regression (Figure 1.24), as well as for other mathematical operations and graphics as shown in the figure. The same applies to the nonlinear version of the model, if needed:

$$b(T) = \text{if } [T \le T_c, 0, k(T - T_c)^m] \tag{1.43}$$

or, explicitly, if $T \le T_c$, $b(T) = 0$; otherwise, $b(T) = k(T - T_c)^{m1}$.

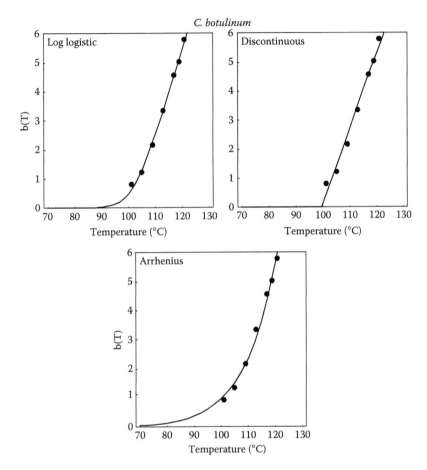

FIGURE 1.24
Demonstration of the use of a discrete secondary model (Equation 1.42) and comparison of its fit from that of the log logistic and Arrhenius models. (From Corradini, M.G. et al., 2005, *J. Food Eng.* 67, 59–69. With permission, courtesy of Elsevier Ltd.)

Other Empirical Models

When the isothermal survival curves are described by models of the kinds discussed in "The Activation Shoulder" and "Sigmoid and Other Kinds of Semilogarithmic Survival Curves," there are no general guidelines as to how to describe the temperature dependence of their coefficients. In the absence of such guidelines, one can choose ad hoc empirical models that best fit the data. Examples of such models can be found in Peleg (2003b) and Peleg et al. (2004). Their usefulness should be judged by how closely they capture the nature of the primary model coefficients' temperature dependence. Because they are totally empirical, they are hard to relate to any specific destructive mechanisms at the cellular level or even to a particular kinetic inactivation mode at the population level, although in principle at least this can be done.

However, being empirical and identified as such, they do not require independent validation because no special meaning is assigned to their parameters from the beginning. The objective of deriving such secondary models, as will be demonstrated in the next chapters, is only to simulate or predict nonisothermal survival patterns. If this is the only purpose, then a mechanistic interpretation of the model's parameters plays no role and therefore is not required. The same can be said about nonthermal inactivation processes in which the secondary models should capture the primary model's parameters' dependence on other factors, such as pressure (Peleg, 2002d), a chemical antimicrobial agent's concentration, etc. (Peleg, 2002c; Corradini and Peleg, 2003a). Such models will be presented and their capabilities demonstrated in Chapter 5 and Chapter 6.

2

Nonisothermal Heat Inactivation

A theory has only the alternatives of being right or wrong. A model has a third possibility: it may be right, but irrelevant.

Jadish Mehra

The Traditional Approach

Commercial thermal processes to destroy microorganisms and/or inactivate spores are rarely, if ever, isothermal. This applies to thermal preservation of foods as well as the sterilization of pharmaceutical products. To be effective, a heat process needs to guarantee that the coldest point in the treated object has been exposed to a sufficiently high temperature for a sufficiently long time to cause the destruction of the targeted organism or spore — regardless of whether the targeted organism is actually present in the product — because the process's aim is to guarantee the products' microbial safety in the worst-case scenario. Also, whatever calculation method is used to assess the heat treatment's requirements, a safety factor is almost always added and other precautions might be implemented for further security. Mandatory storage of sterilized cans in the plant before their release to the market is an example.

Because of heat transfer considerations, reaching the target temperature instantaneously in a conventional heating process is impossible. The come-up and cooling times can be shortened to a very large extent in surface sterilization and in the heat treatment of liquids. Heat exchangers with a large contact area can raise a liquid's temperature within a very short time and a turbulent flow will assure good mixing of the treated product. The same can be said about cooling where the same heat transfer principles apply. Ultrahigh temperature (UHT) equipment accomplishes the rapid heating and cooling by direct injection of live steam into the product and flash evaporation of the water thus added to cool the product almost instantaneously.

Paradoxically, shortening the heat treatment time to a few seconds by raising the product's temperature to a very high temperature (around 135°C, for example) also creates a situation in which the come-up and cooling times can play a significant role in the process lethality. This is because even a very short time between 120 and 135°C can reduce the treated sporal population by a considerable amount.

Another way to raise a product's temperature instantaneously is through the rapid application of ultrahigh pressure of hundreds of megapascals (see Chapter 6). In the resulting adiabatic heating, the temperature will rise all over the affected volume, and heat transfer considerations only apply to heat losses when the pressure vessel is not insulated. This technology, however, is still in the development stage at the time at which this book is written. When it comes to food products thermally processed in their retail containers, like canned solid foods, the heating is not even remotely instantaneous; thus, how to calculate the treatment efficacy correctly is an issue of obvious safety and quality implications.

The F_0 Value and Its Limitations

The starting point of the traditional approach has been to select an organism or spore as the target for elimination. It could be a pathogen actually present in the product that must be inactivated, e.g., *E. coli* 0157:H7, *Staphylococci,* or *Salmonella* in ground meats. However, it can also be a dangerous heat-resistant organism or spore, which *might be* in the product as previously explained. Thus, the spores of *C. botulinum* are usually chosen as the target organism in canned low-acid foods, even when the organism is in most cases not actually present.

The rationale is that if the process is intensive enough to eliminate the most resistant dangerous spores *if they had been present*, it will also be adequate to destroy all other potentially harmful forms of microbial life, which are more heat sensitive. This is by all means a very reasonable and safe strategy. As long as an even more heat-resistant spore (like that of certain bacilli) does not become a threat, the spores of *C. botulinum* will continue to serve as the target of thermal processes and their inactivation kinetics will remain the basis of sterility calculations.

The traditional classic methods to calculate a product's sterility are based on two assumptions:

- The heat inactivation of microorganisms, including that of bacterial spores, always follows first-order kinetics.
- The temperature dependence of the D value is log linear with a slope reciprocal known as the z value.

If the two assumptions are satisfied, one can express the lethality of any nonisothermal heat process in terms of an equivalent time at a reference temperature known to be sufficient to accomplish "commercial sterility." Implementation of this idea has led to the formulation of the F_0 value, which is the established criterion of thermal processes safety of canned low-acid foods to this day. An F_0 value of unity ($F_0 = 1$) is the theoretical minimum to achieve sterility, but industrial thermal processes are designed to reach an F_0 value substantially higher than the theoretical value, of course. Calculation of the F_0 value is done on the basis of a well-known formula, which can be written as (e.g., Jay 1996; Holdsworth 1997; Toledo 1999):

$$F_0(t) = \int_0^t 10^{\frac{T(t)-T_{ref}}{z}} dt \qquad (2.1)$$

where
$F_0(t)$ is the momentary value of the F_0 value after time t
$T(t)$ is the changing temperature, or temperature profile
T_{ref} is a reference temperature
z is the targeted organism's z value

Traditionally, the reference temperature for canned low-acid foods has been 121.1°C (250°F) and the targeted organism has been *C. botulinum* spores, the z value of which is the one incorporated into the formula. For any given temperature profile, $T(t)$, the calculation of the F_0 value can be done graphically or by numerical integration of the right side of Equation 2.1.

Another formula to calculate sterility is based on the assumption that the temperature dependence of the inactivation rate constant, k, the reciprocal of the D value, is governed by the Arrhenius equation, in which case (McKellar and Lu, 2003):

$$PE(t) = \frac{1}{t_0} \int_0^t e^{-\frac{E}{R}\left(\frac{1}{T(t)} - \frac{1}{T_{ref}}\right)} dt \qquad (2.2)$$

where
$PE(t)$ is the integrated lethal effect
t_0 is the inactivation time at the reference temperature T_{ref} (in °K)
$T(t)$ is the process's momentary temperature in °K
E is the inactivation's energy of activation
R is the universal gas constant

Software to do the integration numerically has been developed by several researchers and the method is widely implemented in process calculations in the food industry, including online. The merits of the two formulas (Equation 2.1 and Equation 2.2) and their application have been discussed in numerous publications, which include book chapters, reviews, technical articles and several Websites. They are also accepted by governmental regulating agencies as a way to evaluate the safety of thermal processes in the food and pharmaceutical industries.

However, consider the following. *If* the isothermal survival curves of the targeted organism are all log linear in the pertinent temperature range, and *if* the temperature dependence of the D value is also log linear, then the F_0 value and survival ratio are closely related, i.e.,

$$\log_{10} S(t) = \frac{F_0(t)}{D_{@T=T_{ref}}} \tag{2.3}$$

where

$S(t)$ is the momentary survival ratio — i.e., $N(t)/N_0$, where $N(t)$ and N_0 are the momentary and initial numbers, respectively (see Chapter 1)

$D_{@T = T_{ref}}$ is the D value at the reference temperature

In other words, if the stated conditions are satisfied, the F_0 value would be a direct representation of the actual momentary survival ratio. Also, when the temperature dependence of the D value is log linear, then the selection of the reference temperature is unimportant. This can be demonstrated in the following way.* Consider the survival ratio at a given temperature, T, calculated with the reference temperatures T_{ref1} and T_{ref2}. Because by definition of the log linear model, $\log_{10}(D_{@T_{ref2}}/D_{@T_{ref1}}) = (T_{ref1} - T_{ref2})z$ or $D_{@T_{ref2}} = D_{@T_{ref1}} 10^{(T_{ref1} - T_{ref2})/z}$:

$$\frac{10^{\frac{T-T_{ref2}}{z}}}{D_{@T_{ref2}}} = \frac{10^{\frac{T-T_{ref2}}{z}}}{D_{@T_{ref1}} 10^{\frac{T_{ref1}-T_{ref2}}{z}}} = \frac{10^{\frac{T-T_{ref1}}{z}}}{D_{@T_{ref1}}} \tag{2.4}$$

Consequently, the survival ratio, calculated by integrating the log linear model will be exactly the same survival ratio, $S(t)$, if it is expressed in terms of T_{ref2} and $D_{@T_{ref2}}$ or T_{ref1} and $D_{@T_{ref1}}$. In other words, the selection of a reference temperature and its corresponding D value will have no effect on the calculated F_0 value and the corresponding survival ratio (Equation 2.3). However, this argument will not hold if the temperature dependence of k, the reciprocal of the D value, follows the Arrhenius model, for example. In such a case, there will be a discrepancy that will grow with

---

* I owe this observation to Dr. John Larkin of the FDA.

the difference between the process and the reference temperatures, as was shown by Datta (1993).

The relationship among the F_0 value, the survival ratio, and the reference temperature becomes even more obscure whenever the isothermal survival curves of the targeted organism or spore are not log linear to begin with, in which case neither the D value nor the rate constant k has a clear meaning. This is a very serious flaw of the standard methods to calculate sterility. Any theory that implies that the survival ratio can depend on the type of expression chosen to calculate it and/or on an arbitrary reference temperature cannot be considered valid and ought to be dismissed offhand. The number of survivors and thus the survival ratio are a real measurable quantity and cannot be altered by choosing the parameters of the inactivation model. The actual number of surviving cells is determined by the organism's heat resistance, its growth stage when subjected to the treatment, the medium in which it is treated, and the heat process's temperature history. The same can be said about bacterial spores, except that they do not have a growth stage.

Consequently, any valid inactivation criterion must be based on the actual number of surviving organisms, or spores, or the corresponding survival ratio and not on a hypothetical, model-dependent equivalency to a reference treatment at an arbitrary temperature. Only when the traditional standard model's two underlying assumptions can be verified experimentally for the whole lethal temperature range will the F_0 value be considered a theoretically valid sterility criterion.

However, as recent data clearly show, one cannot assume *a priori* that either condition is satisfied. For historic reasons, industrial microbiologists dealing with heat treatments feel more comfortable relating the efficacy of a given process to that of an isothermal treatment at a reference temperature. If so, then it will be more reasonable to compare the two processes in terms of the same accomplished theoretical *survival ratio* rather than of an F_0 value, whose foundations are questionable at best.

Thus, one could safely say that a heat process that theoretically produces eight orders of magnitude reduction in the sporal population of *C. botulinum*, for example, is equivalent to about 4 min at 121.1°C, which also produces the same survival ratio. This would be a statement whose validity is independent of that of any preconceived kinetic model, in contrast with expressing the equivalency in terms of an F_0 value. [For more on this topic see Chapter 3.]

The Proposed Alternative

Consider the following:

- On the pertinent time scale, a microbial cell once destroyed or a spore inactivated cannot be revived. It also has no time to grow during the process or to adapt biochemically to the treatment.
- All the isothermal survival curves in the pertinent temperature range can be described by the same primary model. No restrictions are imposed on the mathematical structure of this model and the number of its adjustable parameters. It also need not be unique. Obviously, an attempt should always be made to choose the simplest model, using Ockham's razor as a guideline, or the most convenient for other reasons (see below).
- The temperature dependence of the primary model's parameters can be described mathematically by what's known as 'secondary models.' These models can be continuous algebraic functions, e.g., the log logistic equation, or discontinuous — that is, containing one or more "if statements" (see Chapter 1).
- The temperature history of the coldest point in the treated food, or "profile," $T(t)$, can be described mathematically. Again, the expression can be algebraic or discontinuous — that is, containing one or more 'if statements,' depending on the actual process at hand. (The case in which the profile is entered as discrete temperature readings, simulated or real, will be discussed separately in Chapter 3.)
- Factors like pH, oxygen tension, water activity, and the like remain practically unchanged during the process.

In general, the inactivation rate even under isothermal conditions can be time dependent (see Figure 2.1). Thus, a valid survival model must take it into account that, under nonisothermal conditions, the momentary rate does not depend only on the momentary temperature, as the traditional models require, but also on the treated population's *momentary state*. To demonstrate the point, let us examine the three hypothetical scenarios shown in Figure 2.2. This figure depicts the temperature histories (profiles) of three parts of the same suspension of bacterial spores — i.e., all three are treated in the same medium, whether it is a real food or nutrient broth.

- One part of the suspension (A) is heated to 115°C in 1 min, held at that temperature for 8 min, and then cooled to 110°C within 1 min.
- The second part of the suspension (B) is heated slowly to 110°C and reaches this temperature after 10 min.
- The third part of the suspension (C) is held at a nonlethal temperature for 9 min and then heated to 110°C within 1 min.

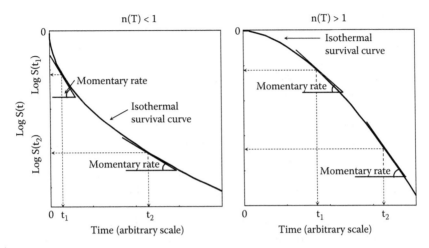

FIGURE 2.1
Schematic view of curvilinear isothermal semilogarithmic survival curves with upward and downward concavity. Notice that even under isothermal conditions, the momentary inactivation rate, the local slope of the survival curve, depends on the momentary survival ratio and thus is a function of time.

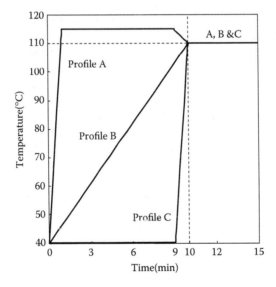

FIGURE 2.2
Schematic view of three hypothetical temperature profiles A, B, and C. Notice that after 10 min, all three parts end up in the constant temperature of 110°C. If the traditional inactivation models based on the log linearity of the D value's temperature dependence or Arrhenius equation had been correct, the logarithmic rate after 10 min would be exactly the same in all three scenarios. This is because both models entail that the logarithmic inactivation rate is only a function of temperature but not of time.

Initially, all three parts of the suspension had exactly the same number of spores. After 10 min, all three are at 110°C after being heat treated for the same time except that the temperature profiles were different. From that moment the 110°C temperature is maintained constant (Figure 2.2). Obviously, the number of viable spores in the three suspensions after the ten minutes would be very different — there is no question about that. However, the suggestion that the spore's *logarithmic inactivation rate* in the three suspensions at that *moment* and forever after (i.e., beyond the 10 min of the original treatment) *must be exactly the same* would certainly be received with some skepticism, to say the least.

Yet, the hypothesis that the momentary logarithmic inactivation rate is only a function of the momentary temperature and therefore totally independent of the spores' thermal history is the basis of all the standard methods to calculate sterility in the food and other industries. If heat inactivation of all microbial cells and bacterial spores had really followed the first-order kinetics, the momentary rate would indeed be determined by the momentary temperature only as the log linear model and the Arrhenius equation imply.

The same reasoning applies to models based on conventional higher order kinetics, in which case the isothermal rate constant, k, would be differently defined, e.g., as $dN(t)/N(t)^m = -kt$. This rate constant would still be a function of the momentary temperature, but not of the exposure time. Unfortunately, ample evidence indicates that the logarithmic inactivation rate *is* time dependent even under isothermal conditions (see Chapter 1 and Figure 2.1) and thus that there is a good reason to reassess the utility of the standard models to calculate heat sterility and predict the efficacy of heat treatments in general.

This leads us to the proposal that the momentary logarithmic inactivation rate in a nonisothermal process is the isothermal logarithmic inactivation rate at the momentary temperature, *at a time that corresponds to the momentary survival ratio* (see Figure 2.1.) This is a testable hypothesis. If correct, then one can use a set of isothermal survival data obtained at different lethal temperatures to *predict* the survival curve that will be observed in any number of nonisothermal treatments, as will be shown later.

An alternative hypothesis is that the momentary logarithmic inactivation rate depends not only on the momentary temperature and survival ratio, but also on the *path* to reach this survival ratio. If so, attempts to predict the outcome of nonisothermal treatment by the proposed alternative models will fail because the temperature history is totally unrelated to the isothermal inactivation data and can be varied at will. In fact, if the first hypothesis is correct, then, at least in some cases, it would also be possible to predict the outcome of nonisothermal heat treatments from survival curves determined under *other nonisothermal temperature histories* (as will be shown in Chapter 4).

Nonisothermal Weibuillian Survival

The Rate Model

Consider a microbial population whose isothermal survival curves in the lethal temperature range can all be adequately described by the Weibullian–power law model (Equation 1.1):

$$\log_{10}S(t) = -b(T)t^{n(T)} \tag{2.5}$$

The momentary time-dependent isothermal logarithmic inactivation rate in this case is

$$\left|\frac{d\log_{10}S(t)}{dt}\right|_{T=\text{const}} = -b(T)n(T)t^{n(T)-1} \tag{2.6}$$

According to this model, the time that corresponds to any given survival ratio, t^*, is the inverse of Equation 2.5:

$$t^* = \left|\frac{-\log_{10}S(t)}{b(T)}\right|^{\frac{1}{n(T)}} \tag{2.7}$$

Our hypothesis, recall, is that the momentary inactivation rate under nonisothermal conditions is that which corresponds to the momentary temperature, $T(t)$, at the time corresponding to the momentary survival ratio, $\log_{10}S(t)$, as shown in Figure 2.3. This time, though, is t^* as defined by Equation 2.3. We can therefore combine Equation 2.6 and Equation 2.7 to produce the nonisothermal inactivation rate equation (Peleg and Penchina, 2000):

$$\frac{d\log_{10}S(t)}{dt} = -b[T(t)]n[T(t)]t^{*\frac{n[T(t)]-1}{n[T(t)]}} \tag{2.8}$$

or

$$\frac{d\log_{10}S(t)}{dt} = -b[T(t)]n[T(t)]\left\{\frac{-\log_{10}S(t)}{b[T(t)]}\right\}^{\frac{n(T)-1}{n(T)}} \tag{2.9}$$

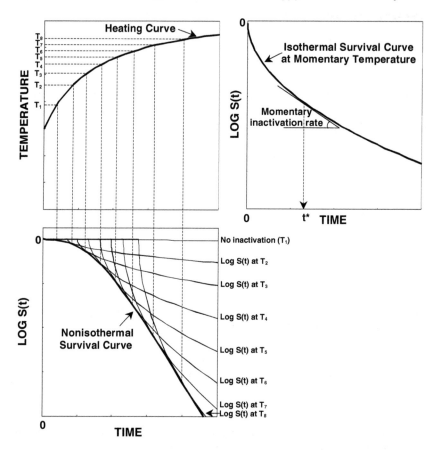

FIGURE 2.3
Schematic view of the construction of a nonisothermal survival curve. (Adapted from
Corradini, M.G. et al., 2005, *J. Food Eng.*, 67, 59–69, and Corradini, M.G. and Peleg, M., 2005,
J. Appl. Microbiol., 99, 187–200.)

This differential equation can be solved numerically using a program
like Mathematica® (Wolfram Research, Champaign, IL) and other mathematical software for almost any conceivable temperature profile, $T(t)$, as
will be shown later. (Equation 2.9 can also be converted into a *difference
equation* — see Chapter 3 — in which case, it can be solved by widely
used general purpose software like Microsoft Excel®.)

Let us examine some potential uses of the model. For simplicity, we will
assume that $n(T) = $ constant $= n$ and that $b(T)$ can be described by the log
logistic equation:

$$b(t) = b[T(t)] = \log_e[1 + \exp\{k[T(t) - T_c]\}] \qquad (2.10)$$

(Notice that nesting, i.e., $b(t) = b[T(t)]$ in our case, is an almost trivial step in Mathematica.)

Inserting a constant n and $b[T(t)]$ as defined by Equation 2.10 into Equation 2.9 produces the survival rate model:

$$\frac{d\log_{10} S(t)}{dt} = -\log_e\left[1 + \exp\{k[T(t) - T_c]\}\right] \cdot n$$

$$\cdot \left\{\frac{-\log_{10} S(t)}{\log_e[1 + \exp\{k[T(t) - T_c]\}]}\right\}^{\frac{n-1}{n}} \qquad (2.11)$$

With any given survival parameters, n, k, and T_c, and a temperature profile, $T(t)$, the differential equation can be solved numerically to generate the corresponding survival curve, i.e., the sought $\log_{10} S(t)$ vs. t relationship. Once the solution is found, it can be plotted and compared with other such curves calculated for different scenarios. These can be survival curves of different organisms or spores treated under the same profile and/or of the same organism or spore subjected to different heat treatments. In the first case, the different organisms will be described by their different survival parameters — namely, n, k, and T_c — under the same $T(t)$s; in the second, n, k, and T_c will remain constant but the $T(t)$s will be varied and compared.

Heating and Cooling

Simulation of Heating Curves by Empirical Models

Because the objective here is to demonstrate the methodology rather than to account for any particular organism's response to any particular thermal process, we will use empirical models to describe the temperature profiles and not heat transfer-based models. A variety of empirical models can be used to simulate industrial temperature profiles. Nevertheless, for those interested in a general and flexible mathematical model that almost always produces realistic heating and cooling curves, our suggestion is the combined logistic and Fermi expression (Peleg, 2004), i.e.,

$$T(t) = \left\{T_0 + \frac{T_{\text{target}} - T_0}{1 + \exp[k_{\text{heating}}(t_{\text{heating}} - t)]}\right\} \cdot \frac{1}{1 + \exp[k_{\text{cooling}}(t - t_{\text{cooling}})]} \qquad (2.12)$$

where

T_0 is the initial temperature

T_{target} is the process's target (asymptotic) temperature

$k_{heating}$ is a heating rate parameter

$t_{heating}$ is a location parameter that marks the inflection point of the temperature's rise and thus to its lag

$k_{cooling}$ is a cooling rate parameter

$t_{cooling}$ is a second location parameter that marks the time at which the cooling stage replaces the heating

Examples of heating–cooling curves produced by this model are shown in Figure 2.4. A holding time can be produced by selecting $t_{cooling}$ and $t_{heating}$ in such a way that $t_{cooling} - t_{heating}$ will be sufficiently large. (An MS Excel® program to generate heating and cooling curves using this model is part of freeware posted on the Web to generate survival curves in pasteurization and sterilization. See www.umass.edu/foodsci as well as Chapter 3.)

In some thermal processes, the cooling stage appears to be approximately linear. In such a case, the temperature profile can be described by combining the logistic heating part with a log logistic cooling term, i.e.,

$$T(t) = \left\{ T_0 + \frac{T_{target} - T_0}{1 + \exp[k_{heating}(t_{heating} - t)]} \right\}$$

$$\cdot \left(1 - \log_e \{1 + \exp[k_{cooling}(t - t'_{cooling})]\} \right) \tag{2.13}$$

(Notice that the cooling must be stopped before the temperature dives into a negative territory.)

An alternative simpler model is:

$$T(t) = \left\{ T_0 + \frac{(T_{target} - T_0)t}{k''_{heating} + t} \right\} (1 - \log_e \{1 + \exp[k''_{cooling}(t - t''_{cooling})]\}) \tag{2.14}$$

If only the heating regime is of interest, the second term of the right part of each of the three profile equations can be simply dropped. This will simplify the model to a great extent by reducing the number of control parameters. Examples are shown in Figure 2.4 (In case one tries to fit an actual experimental temperature profile with any of these three models, it would be advisable to start with the heating and holding data only and use only the first terms as a regression model. Ignoring the cooling stage initially and dropping the second term will eliminate two adjustable parameters, thus facilitating the choice of the initial guesses of $k_{heating}$ and

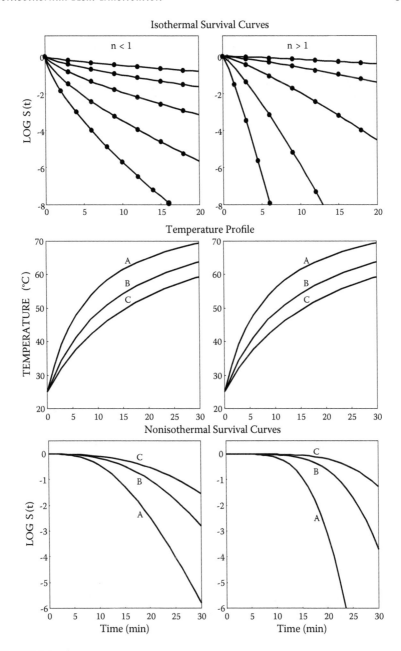

FIGURE 2.4
Schematic view of Weibullian nonisothermal survival curves during monotonic heating. Notice that, although all the isothermal survival curves of the hypothetical organism on the left have upward concavity, the nonisothermal survival curve during heating could have a downward concavity. Courtesy of Dr. Maria G. Corradini.

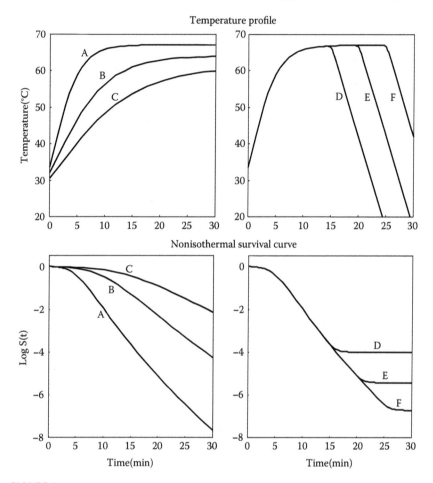

FIGURE 2.5
Simulated temperature profiles produced with Equation 2.12 as a model and corresponding
survival curves generated with Equation 2.11 as a model.

$t_{heating}$ or $k_{heating}$. The cooling stage data can then be added and the restored
whole curve be fitted with the complete model.)

Simulated Survival Curves for Processes with Different Target
Temperature and Holding Durations

The effect of the target temperature on the theoretical survival curves of *C.
botulinum* like spores is shown at the bottom left of Figure 2.5. They were
produced by Equation 2.11 as the survival model and Equation 2.12 as the
temperature profile model. The survival parameters $n = 0.35$, $k = 0.31°C^{-1}$,

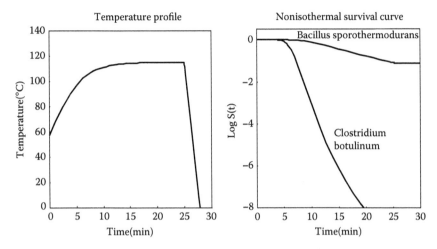

FIGURE 2.6
Simulated view of how the survival curves of two spores can be compared when subjected to the same heat treatment. This was created with Equation 2.12 as the heating model and Equation 2.11 as the inactivation model. Notice that a process effective in the elimination of the spores of *C. botulinum* can be ineffective against those of *B. sporothermodurans*.

and $T_c = 102°C$ are approximately those of C. *botulinum* spores determined from the isothermal experimental inactivation data reported by Anderson et al. (1996). Similar simulated survival curves are shown at the bottom right of the figure. These were produced for the same hypothetical spores for processes with one target temperature but where the cooling starts at different times.

Naturally, the user of the model can select any number of target temperatures and processing time combinations in order to find a suitable process. The same can be done by changing the survival parameters (see Figure 2.6), in which case a contemplated process can be examined as to the effect it might have on a resistant strain, for example, or a new emerging organism that has become a potential hazard. With very minor modification, the program can produce the survival curves of any number of different organisms or spores simultaneously and on the same graph for comparison as demonstrated in the figure.

The same procedure can be also used to examine the potential effects of either variation in the process parameters, e.g., of different profiles in various locations of a stationary retort or changes in the resistance of the targeted organism or organisms as a result of changes in the product's formulation and properties. Notable among these are added or eliminated compounds with antimicrobial activity, altered pH or the concentration of salts, etc.

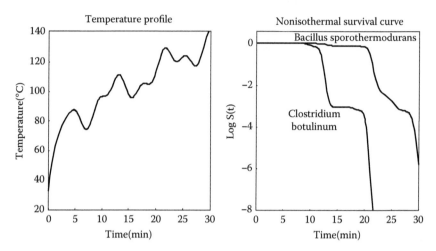

FIGURE 2.7
Simulated heat treatments with oscillating temperature and corresponding survival curves generated with Equation 2.11 as a model. Notice that the temperature profile's complexity has hardly any effect on the ability of a program like Mathematica® to solve the resulting differential equation.

Temperate Oscillations

Although the heating curve in the vast majority of industrial processes is smooth, a faulty controller can cause significant temperature oscillations. These can have regular periodicity or, if the problem is more serious, exhibit an irregular or totally chaotic fluctuating pattern, especially if coupled with an unsteady steam supply. Examples of simulated temperature profiles with irregular oscillations are shown in Figure 2.7 together with the corresponding survival curves. The temperature fluctuations can be produced by superimposing a periodic (sinusoidal) function with one or more periodicities on a regular smooth profile. Random temperature oscillations can be produced by replacing the fixed coefficients of a periodic function with random entries so that their amplitudes and/or frequencies will vary within the expected range.

Regardless of whether the introduced oscillation is periodic or random, the mathematical expression of the temperature profile, $T(t)$, can become rather elaborate. Consequently, the corresponding model's rate equation (e.g., Equation 2.11) into which such an expression is incorporated will become cumbersome. Yet, the numerical solution of the differential equation by a program like Mathematica® hardly takes longer than that of a rate equation of a process with a smooth temperature profile.

Simulations of this kind can be used to assess the effect of runaway processes on the safety of the treated products. They can also help in the decision on what to do when such a process has occurred and whether a remedial treatment will be necessary (Peleg, 2004). By adjusting the oscillations' amplitudes in a set of computer simulations, one can visualize the potential safety consequences of equipment malfunction. These in turn can be used to instruct or train the operator or person in charge how to proceed in such an event.

Discontinuous Temperature Profiles

Sudden temperature changes, like those produced by an interruption in the steam supply, for example, followed by cooling can be described by discontinuous models. These include if statements and can have a form like:

$$T(t) = \text{if } [t \leq t_1, T_1(t), \text{ if } [t \leq t_2, T_2(t - t_1), \text{ if } [t \leq t_3, T_3(t - t_2),$$
$$\text{if } [t \leq t_4, T_4(t - t_3), \ldots]]]] \tag{2.15}$$

In plain language, before time t_1, the temperature profile obeys the expression $T_1(t)$. Then, between t_1 and t_2, it obeys the expression $T_2(t - t_1)$. Between t_2 and t_3, it obeys the expression $T_3(t - t_2)$, between t_3 and t_4, it obeys the expression $T_4(t - t_3)$, and so forth. In software like Mathematica®, as already stated several times, a function $T(t)$ defined in this manner is treated by the program in the same manner as a regular continuous function is treated. That is, $T(t)$ described by Equation 2.12 or a similar expression can be used as a model for nonlinear regression and as a legitimate term in a differential equation that requires a numerical solution.

In our case, therefore, a temperature profile characterized by a series of if statements can be incorporated into the survival model's rate equation parameters as such to produce the term $b[T(t)]$. Subsequently the equation can be solved using the standard NDSolve command as if it were a "normal" differential equation with continuous coefficients. In fact, the term into which the discontinuous temperature profile expression is incorporated can itself have an 'if statement,' e.g., $b(T) = \text{if } [T \leq T_c, 0, k(T - T_c)]$. Still, the resulting model's rate equation can be solved by Mathematica®, despite the double nested 'if statements' in its coefficients (Corradini et al., 2005).

Simulated discontinuous temperature profiles and their corresponding survival curves are shown in Figure 2.8. They were produced to demonstrate that discontinuities in the temperature profile, the Weibullian rate parameter, or both, are no hindrance to the differential equation solution. Models of this kind could be particularly useful in the assessment of the impact of accidents and unplanned interruptions in the process on the residual survival ratio of the targeted organism or spore (Peleg, 2004). Once the survival ratio has been estimated, the simulation can help to evaluate the efficacy of remedial measures, like prolonging the holding time, elevating the process temperature, or both.

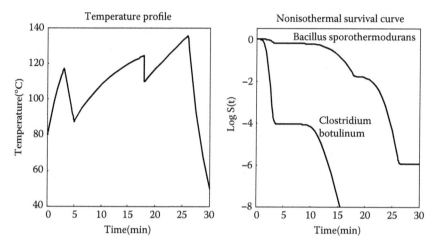

FIGURE 2.8

Simulated discontinuous temperature profiles and corresponding survival curves produced with Equation 2.11 as a model. Notice that the "if statements" in the temperature profile's equation (and consequently in the inactivation differential rate equation's coefficients) are no hindrance to its numerical solution with a program like Mathematica®.

The Special Case of Log Linear Isothermal Survival

According to the described concept, if all the isothermal survival curves of the investigated organism happen to be log linear, the Weibullian model describing them (Equation 2.1) will have a shape factor, or power, of 1. However, when $n(T) = 1$, $[n(T) - 1]/n(T) = 0$ and the model's rate equation is reduced to:

$$\frac{d\log_{10} S(t)}{dt} = -\log_e[1 + \exp\{k[T(t) - T_c]\}] \tag{2.16}$$

in which case the momentary logarithmic inactivation rate would indeed be only a function of the momentary temperature, but not of the momentary survival ratio. (The difference between Equation 2.16 and the conventional models based on the D and z values or Arrhenius equation is that the lethal region is marked by T_c and that the unnecessary logarithmic transformation of k is avoided.)

The inactivation model expressed by Equation 2.16 can be presented in an integral form:

$$\log_{10} S(t) = -\int_0^t \log_e[1 + \exp\{k[T(t) - T_c]\}]dt \tag{2.17}$$

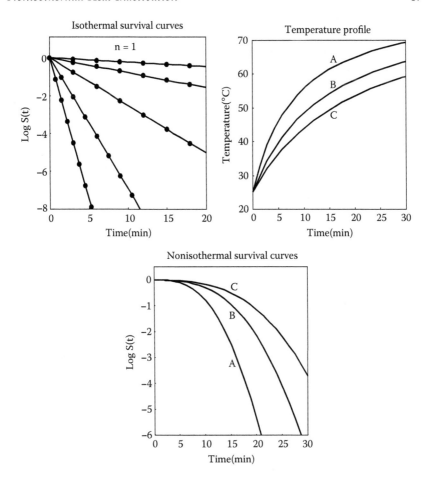

FIGURE 2.9
Examples of nonisothermal survival curves of a hypothetical organism whose inactivation follow the first-order kinetics produced with Equation 2.17 as a model. Notice the downward concavity of the curves and that the temperature dependence of the rate constant in this simulation is not log linear, but rather log logistic. (Courtesy of Dr. Maria G. Corradini.)

For certain temperature profiles, $T(t)$, Equation 2.17 can be integrated analytically; otherwise, it must be solved numerically in order to produce the nonisothermal survival curve, $\log_{10}S(t)$ vs. t. The conditions under which the equation has an analytical solution are discussed in more detail in Chapter 3.

Examples of simulated survival curves using Equation 2.17 as a model are shown in Figure 2.9. They demonstrate that in any heating process in which the temperature increases monotonically, the semilogarithmic survival curve of a "log linear organism" *will always have a downward concavity.* Similar downward concavity is produced when a Weibullian organism whose isothermal survival curves have $n < 1$ is exposed to an accelerated heating regime (compare with Figure 2.4). Needless to say, if an organism's Weibullian isothermal semilogarithmic survival curves are all concave downward (Equation 2.1 with $n(T) > 1$), so will be the nonisothermal curve during the heating stage be as shown in Figure 2.4.

Thus, a visual inspection of the shape of a nonisothermal survival curve recorded during heating will rarely reveal the concavity direction of the organism's or spore's isothermal survival curves. What can be safely said is that if the nonisothermal semilogarithmic survival curve during a monotonic heating regime *appears linear*, then the isothermal curves of the particular cells or spores must have *upward concavity* or, in the Weibullian terminology, $n(T) < 1$. If the nonisothermal survival curves have upward concavity during the heating stage, then the isothermal survival curves of the particular cells or spores must all have upward concavity — that is, they exhibit noticeable tailing.

Non-Weibullian Survival Models

The Weibull–power law model of isothermal survival curve has two main advantages:

- Although formulated as a simple algebraic expression, it is a very flexible mathematical model that can describe the inactivation patterns of a large class of organisms and spores (perhaps their majority).
- It is also a very convenient primary model for calculating nonisothermal survival curves, as shown earlier, because t^* as a function of $\log_{10}S(t)$ can be expressed algebraically (Equation 2.7). (Its derivative, $d\log_{10}S(t)/dt$, is also a simple algebraic expression. However, this is only a minor advantage. In a program like Mathematica®, the derivative need not be expressed explicitly. Thus, if we call the momentary logarithmic survival ratio $LS(t)$, for example, then all we need for defining the derivative, $dLS(t)/dt$, is to write $dLS[t\_] = D[LS[t],t]$. This new function, $dLS(t)$, is now fully defined regardless of how elaborate the expression of $LS(t)$ might be — see below).

The main limitation of the Weibull–power law model is that it can only describe, effectively, isothermal survival curves whose slope (in absolute terms) monotonically increases ($n > 1$), decreases ($n < 1$) or remains constant ($n = 1$). In any event, the model entails that, under isothermal conditions, as $t \to \infty$, $S(t) \to 0$ and $\log_{10} S(t) \to -\infty$. In other words, the Weibullian model cannot describe isothermal survival curves having a finite residual survival ratio, which might be encountered in extreme cases of tailing, especially at relatively low treatment temperatures. Similarly, the Weibull–power law model cannot describe isothermal survival curves that might indicate the existence of a real thermal death time (see Chapter 1), i.e., that the survival ratio will drop to zero at a finite time.

Obviously, the Weibull–power law model cannot describe sigmoid isothermal survival curves of both kinds — i.e., when the concavity direction changes from downward to upward or vice versa, curves with a flat shoulder whose continuation into the lethal regime is clearly log linear and curves exhibiting an activation shoulder. These, as shown in the previous chapter, can be described by an assortment of alternative empirical survival models. As far as the nonisothermal survival patterns are concerned, these alternative models can be divided into two groups: those whose isothermal survival curves' model equation has an analytic inverse and those whose model's equation does not have an analytic inverse.

More specifically, the distinction is between isothermal survival models that enable to express t^*, the time that corresponds to the momentary survival ratio, algebraically (i.e., as an explicit function of the survival ratio and time) and models where t^* can only be expressed as a numerical solution of the isothermal survival equation. Both kinds can be handled by a program like Mathematica®, as will be shown later. Yet, finding a primary model that has an analytic inverse will greatly facilitate the programming, especially in case one wants to calculate nonisothermal survival curves' general purpose software, like Microsoft Excel®.

Logistic (Fermian) Survival

Consider an organism whose isothermal survival curves can be described by the logistic (Fermi) model already introduced in Chapter 1:

$$S(t) = \frac{1}{1 + \exp[k(T)(t - t_c(T))]} \tag{2.18}$$

where $k(T)$ and $t_c(T)$ are the temperature-dependent survival parameters (or, to keep with the convention of expressing the survival ratio by its base-10 logarithm):

$$\log_{10} S(t) = -\frac{\log_e\{1 + \exp\{k(T)[t - t_c(T)]\}}{\log_e 10} \tag{2.19}$$

The momentary isothermal inactivation rate according to this model is:

$$\left|\frac{dS(t)}{dt}\right|_{T=const} = -\frac{k(T)\exp[t-t_c(T)]}{(1+\exp\{k(T)[t-t_c(T)]\})^2} \tag{2.20}$$

and the time t^*, which corresponds to any given survival ratio $S(t)$, is:

$$t^* = \frac{t_c + \log_e[1/S(t)-1]}{k(T)} \tag{2.21}$$

Combining Equation 2.20 and Equation 2.21 yields the rate model:

$$\frac{dS(t)}{dt} = \frac{-k[T(t)]\exp[t^*-t_c[T(t)]]}{[1+\exp(k[T(t)]\{t^*-t_c[T(t)]\})]^2} \tag{2.22}$$

where, again, $T(t)$ is the temperature profile and t^* is defined by Equation 2.21.

Simulated heating curves and corresponding survival curves using Equation 2.22 as a model are shown in Figure 2.10. The reader will be reminded that the logistic (Fermi) model would be particularly suitable for survival patterns characterized by a noticeable flat shoulder at times $t < t_c$ and linear semilogarithmic continuation at $t > t_c$. Thus, the apparent shoulder in the *nonisothermal survival* curves is a manifestation of the logistic survival pattern of the investigated organisms as well as the fact that, at low temperatures, hardly any inactivation takes place, irrespective of whether the organism's isothermal survival curves exhibit a shoulder. (Compare Figure 2.10 with Figure 2.4 and Figure 2.5, which show an apparent shoulder although the isothermal curves have none.)

Extreme Tailing

Extreme tailing, especially at low processing temperatures, can be manifested in a substantial residual survival, which hardly changes with time (on the pertinent time scale, that is). Isothermal survival curves of this kind (see Chapter 1) can be described by the empirical model (Peleg et al., 2004):

$$\log_{10} S(t) = -\frac{k_1(T)t}{k_2(T)+t} \tag{2.23}$$

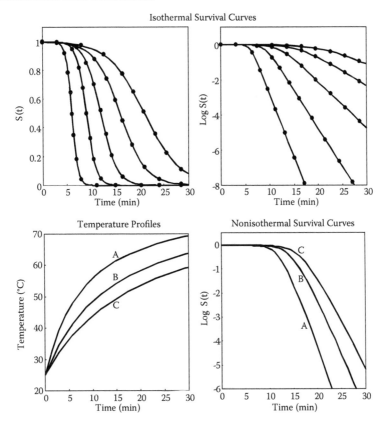

FIGURE 2.10
Simulated heating profiles and corresponding logistic survival curves produced with Equation 2.22 as a model. Notice that an isothermal logistic survival can be manifested in a flat shoulder as well. (Courtesy of Dr. Maria G. Corradini.)

where $k_1(T)$ is the asymptotic residual and $k_2(T)$ a of sort of rate parameter, which represents the steepness of the survival curve. (In fact, the initial decay rate is $k_1(T)/k_2(T)$, i.e., $k_1(T)/k_2(T) = -d\log_{10}S(t)/dt$ @ $t = 0$.)

If, after treatments at high temperatures, the residual survival is eliminated, the model can still be used, except that $k_1(T)$ will be set to a value well below the detectable level. For example, if the detectable survival ratio level is 10^{-8}, then a model with $k_1(T) = -14$ or -16, for example, will describe survival curves whose initial part might be practically indistinguishable from those produced by the Weibullian model with $n < 1$ (Peleg and Penchina, 2000; Peleg et al., 2005). Thus, this model can be quite useful for describing inactivation patterns in which the residual "asymptotic" survival ratio progressively diminishes and then totally disappears when the temperature is increased (Figure 2.11). According to this model (Peleg et al., 2005), the momentary isothermal inactivation rate is:

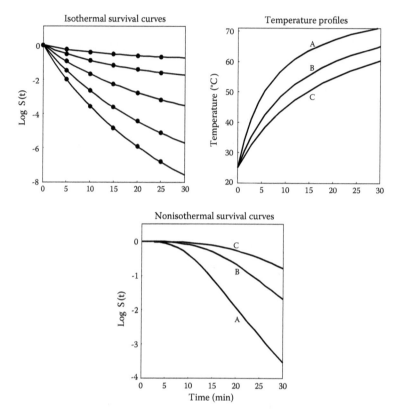

FIGURE 2.11
Simulated heating profiles and corresponding survival curves produced with Equation 2.26 as a model. Notice that the extreme tailing manifested in a residual survival ratio under isothermal conditions is hardly evident in the shape of the nonisothermal survival curves. (Courtesy of Dr. Maria G. Corradini.)

$$\left| \frac{d\log_{10} S(t)}{dt} \right|_{T=const} = -\frac{k_1(T)k_2(T)}{[k_2(T)+t]^2} \qquad (2.24)$$

and the time that corresponds to the momentary survival ratio, t^*, is:

$$t^* = \frac{-k_2(T)\log_{10} S(t)}{k_1(T)-\log_{10} S(t)} \qquad (2.25)$$

Combining the two produces the model rate equation:

$$\frac{d\log_{10} S(t)}{dt} = -\frac{k_1[T(t)]k_2[T(t)]}{\left\{k_2[T(t)] - \frac{k_2[T(t)]\log_{10} S(t)}{k_1[T(t)] - \log_{10} S(t)}\right\}^2} \quad (2.26)$$

Again, Equation 2.26 is a rather cumbersome looking differential equation, but it can easily be solved by a program like Mathematica®. Examples of survival curves generated with this model are shown in Figure 2.10.

Sigmoid Survival Curves

Sigmoid isothermal survival curves can be described by a variety of models (see Chapter 1), among them:

$$\log_{10} S(t) = -\frac{k_1(T)t}{[1 + k_2(T)t][k_3(T) - t]} \quad (2.27)$$

or

$$\log_{10} S(t) = -\frac{k_1(T)t}{k_2(T) + t} - \frac{k_3(T)t}{k_4(T) - t} \quad (2.28)$$

where the $k(T)$ is a temperature-dependent coefficient.

Both equations have an analytic inverse: one of two solutions of a quadratic equation (Peleg, 2003a). Thus, t^*, the time that corresponds to the momentary survival ratio $\log_{10} S(t)$, can be expressed algebraically. Because an analytic derivative always exists to the function $\log_{10} S(t)$, one can combine the two to produce the rate equation in the same manner as in the previous cases, where the isothermal survival parameters had only two temperature-dependent parameters. The resulting differential equations however, might be too cumbersome to be copied into this chapter and therefore will not be shown. Suffice it is to say, again, that once the model's terms are defined in the syntax of a program like Mathematica®, they can be incorporated as such in the differential and equation. Thus, writing the model's rate equation explicitly becomes unnecessary.

Simulated nonisothermal survival curves, when the underlying isothermal curves are described by Equation 2.28, are shown in Figure 2.12 and Figure 2.13. It should be mentioned once more that determining the four parameters of Equation 2.28 from isothermal experimental data by regression need not be difficult (Peleg, 2003a). Nevertheless, finding meaningful secondary models to describe these parameters' temperature dependence

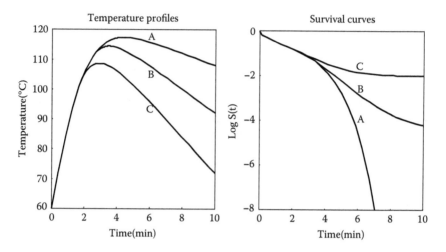

FIGURE 2.12
Simulated heating and cooling temperature profiles and corresponding survival curves of an organism whose isothermal survival curves have a sigmoid shape governed by Equation 2.28, which has an analytical inverse, as a model. (From Peleg, M., 2003, *Crit Rev. Food Sci. Nutr.*, 43, 645–658. With permission, courtesy of CRC Press.)

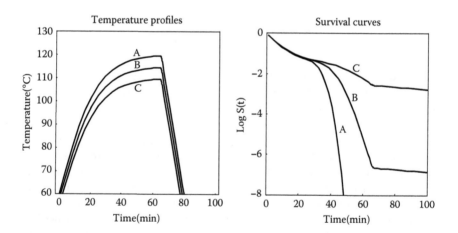

FIGURE 2.13
Simulated oscillating heating profile and corresponding survival curves of an organism whose isothermal survival curves have a sigmoid shape governed by a model that has no analytical inverse. Notice that the combined complexity of the rate model's equation is not a hindrance to its solution by a program like Mathematica. (From Peleg, M., 2003, *Crit. Rev. Food Sci. Nutr.*, 43, 645–658. With permission, courtesy of CRC Press.)

can become a serious problem if the experimental data are few and scattered. Thus, the potential value of such models is primarily in simulating nonisothermal survival patterns.

Isothermal Survival Model's Equation with No Analytic Inverse

Some isothermal survival curves have no analytical inverse, i.e., the time, t^*, cannot be expressed as an explicit algebraic function of $\log_{10}S(t)$. Examples are certain kinds of sigmoid survival models and the empirical model that describes survival curves with an activation shoulder. One such model has the form (see Chapter 1):

$$\log_{10}S(t) = \frac{t(1 - \log_e[1 + \exp[k_1(T)[t - t_c(T)]]]^m)}{k_2(T) + k_3(T)t} \qquad (2.29)$$

where the $k(T)$'s and $t_c(T)$ are the temperature-dependent, activation–inactivation parameters.

With $\log_{10}S(t)$ defined by Equation 2.26, $t^* = t^*(\log_{10}S(t))$ cannot be expressed algebraically. Yet, t^* *can be written as the numerical solution of the equation with respect to t*, or, in the syntax of Mathematica®:

$$tstar = t/.x \rightarrow \text{first } [\text{NSolve}[LS[t] == LS[x],x]] \qquad (2.30)$$

where
  $LS[t]$ represents the momentary value of $\log_{10}S(t)$
  $LS[x]$ is the completed expression of Equation 2.26, except that the independent variable t is replaced by a new dummy variable x
  The coefficients, e.g., $k_1[T(t)]$, $k_2[T(t)]$, etc., remain functions of time as before

This command tells the program that t becomes *tstar* by receiving the value of x, which is the numerical solution of the equation $LS(t) = LS(x)$ when x is the equation's unknown and $LS(t)$ is the momentary value of $\log_{10}S(t)$. This expression defines the momentary value of t^* at any momentary value of $\log_{10}S(t)$. As far as Mathematica® is concerned, *tstar* thus defined is a term to be treated just as any other term, except that its value is calculated as the numerical solution of an equation rather than by a standard built-in procedure of the kind used in the execution of more routine commands such as Log[], Exp[], and the like.

As previously mentioned, the isothermal logarithmic inactivation rate at any given temperature, $-d\log_{10}S(t)/dt$, can always be expressed algebraically. Yet, writing the time derivative of $\log_{10}S(t)$ when defined by Equation 2.29 will produce an extremely cumbersome mathematical expression, especially once the time, t, is replaced by t^* as defined by Equation 2.30. However, again, writing the derivative as an explicit algebraic

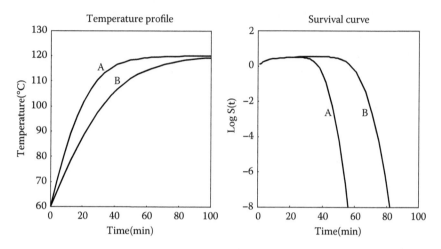

FIGURE 2.14

Simulated heating–cooling curves and the corresponding survival curves of spores with an activation shoulder. Notice that, although the primary model's equation (Equation 2.29) has no analytical inverse, the nonisothermal survival curve can still be produced by incorporating t^* as an iterative numerical solution (Equation 2.30) into the rate equation's coefficients. (From Peleg, M. 2002. *J. Food Sci.* 67, 2438–2443. With permission, courtesy of the Institute of Food Technologists.)

term is totally unnecessary in Mathematica. As before, all one must do is to define the derivative as a new function — $dLS[t\_] = D[LS[t],t]$, for example. What it means is that the new function, $dLS[t]$, is the first derivative of the function $LS[t]$ with respect to time, which in our case is $d\log_{10}S(t)/dt$. Once $dLS[t]$ is determined in this way, it can be converted into a function of *tstar* (t) with $T(t)$ incorporated into corresponding coefficients to produce the nonisothermal survival model.

Although at first glance, the procedure seems extremely complicated, in reality it is not. Once the user becomes familiar with the syntax of Mathematica® changes in the model, its parameters magnitude and the temperature profile equation become extremely easy to implement. The actual rate model *is* complicated because it requires the numerical solution of a differential equation whose coefficients are a numerical solution of an algebraic equation performed at each iteration. Yet, even for elaborate temperature profiles, the calculation time is only on the order of minutes at worst, depending on the computer's processor speed. Examples of survival curves produced by the preceding procedure are shown in Figure 12.14 and Figure 12.15. They demonstrate that the method works even for temperature profiles that contain random temperature oscillations.

The described procedure can be also employed when the organism's isothermal survival curves do have an analytical inverse. For example, a

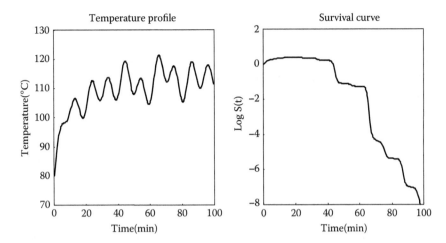

FIGURE 2.15

A simulated irregular oscillating temperature profile and corresponding survival curves of spores with an activation shoulder. Notice that, despite of that t^* is defined as the iterative numerical solution of the primary model equation (Equation 2.30) and the temperature profile's complexity, the rate equation can still be solved to render the sought survival curves. (From Peleg, M. 2002. *J. Food Sci.* 67, 2438–2443. With permission, courtesy of the Institute of Food Technologists.)

user might want to examine, simultaneously, the response of several organisms, some of whose models' equations have analytic inverse and some of which do not. In such a case, one can set the numerical calculation of t^* as the routine procedure and use a single program to calculate the survival curves. If any of the models happens to have an analytical inverse, it will probably be found by Mathematica®, calculated symbolically, and then used as such in the numerical solution of the rate equation (Peleg, 2003a).

Independence of the Calculated Nonisothermal Survival Curve of the Chosen Survival Model

The nonisothermal survival curve of heat-treated microbial cells or bacterial spores is determined by the organism's heat resistance, its growth conditions and history, the medium in which the process takes place, and the temperature profile. However, the survival curve cannot be influenced by the model chosen to calculate it. Although this point has already been stated several times, it cannot be overemphasized. Therefore, if the described procedure to calculate nonisothermal survival curves is self-consistent, it should yield the same survival curve, irrespective of the primary and secondary models used to construct the rate equation.

To demonstrate the point, we have fitted a set of published isothermal survival data of *Salmonella* with the Weibullian–power law model using Equation 2.5 and again with Equation 2.23 as a model. Thus, we have two distinct primary models to describe the same experimental isothermal survival data (see Figure 2.6). The corresponding secondary models were, of course, also very different. The Weibullian–power law model had a constant shape factor-power, n, and a rate parameter $b(T)$ with a log logistic temperature dependence. The non-Weibullian model had its parameter, $k_1(T)$ constant, and its $k_2(T)$, described by an empirical single term exponential model. This has produced two survival rate equations. One is based on the three Weibullian survival parameters n, k, and T_c, where k and T_c define $b(T)$ and thus $b[T(t)]$. The other, based on Equation 2.23 as a model, also has three survival parameters: $k_1(T)$ and the two coefficients of $k_2(T)$ vs. T relationship calculated by regression.

The mathematical structure of the resulting two rate models, which had been based on the derivatives of the two isothermal survival models and on their corresponding t^*, was quite different. In the first case, it was in the form of Equation 2.9 and in the second in the form of Equation 2.26. However, when nonisothermal survival curves produced by the two models for the same temperature profile are compared, they are almost identical, as can be seen in Figure 2.16. The slight discrepancies reflect the imperfect fit of the two primary models.

In light of the inevitable scatter in experimental microbial counts (usually on the order of half a log [base 10] unit), if such curves had been predictions of the outcome of a real heat treatment, they would be considered as identical for all practical purposes. The same can be said about *any* isothermal survival model that fits the data well, irrespective of its mathematical structure and the number of its adjustable parameters. However, this would be true only if the temperature range, the treatment duration, and the corresponding survival ratios of the nonisothermal treatment are all covered by the original experimental isothermal data.

Experimental Verification of the Model

The true test of a theory's or model's validity is its ability to predict the outcome of experiments whose results were not used in its formulation. In our case, if the described concept is correct and the assumptions on which it is based are sound, then one could use isothermal survival data and predict the survival curve(s) of an organism in the same or comparable medium when subjected to a nonisothermal heat treatment. Surprising as this might sound, suitable data for such a test are rare in the food

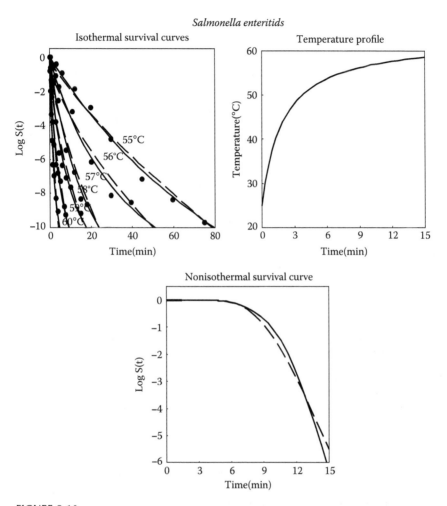

FIGURE 2.16
Simulated demonstration that the prediction of the nonisothermal survival curve does not depend on the primary model's selection, as long as it fits the isothermal data well and these cover the time and temperature range of the contemplated heat treatment. Solid lines: curves produced with the Weibullian–log logistic model (Equation 2.11). Dashed lines: curves produced with a non-Weibullian model (Equation 2.26). (Courtesy of Dr. Maria G. Corradini.)

literature. Our attempt to produce such data, except in one case — when we got help from the then Nabisco Corporation (now Kraft) — were unsuccessful for lack of funding. The following is almost all the evidence that we have at the time at which this book is written. It concerns the inactivation of three microbial vegetative cells and the spores of a resistant bacillus. The four cases will be presented in the chronological order of their analysis and publication.

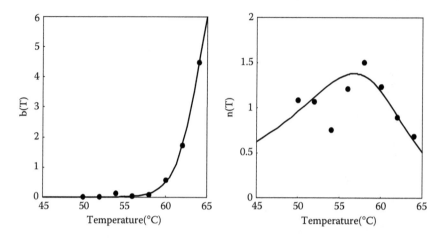

FIGURE 2.17

The temperature dependence of the Weibullian survival parameters of *Listeria monocytogenes*, based on the isothermal inactivation data of Stephens, P.J. et al., 1994, *J. Appl. Microbiol.*, 77, 702–710. (From Peleg, M. et al., 2001, *Food Res. Int.*, 34, 383–388. With permission, courtesy of Elsevier Ltd.)

The Isothermal and Nonisothermal Inactivation Patterns of *L. monocytogenes*

Published isothermal survival curves in the range of 50 to 64°C (Stephens et al., 1994) were fitted by the Weibullian–power law model (Equation 2.5). [The regression coefficients, r^2, were in the range of 0.926 to 0.993.] The $b(T)$ and $n(T)$ vs. T plots are shown in Figure 2.17. Both were described by empirical models whose fit is also shown in the figure. Nonisothermal survival data of *L. monocytogenes* under a linear temperature profile compiled from three different experiments reported by Coote et al. (1991) are shown in Figure 2.18 as filled circles. The arrows mark survival ratios too small to be experimentally determined. The solid curve is the prediction using Equation 2.9 as a model with the parameters calculated from the isothermal data of Stephens et al. (1994).

Although the agreement is far from perfect, the predicted curve can be still considered as in general agreement with the actually observed data. Part of the discrepancy can be explained by the imperfections and "noise" of the reported data. It is obvious, for example, that *Listeria* cells are not reduced by 1.5 to 2 orders of magnitude by a short exposure to temperatures around 30°C, as some of the data points indicate. Apparently, some of the initial data points reflect nonuniform heating or less than perfect recovery of the cells (Peleg et al., 2001). However, the burden of proof is on the one who proposes the alternative model, so any skepticism concerning whether what has been shown really confirms the model is understandable.

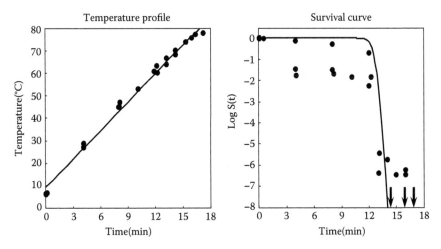

FIGURE 2.18
Prediction of the nonisothermal survival of *Listeria monocytogenes* from isothermal inactivation data. Filled circles: compiled data reported by Coote, P.J. et al., 1991, *Appl. Microbiol.*,
70, 489–494. Solid line: the predicted survival curve based on the isothermal data of Stephens,
P.J. et al., 1994, *J. Appl. Microbiol.*, 77, 702–710. Constructed by superimposed figures originally published by Peleg, M. et al., 2001, *Food Res. Int.*, 34, 383–388. Arrows indicate survival
ratios below the detection threshold.

The Isothermal and Nonisothermal Inactivation of *Salmonella*

Figure 2.19 shows isothermal survival curves of *Salmonella* obtained in
the U.K. by Dr. Karen Mattick and fitted with the Weibullian–power law
model (Equation 2.5) and the corresponding temperature dependence of
the model's parameters $b(T)$ and $n(T)$. Also shown in the figure is the fit
of the two empirical models used to describe the relationship between
$b(T)$ and $n(T)$ and T (Mattick et al., 2001). (The same data could also be
fitted by the same model, but with a fixed power, i.e., $n(T)$ = constant =
n, [Peleg and Normand, 2004]; see Chapter 3.)

The empirical $b(T)$ and $n(T)$ models were combined with the temperature profile models of various nonisothermal treatments to produce the
corresponding $b[T(t)]$ and $n[T(t)]$. These in turn were incorporated into
the model rate equation (Equation 2.9), which was then solved numerically by Mathematica® to generate the corresponding survival curve for
each temperature profile. These theoretical curves, calculated solely from
the isothermal data, were superimposed on the experimental survival
ratios, which were recorded by Dr. K.L. Mattick in New Jersey the following year.

Four examples of the predicted vs. actually observed survival patterns,
taken from Mattick et al. (2001), are shown in Figure 2.20. They clearly

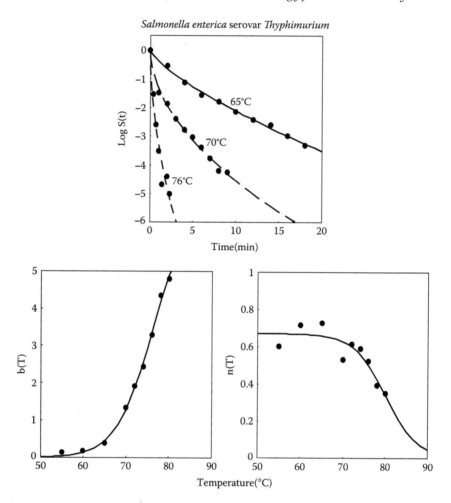

FIGURE 2.19

Isothermal survival curves of *Salmonella* fitted with the Weibullian model (Equation 2.5) and the temperature dependence of the survival parameters $b(T)$ and $n(T)$. (From Mattick, K.L. et al., 2001, *J. Food Prot.*, 64, 606–613. Courtesy of the International Association for Food Protection.)

show that the model (Equation 2.9) was adequate to describe nonisothermal survival curves and could also be used to predict them with fairly good accuracy. It should be reemphasized that the figure shows predictions in the scientific sense of the term; i.e., none of the survival data recorded in the nonisothermal treatments had been used to formulate the model.

The successful predictions also confirm the validity of the assumptions on which the model is based — that is, that the momentary inactivation rate primarily depends on the momentary temperature and survival ratio,

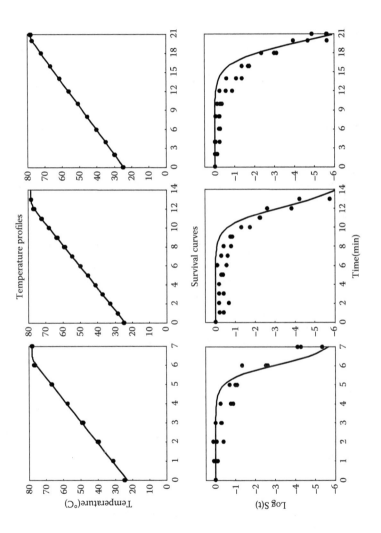

FIGURE 2.20
Prediction of the nonisothermal survival patterns of *Salmonella* from isothermal inactivation data. The filled circles in the plot at the bottom are the experimental data and the solid line is the predicted curves using Equation 2.9 as a model. (From Mattick, K.L. et al., 2001, *J. Food Prot.*, 64, 606–613. Courtesy of the International Association for Food Protection.)

but not on the path at which the momentary survival ratio has been reached. If the rate had depended significantly on the path, the quality of the predictions would have been influenced by the heating rate. However, no such trend has been observed, so one must conclude that, at least for the *Salmonella* cells examined and the particular medium in which they were treated, the assumptions underlying the model were basically correct.

As shown by Mattick et al. (2001) and Periago et al. (2004) (see below), the model should not be used to estimate the outcome of marginal heat treatments that accomplish only one to two orders of magnitude reduction in the survival ratio. This is because the data scatter, sometimes on the order of one log cycle, may create the impression of poor prediction.

Isothermal and Nonisothermal Survival Curves of *B. sporothermodurans* Spores in Soups

Isothermal experimental survival curves recorded in The Netherlands of *B. sporothermodurans* spores in three soups (Periago et al., 2004) are shown in Figure 2.21. The figure also shows the fit of the Weibullian–power law model (Equation 2.1) with a fixed $n(T)$ and the temperature dependence of the rate parameter, $b(T)$, fitted by the log logistic model (Equation 2.10). The regression parameters are given in Table 2.1. (Notice the mention of the "assumed come-up time 30s." Because the actual come-up time had not been determined in the original experiments, the calculations were repeated with assumed come-up times of 45 and 60s. However, it has been shown that, within the realistic range of 30 to 60s, the results were practically unaffected by the assumed come-up time [Periago et al., 2004].)

The organism's survival parameters listed in the table — namely, n, T_c, and k — were used to predict the outcome of nonisothermal heat treatments in experiments performed in Spain, using Equation 2.11 as a model. The results are shown in Figure 2.22. Again, although the predicted survival curves (solid lines) were far from perfect (the dashed lines are the data *fitted* with the model, with k and T_c the adjustable parameters), they were also not too far off, considering that the isothermal and nonisothermal data were obtained by different experimental procedures and different instruments (see Periago et al., 2004.)

The Isothermal and Nonisothermal Inactivation of *E. coli*

Valdramidis et al. (2004) reported survival curves of *E. coli* K12 MG1655, a surrogate to the pathogenic *E. coli* 0157:H7, under several isothermal conditions and during two nonisothermal heat treatments. The isothermal survival data were fitted with Equation 2.5 as a model with variable and fixed $n(T)$ and the results are shown in Figure 2.23. (The fitted curves

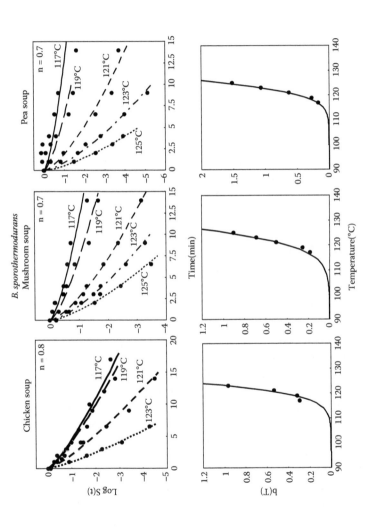

FIGURE 2.21

Isothermal survival curves of *Bacillus sporothermodurans* spores in soup fitted with the Weibullian model (Equation 2.5) with a fixed power, $n(T) = n =$ const, and the temperature dependence of $b(T)$. Notice the almost perfect fit of the log logistic model (Equation 2.10). The regression parameters are reported in Table 2.1. (From Periago, P.M. et al., 2004, *Int. J. Food Microbiol.*, 95, 205–218. With permission, courtesy of Elsevier Ltd.)

TABLE 2.1

Inactivation and Regression Parameters of *B. sporothermodurans* Spores in Three Soups[a]

Soup	Temp (°C)	Isothermal survival parameters			Log logistic $b(T)$ parameters		
		$b(T)$	$n(T)$	χ^2	T_c (C)	k (C^{-1})	χ^2
Chicken	117	0.30	0.8	0.033	122	0.32	0.007
	119	0.32		0.013			
	121	0.53		0.103			
	123	0.97		0.011			
Calculated directly from the nonisothermal data			0.8		124	0.43	0.021[b]
Mushroom	117	0.18	0.7	0.010	123	0.24	0.002
	119	0.25		0.026			
	121	0.50		0.002			
	123	0.69		0.028			
	125	0.91		0.034			
Calculated directly from the nonisothermal data			0.7		123	0.41	0.190[b]
Pea	117	0.17	0.7	0.115	122	0.37	0.003
	119	0.27		0.049			
	121	0.64		0.083			
	123	1.07		0.027			
	125	1.52		0.006			
Calculated directly from the nonisothermal data			0.7		122	0.36	0.097[b]

[a] Assumed come-up time is 60 s.
[b] See Chapter 4.

Source: Periago, P.M. et al., 2004, *Int. J. Food Microbiol.*, 95, 205–218. With permission, courtesy of Elsevier Ltd.

when $n(T)$ was fixed or allowed to vary are shown as dashed and solid lines, respectively.) Both models had a comparable fit as judged by the mean square error and neither was found to be consistently superior.

As expected, when $n(T)$ was an adjustable parameter, the models' fit was slightly better in most cases but not in all. Nevertheless, the adjusted values of $n(T)$ showed no clear trend of increasing or decreasing with temperature. Consequently, the regression was repeated with the averaged value of $n(T)$, which was about 1.5. This value was considered a representative of the Weibullian model exponent. The corresponding $b(T)$ values, also shown in Figure 2.23, were all calculated with the fixed $n = 1.5$. The temperature dependence of $b(T)$ could be described by the log logistic model (Equation 2.10), which had an almost perfect fit. With n fixed by averaging and k and T_c determined by regression, the survival parameters of the organism in the given medium were $n = 1.5$, $k = 0.88°C^{-1}$, and $T_c = 60.5°C$.

The two nonisothermal temperature profiles, A and B, reported by Valdramidis et al. (2004) are shown in Figure 2.24. They are characterized

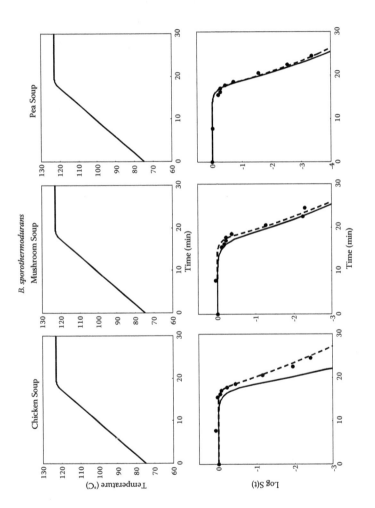

FIGURE 2.22
Nonisothermal survival curves of *Bacillus sporothermodurans* spores in chicken, mushroom, and pea soups predicted from their isothermal inactivation data. Solid lines: the prediction of the Weibullian–log logistic model (Equation 2.11) with coefficients listed in Table 2.1. Dashed line: the fit of the model when n, k, and T_c are adjustable parameters. (From Periago, P.M. et al., 2004, *Int. J. Food Microbiol.*, 95, 205–218. With permission, courtesy of Elsevier Ltd.)

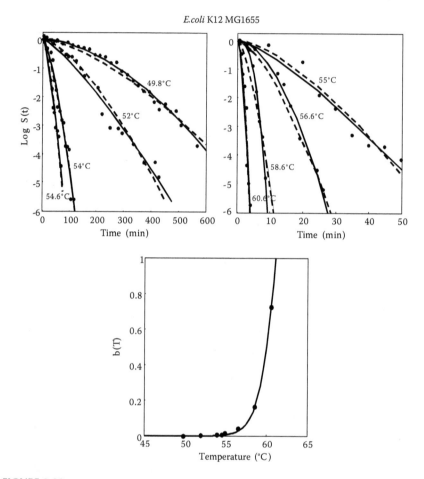

FIGURE 2.23

Isothermal survival curves of *E. coli* K12 MG1655 fitted with the Weibullian model (Equation 2.5) with variable $n(T)$ (solid lines) and a fixed $n = 1.5$ (dashed lines) (top). The log logistic temperature dependence of $b(T)$ is shown at the bottom. Notice the almost perfect fit of Equation 2.10. The original experimental data are from Valdramidis, V.P. et al., 2004, *Proceedings of the 9th International Conference of Engineering and Food ICEF 9*, Montpellier, France. (From Corradini, M.G. and Peleg, M. 2004, *J. Food Prot.*, 67, 2617–2621. With permission, courtesy of the International Association for Food Protection.)

by a region of constant rate heating starting at 30°C followed by a constant temperature region at 55°C. This kind of a profile could be described by the empirical models:

$$\text{Profile A: } T(t) = 55 - \log_e\{1 + \exp[0.798(31.0 - t)]\} \qquad (2.31)$$

$$\text{Profile B: } T(t) = 55 - \log_e\{1 + \exp[1.66(14.5 - t)]\} \qquad (2.32)$$

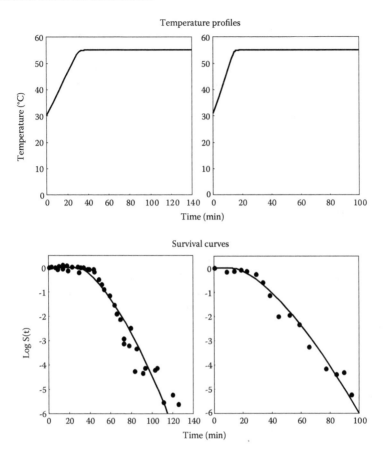

FIGURE 2.24
Two nonisothermal survival curves of *E. coli* K12MG1655 predicted from the isothermal inactivation data shown in Figure 2.23. Filled circles: the experimental data reported by Valdramidis, V.P. et al., 2004, *Proceedings of the 9th International Conference of Engineering and Food ICEF 9*, Montpellier, France. Solid lines: the curves predicted with the Weibullian–log logistic model (Equation 2.11). (From Corradini, M.G. and Peleg, M., 2004, *J. Food Prot.*, 67, 2617–2621. With permission, courtesy of the International Association for Food Protection.)

The two profiles were incorporated into the log logistic expression to produce the corresponding $b[T(t)]$ terms, which were then inserted into the Weibullian model's rate equations (Equation 2.9). These were then solved numerically with Mathematica® to generate the organism's estimated survival curves under the two heating profiles.

The survival curves thus calculated are shown as solid lines in the plots at the bottom of Figure 2.24. Also shown in these plots are the experimentally determined survival ratios (filled dots) as reported by Valdramidis et al. (2004). The figure demonstrates the agreement between the estimated and actual survival curves and provides another example that the Weibull–log logistic model combination can be used not only to characterize the isothermal inactivation curves of *E. coli*, but also to predict correctly its survival patterns during nonisothermal heat treatments.

Nevertheless, recall that, in the previous demonstration of the models' capabilities (the one that referred to the *B. sporothermodurans* spores' survival), the estimated survival ratios were consistently underestimated, particularly in the case of the chicken soup. This should serve as a reminder that if the model's estimates are to be implemented in thermal preservation of food and pharmaceuticals, an added safety factor would be not only prudent, but necessary.

It should be mentioned, though, that the limited experimental data available to us for the analysis of the *B. sporothermodurans* spores inactivation had not been originally intended to test the proposed model or any other survival model for that matter. Consequently, the temperature profiles of the three soups' heat treatments, the number of data points, and their spacing were far from ideal for the purpose. This is in contrast with those in the *Salmonella* experiments reported by Mattick et al. (2001), which have been discussed earlier, and the *E. coli* experiments of Valdramidis et al. (2004), which were originally designed to test a kinetic model, albeit of a different kind.

Heat-Induced Chemical and Physical Changes

Heat not only destroys microorganisms. Exposure of a food or a pharmaceutical product to a high enough temperature usually accelerates other changes, whether they are textural as in processed foods and/or chemical in both. The chemical changes can have an adverse effect on the quality of the processed product in two ways:

- Loss of a desirable component, e.g., a vitamin, pigment, flavor compounds, etc.
- Formation of an undesirable chemical compound or compounds, as in the case of nonenzymatic browning reactions.

The products of such chemical reactions may merely impart an unpleasant off-flavor or adversely affect the product's appearance; however, they may also become a health risk (e.g., Arnoldi, 2001). As a result, the design of an optimal thermal process of a food or a heat-sensitive pharmaceutical product may require examination of the heat's lethal effect on microbial cells and spores and how it might influence the product's chemistry. Examination of this kind requires that the kinetics of microbial inactivation be linked to that of chemical and biochemical reactions and to that of biophysical processes, which can also affect the product's quality and safety.

Some of the problems found in the traditional microbial inactivation kinetic models also exist in the traditional kinetic models of complex biochemical reactions and biophysical processes (Peleg et al., 2004). The principal one is the application of the Arrhenius equation to account for the temperature dependence of complex chemical reactions' rate constant. Again, what the rate constant means in the context of nonlinear reaction kinetics is unclear. Why must all biochemical reactions, regardless of the number and complexity of their interacting pathways, have a single energy of activation is an idea yet to be satisfactorily explained.

It seems that it will be more reasonable to assume that the rate of complex biochemical reactions and biophysical processes depends on the momentary temperature and the system's *state* (Corradini and Peleg, 2004a; Peleg et al., 2004). If so, then we can use the same modeling approach that we applied to microbial inactivation (and growth). This concept has been demonstrated in the isothermal degradation patterns of an enzyme, a vitamin, and a pigment. All three could be described by the Weibullian–power law model with a fixed shape factor of power, n, as shown in Figure 2.25. This observation is not at all surprising because an enzyme's inactivation curve and a pigment's or vitamin's chemical degradation curves are in essence survival curves too. The survivors in this case, however, are not microbial cells or spores, but rather molecules that remain active or intact after their exposure to heat. Thus, the survival curves of microorganisms and complex molecules seem to be a manifestation of a progressive "failure phenomenon," of which the Weibullian rather than the first-order kinetics model might be a more faithful representative.

Once described in this manner, the same mathematical procedures to calculate the nonisothermal microbial inactivation could be used to describe and predict biochemical degradation patterns too. As shown in Figure 2.25, the Weibullian model with a constant power, n, fitted the experimental data very well and the rate parameter, $b(T)$, of all three molecular species indeed had a log logistic temperature dependence. Thus, the rate models and calculation procedures described in this chapter could probably be used to estimate the fate of these compounds under any temperature profile, at least in principle.

Initial results (Corradini and Peleg, 2006b) indicate that it can be true in practice as well — i.e., the outcome of nonisothermal degradation processes can be predicted from isothermal degradation data. In fact, the degradation curves of nutrients or pigments can be calculated and plotted

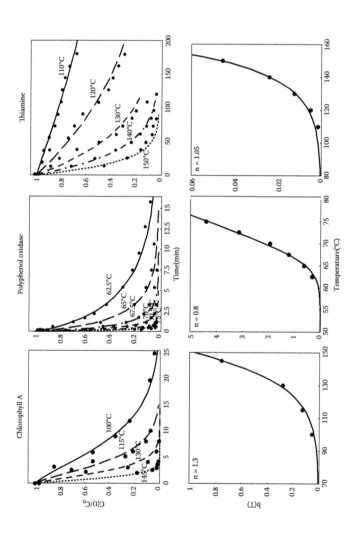

FIGURE 2.25

Isothermal degradation curves of chlorophyll A and thiamin and the inactivation of polyphenol oxidase, fitted with the Weibullian model (Equation 2.5) with a fixed power n. Notice that the degradation/inactivation did not follow the first-order kinetics that has been traditionally assumed ($n = 1.3$, 1.05, and 0.8, respectively). The temperature dependence of $b(T)$ fitted with the log logistic model (Equation 2.10) is shown at the bottom of the figure. Notice that there is no reason to assume that the degradation/inactivation process must have a single energy of activation, as has been traditionally done. (From Corradini, M.G. and Peleg, M., 2004, *J. Sci. Food Agric.*, 84, 217–226. With permission, courtesy of Wiley Publishing.)

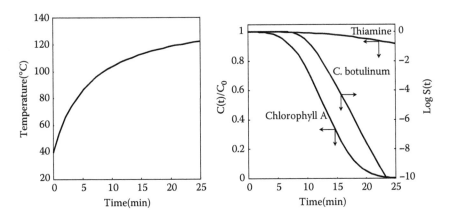

FIGURE 2.26
Simultaneous presentation of microbial inactivation and nutrient/pigment degradation in a heat treatment using the Weibullian–log logistic model (Equation 2.11). Notice that such curves can be produced in real time during an industrial process (see Chapter 3) as well as in computer simulations. (From Corradini, M.G. et al., 2005, *J. Food Eng.*, 67, 59–69. With permission, courtesy of Elsevier Ltd.)

simultaneously with the survival curves of the targeted organism, as shown in Figure 2.26 (Corradini et al., 2005). Such plots can help to design heat processes that would guarantee not only the product's microbial safety, but also maximal retention of nutrients and other desirable ingredients. In principle at least, the same can be done for purely physical attributes, such as texture or color. However, because of heat transfer considerations, isothermal softening data might be difficult if not impossible to obtain experimentally because the product's texture changes rapidly. (There is simply no capillary equivalent for mechanical properties' determination.)

However, whenever mechanical changes could be reasonably monitored isothermally, one would be able to estimate the effect of heat treatments on the targeted microbial population and the product's nutritional quality and appearance, as well on its texture. There is no reason why the same mathematical procedures described in this chapter could not be applied to biochemical reactions whose outcome is growth or accumulation rather than degradation or decay (Peleg et al., 2004). Nonenzymatic browning and the formation of oxidized volatiles during storage, would be natural candidate systems to test this hypothesis. (As to predicting nonisothermal *microbial growth*, the methodology has already been successfully applied, as will be shown in Chapter 8).

3

Generating Nonisothermal Heat Inactivation Curves with Difference Equations in Real Time (Incremental Method)

> God does not care about our mathematical difficulties. He integrates empirically.
>
> **Albert Einstein**

The deficiencies of the current methods to calculate the efficacy of thermal preservation processes and the theories on which they are based have been known to many researchers in the field for many years. Yet, the literature on the subject has had very little impact, if any at all, on the way in which sterility is calculated in the food industry and probably in other industries as well. The same can be said about the way in which microbial inactivation kinetics is taught at universities in the U.S. and around the world. Conservatism and the widespread belief that microbial mortality really follows the first-order kinetics model are probably the main causes.

Many practicing microbiologists and food technologists, perhaps the vast majority, think that the nonlinearity of survival curves is a marginal issue of interest to academicians only, with no relevance to the "real world" of commercial food processing and sterilization in the pharmaceutical industry. However, the overt objection by many to any attempt to reassess the validity of the traditional methods to calculate sterility has two other causes.

The first is the fear that any change in the calculation procedures currently in use will inevitably lead to relaxation of safety standards — a risk no responsible person is willing to take. The argument is that the traditional methods based on D, z, and F_0 values have worked fine for decades and therefore there is no reason to change anything. ("If it ain't broke, don't fix it.") The response to this argument, as already mentioned, is that systematic overprocessing of foods, at least, need not be beneficial in the long run, not to mention the quality and nutritional losses that might accompany it. Alternatively, in some situations, at least theoretically, the standard method of calculation can lead to underprocessing and thus to an increased risk.

The second reason is that an easy-to-learn *practical alternative* that could readily replace the methods now in use has yet to be developed. Not surprisingly, the models and methods discussed in the previous chapters, which are described in several technical publications, have made little impression on the food industry, with the exception of a few research groups. (The same can be said about academia, in which nonlinear inactivation kinetics is not a topic emphasized in the curriculum.) One reason is that, to implement the described models, one needs to solve a differential equation numerically, a task that requires special programming or the use of advanced software like Mathematica®. Neither, it seems, is an attractive option to industrial and academic microbiologists, unless they have special training or interest in advanced methods of computation.

It would therefore be beneficial to the food industry and probably to the pharmaceutical and other industries as well if the proposed alternative models and methods could be implemented in a general-purpose software like Microsoft Excel®. Once available in the form of a spreadsheet or the format of any other familiar program, the method could be used for:

- Following the progress of thermal processes in the form of an evolving nonisothermal survival curve or several survival curves recorded simultaneously *in real time.*
- Incorporating the program in the control system of industrial processes to assure their safety. (Microbial safety will be defined in terms of a final theoretical survival ratio deemed safe and an added safety factor.)
- Simulation and analysis of the performance of existing and planned thermal processes (in industrial research laboratories and regulating agencies).
- Teaching and instruction in industrial professional development and academic courses on thermal processing.

The Difference Equation of the Weibullian–Log Logistic Nonisothermal Survival Model

Consider a typical or bacterial spore organism whose isothermal survival curves obey the Weibullian–power law model with a fixed (or representative) power, n:

$$\log_{10}S(t) = -b(T)t^n \tag{3.1}$$

and where the temperature dependence of $b(T)$ follows the log logistic model:

$$\log_{10} b(T) = \log_e\{1 + \exp[k(T - T_c)]\} \qquad (3.2)$$

As previously explained (see Chapter 2), the nonisothermal rate equations of the organism under a temperature profile, $T(t)$, will be:

$$\frac{d \log_{10} S(t)}{dt} = -b[T(t)]n\left\{\frac{-\log_{10} S(t)}{b[T(t)]}\right\}^{\frac{n-1}{n}} \qquad (3.3)$$

where $b[T(t)]$ is defined by replacing T in Equation 3.2 by $T(t)$.

As has already been shown, Equation 3.3 can be solved numerically by a program like Mathematica® to produce the $\log_{10} S(t)$ vs. t relationship — i.e., the survival curve that corresponds to the particular heat treatment as described by the temperature profile $T(t)$.

Suppose now that we have temperature readings at time intervals short enough so that the segments of the survival curve within each of these intervals are approximately linear. In such a case, the slope of the survival curve or the average inactivation rate within the interval can be approximated by:

$$\frac{d \log_{10} S(t)}{dt} \approx \frac{\log_{10} S(t_i) - \log_{10} S(t_{i-1})}{t_i - t_{i-1}} \qquad (3.4)$$

where

t_i and t_{i-1} are the times at the beginning and end of the interval, respectively

$\log_{10} S(t_i)$ and $\log_{10} S(t_{i-1})$ are the corresponding survival ratios, respectively

The average survival ratio within the time interval:

$$\overline{\log_{10} S(t)} = \frac{\log_{10} S(t_i) + \log_{10} S(t_{i-1})}{2} \qquad (3.5)$$

and the mean value of the Weibullian rate parameter, $b[T(t)]$, is:

$$\frac{\overline{b[T(t)]}}{b[T(t)]} = \frac{b[T(t_i)] + b[T(t_{i-1})]}{2} \qquad (3.6)$$

Replacing the $\log_{10} S(t)$ and $b[T(t)]$ terms in the survival rate model equation (Equation 3.3) by the preceding expressions yields the *difference equation*:

$$\frac{\log_{10} S(t_i) - \log_{10} S(t_{i-1})}{t_i - t_{i-1}} = -\frac{b[T(t_i)] + b[T(t_{i-1})]}{2} \cdot n$$

$$\cdot \left\{ \frac{-[\log_{10} S(t_i) + \log_{10} S(t_{i-1})]}{b[T(t_i)] + b[T(t_{i-1})]} \right\}^{\frac{n-1}{n}} \qquad (3.7)$$

which now becomes the discrete replacement for the continuous survival of the model.

If n is not a constant, but a function of temperature, too — i.e., $n = n[T(t)]$ — one can write:

$$N[T(t)] \approx \frac{n[T(t_i)] + n[T(t_{i-1})]}{2} \qquad (3.8)$$

and:

$$\frac{n[T(t)] - 1}{n[T(t)]} \approx \frac{n[T(t_i)] + n[T(t_{i-1})] - 2}{n[T(t_i)] + n[T(t_{i-1})]} \qquad (3.9)$$

which will replace the fixed $(n - 1)/n$ in the model's less elaborate form (Equation 3.7).

For the sake of simplicity, though, what follows will deal only with the case of a constant n, i.e., $n(T) = n$. However, it will become obvious that the general calculation procedure applies to temperature-dependent and -independent n. Because n can be smaller, equal to, or larger than one, all the preceding and what follows also apply to organisms whose isothermal survival curves are log linear ($n = 1$). The departure from the conventional model here is that the temperature dependence of $b(T)$ follows the log logistic rather than the log linear or Arrhenius model.

Suppose that the starting point of the thermal process at hand is specified by $t_0 = 0$, $T(0) = T_0$ (the initial temperature), $\log_{10} S(0) = 0$, and $b[T(0)] = b[T_0]$. If so, then after a short time, t_1, the temperature reaches $T(t_1)$, the Weibullian rate parameter becomes $b[T(t_1)]$ and the corresponding survival ratio $\log_{10} S(t_1)$. Incorporating these into Equation 3.7 yields:

$$\frac{\log_{10} S(t_1)}{t_1} = -\frac{b[T(t_1)] - b[T_0]}{2} \cdot n \cdot \left\{ \frac{-\log_{10} S(t_1)}{b[T(t_1)] - b[T_0]} \right\}^{\frac{n-1}{n}} \qquad (3.10)$$

Because three survival parameters, n, k, and T_c, are known (determined from experimental isothermal data), n can be inserted into the equation directly. The values of $b[T(t_1)]$ and $b[T_0]$ can be calculated with the log logistic equation (Equation 3.2) and entered into the model equation (Equation 3.10), which now becomes an algebraic equation with only one unknown, namely, $\log_{10} S(t_i)$. This equation can be readily solved analytically or numerically by using Microsoft Excel's "goal seek" command to produce the sought value of $\log_{10} S(t_1)$.

With $\log_{10} S(t_1)$ known, one can calculate the value of $\log_{10} S(t_2)$ in a similar manner by solving Equation 3.7. The first step is to enter $T(t_2)$ after it is measured directly or calculated using the temperature profile model equation, $T(t)$. Once $T(t_2)$ is determined, the corresponding $b[T(t_2)]$ is calculated using Equation 3.2 as before. Incorporating the value of $b[T(t_2)]$ into the model yields the equation:

$$\frac{\log_{10} S(t_2) - \log_{10} S(t_1)}{t_2 - t_1} = -\frac{b[T(t_2)] + b[T(t_1)]}{2} \cdot n$$

$$\cdot \left\{ \frac{-[\log_{10}[S(t_2)] + \log_{10} S(t_1)]}{b[T(t_2)] + b[T(t_1)]} \right\}^{\frac{n-1}{n}} \tag{3.11}$$

This equation, too, is an algebraic equation that has only one unknown this time: $\log_{10}[S(t_2)]$. As before, it can be solved analytically or by using the "goal seek" command of MS Excel®. The procedure can now be repeated to calculate $\log_{10} S(t_3)$, $\log_{10} S(t_4)$, etc., until the whole thermal process's duration is covered. The series of the $\log_{10} S(t_i)$ values vs. the corresponding t_i's is the theoretical survival curve of the targeted organism if it had been exposed to the given temperature profile. The temperature entries, as shown, can be direct measurements, the output of a digital thermometer (logger) or any other temperature-monitoring device, or generated periodically by the program using the temperature profile's equation. Either way, the survival curve will be produced in what for all practical purposes is *real time* — in the first case, that of the process and, in the second, of its simulated progress.

Notice that when $n(T)$ is not a constant, but rather a function of temperature (and thus of time also — see Equation 3.8), its value will also change at each iteration. However, because the temperature $T(t_i)$ at each and every iteration is known, the corresponding value of $n[T(t_i)]$ can be calculated in the same manner that $b[T(t_i)]$ was. Once $n[T(t_i)]$ is calculated in this way, its numeric value is inserted into the model's equation as explained earlier. Although the model's power term will progressively change, like the rate term, still only one unknown will be in the equation

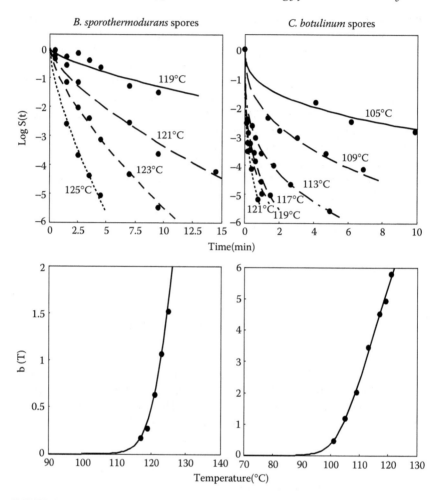

FIGURE 3.1

Isothermal survival curves of *B. sporothermodurans* and *C. botulinum* spores fitted with Equation 3.1 as a primary model (top) and the temperature dependence of the rate parameter, $b(T)$, fitted with the log logistic model (Equation 3.2) as secondary model (bottom). The original experimental data are from Periago, P.M. et al., 2004, *Int. J. Food Microbiol.*, 95, 205–218, and Anderson, W.A. et al., 1996, *J. Appl. Bacteriol.*, 80, 283–290. (From Peleg, M. et al., 2005, *J. Appl. Microbiol.*, 98, 406–417. With permission, courtesy of Blackwell Publishing.)

at each iteration — that is $\log_{10}S(t_i)$, which would be calculated in exactly the same way as before.

Examples of experimental isothermal survival curves fitted with Equation 3.1 as a model are shown in Figure 3.1, together with the fit of the log logistic model to the temperature dependence of $b(T)$. Simulated survival curves that emulate what happens in a realistic heat treatment based on them are shown in Figure 3.2. The plots were produced to show how

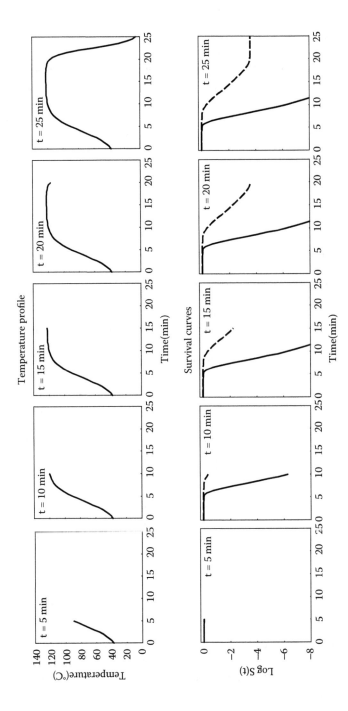

FIGURE 3.2
Nonisothermal survival curves of *C. botulinum* (solid lines) and *B. sporothermodurans* spores (dashed lines) generated with MS Excel® in real time using the incremental model (Equation 3.7). (From Peleg, M. et al., 2005, *J. Appl. Microbiol.*, 98, 406–417. With permission, courtesy of Blackwell Publishing.)

the changing temperature during the process would affect, theoretically, the fate of *C. botulinum* spores and *B. sporothermodurans* if they had been present in the treated hypothetical product.

The survival parameters, n, k, and T_c, were reported by Peleg et al. (2005), based on the experimental data of Anderson et al. (1996) and Periago et al. (2004). The figure also demonstrates that the procedure can be used to generate the survival curves of more than one organism simultaneously and that a process designed to destroy the spores of *C. botulinum* can be ineffective against the much more heat-resistant spores of *B. sporothermodurans*.

Similar curves can be produced by pasting an actual time–temperature data set to determine whether the processed product is safe and whether it would remain stable if the more resistant spores had been present.

Non-Weibullian Survival Curves

The described method of expressing the nonisothermal survival curves in terms of a difference equation that can be solved by MS Excel® or similar software is not limited to the Weibullian–log logistic model combination. In principle at least, it can be applied to organisms whose isothermal survival curves follow *any kind of primary model,* provided that its equation has an analytical inverse and that its parameters' temperature dependence (the secondary models), can be expressed algebraically. (To calculate nonisothermal inactivation patterns of organisms whose primary survival model equation has no analytical inverse will require special programming. The same applies to situations in which the momentary magnitude of the survival parameters is not calculated by an equation but is entered in other ways.)

Here is a demonstration of the method's use with an alternative survival model. Published survival data on *Salmonella* (Mattick et al., 2001) could be fitted not only by the Weibullian but also by the empirical model:

$$\log_{10} S(t) = -\frac{k_1(T)t}{k_2(T)+t} \qquad (3.12)$$

where $k_1(T)$ and $k_2(T)$ are temperature-dependent coefficients.

This model, as has been discussed in previous chapters, is particularly convenient for describing survival curves exhibiting an extreme degree of "tailing" — i.e., of survival curves that end up in a finite number of survivors. However, Equation 3.12 is also a flexible enough model that can fit many "regular" concave upward isothermal curves if the asymptotic survival ratio, $-k_1(T)$, is set to be much smaller than the lowest measurable survival rate. According to this model, the continuous nonisothermal inactivation rate equation, as shown in Chapter 2, is:

$$\frac{d\log_{10} S(t)}{dt} = -\frac{k_1[T(t)]k_2[T(t)]}{(k_2[T(t)]+t^*)^2} \tag{3.13}$$

where

$$t^* = -\frac{k_2[T(t)]\log_{10} S(t)}{k_1[T(t)]-\log_{10} S(t)} \tag{3.14}$$

For small time increments, this differential equation can be transformed into the rather cumbersome difference equation:

$$\frac{\log_{10} S(t_i)-\log_{10} S(t_{i-1})}{t_i-t_{i-1}} = -\frac{\{k_1[T(t_i)]+k_1[T(t_{i-1})]k_2[T(t_i)]+k_2[T(t_{i-1})]\}}{[\{k_2[T(t_i)]+k_2[T(t_{i-1})]\}+2t^*]^2} \tag{3.15}$$

where

$$t^* = -\frac{\{k_2[T(t_i)]+k_2[T(t_{i-1})]\}\cdot[\log_{10} S(t_i)+\log_{10} S(t_{i-1})]}{k_1[T(t_i)]+k_1[T(t_{i-1})]-\log_{10} S(t_i)-\log_{10} S(t_{i-1})} \tag{3.16}$$

Here, again, the momentary survival parameters at each iteration — namely, $k_1[T(t_i)]$, $k_1[T(t_{i-1})]$, $k_2[T(t_i)]$, and $k_2[T(t_{i-1})]$ — can be calculated from their respective models using $T(t_i)$ and $T(t_{i-1})$. Thus, if we start at $t_0 = 0$ and $\log_{10}S(0) = 0$, the only unknown in the equation at the first iteration will be $\log_{10}S(t_1)$, which can be found by solving the equation with the "goal seek" function, as has been done when the isothermal survival curves were described by the Weibullian–power law model. Similarly, the only unknown at the second iteration will be $\log_{10}S(t_2)$ at the third iteration $\log_{10}S(t_3)$ and so forth.

Experimental survival curves of *Salmonella* fitted with Equation 3.12 as a model are shown in Figure 3.3. Also shown in the figure is the temperature dependence of $k_1(T)$ and $k_2(T)$ fitted by ad hoc empirical model. (Because the same isothermal data can also be fitted by the Weibullian model with fixed or variable power (Mattick et al., 2001; Peleg and Normand, 2004), none of the models mentioned here can be considered unique. They are, however, *convenient* models, and their parameters have an intuitive meaning.)

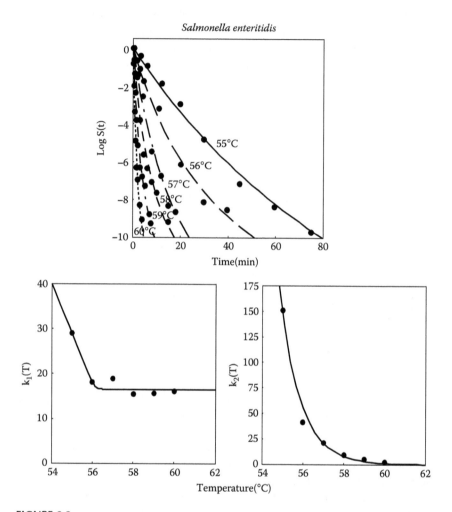

FIGURE 3.3

Isothermal survival curves of *Salmonella* fitted with Equation 3.1 as a model (top) and the temperature dependence of its parameters $k_1(T)$ and $k_2(T)$ fitted with ad hoc empirical secondary models (bottom). The original experimental data are from Mattick, K.L. et al., 2001, *J. Food Prot.*, 64, 606–613. Notice that the same survival data can be also fitted with the Weibullian–power law model, with fixed or variable power. (From Peleg, M. et al., 2005, *J. Appl. Microbiol.*, 98, 406–417. With permission, courtesy of Blackwell Publishing.)

Examples of a temperature profile and the corresponding theoretical survival curves of the organism calculated by the preceding iterative procedure are shown in Figure 3.4. Once more, the purpose of presenting these plots is to demonstrate that the procedure can work not only with the Weibullian–power law model, but also with alternative survival models.

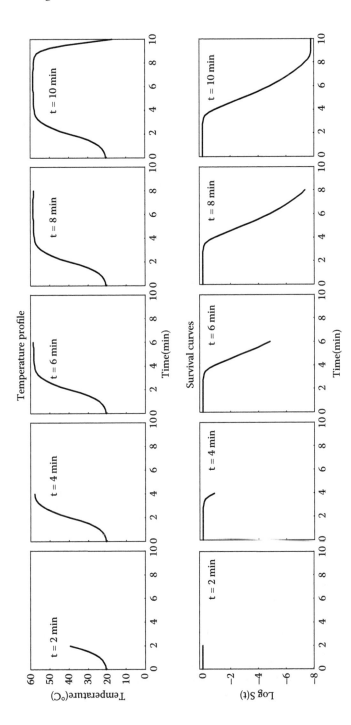

FIGURE 3.4

A nonisothermal survival curve of *Salmonella* generated with MS Excel® in real time using the incremental model (Equation 3.15 and Equation 3.16). (From Peleg, M. et al., 2005, *J. Appl. Microbiol.*, 98, 406–417. With permission, courtesy of Blackwell Publishing.)

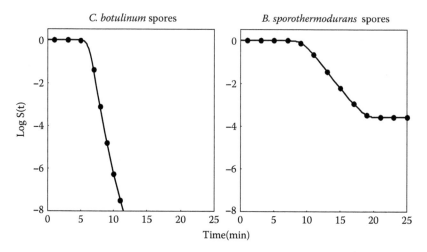

FIGURE 3.5
Comparison of the complete nonisothermal survival curves of *C. botulinum* spores and *B. sporothermodurans* generated with the incremental method (solid circles) and the continuous model. (From Peleg, M. et al., 2005, *J. Appl. Microbiol.*, 98, 406–417. With permission, courtesy of Blackwell Publishing.)

Comparison between the Continuous and Incremental Models

Examples of nonisothermal survival curves generated with the continuous models (Equation 3.3 and Equation 3.13) and their incremental versions, Equation 3.7 and Equation 3.15, respectively, are shown in Figure 3.5 and Figure 3.6. The solid lines in the plots are the outputs of the Mathematica® program based on the continuous models and the filled circles those of the discrete versions of the model. The agreement between the two models is almost perfect and demonstrates that the continuous and incremental versions of these models can be used interchangeably.

In fact, by setting the time increments and accuracy goal of the equation's numerical solution, one can reach almost any degree of agreement between the survival curves produced by the differential and difference equations. This is, of course, not at all surprising because the numerical solution of the continuous version of the survival model's equation by a program like Mathematica® is also an iterative incremental process, except that the discrete intervals are much shorter than those chosen for the demonstration. Also, the algorithms used by Mathematica® to render the solution are probably much more effective and faster than the one we have chosen.

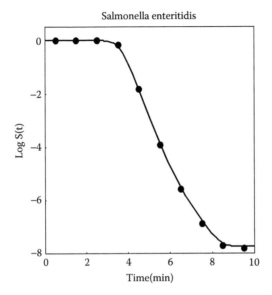

FIGURE 3.6
Comparison of the complete nonisothermal survival curves of *Salmonella* generated with the incremental method (solid circles) and the continuous model. (From Peleg, M. et al., 2005, *J. Appl. Microbiol.*, 98, 406–417. With permission, courtesy of Blackwell Publishing.)

As already mentioned, the incremental method as described is limited to primary survival model equations that have an analytical inverse — that is, t^*, the time corresponding to the momentary survival ratio, $\log_{10}S(t)$, can be expressed algebraically. In principle, this limitation can be overcome by programming a procedure to calculate t^* numerically. However, one can argue that most of the *commonly encountered survival patterns*, including sigmoid survival curves, can be described by primary model equations that do have an analytical inverse and, therefore, that the effort to program the t^* extraction would not be worthwhile.

Nevertheless, in the future, one might consider the development of general purpose software that could be used with any primary survival model without the need to check whether its equation has an analytical inverse or not. In such a case, t^* will always be calculated numerically. If the experience with Mathematica® is any indication, this would have only a minor effect on the time to generate the survival ratios.

Because the described incremental procedure yields results practically indistinguishable from those obtained by solving the original differential equation, but requires less specialized software, it would be most likely the preferable choice of the majority of users, especially in industry. It is to be hoped that its availability in the form of free, user-friendly software will help the food and other industries discontinue the use of the F_0 value as a measure of thermal process efficacy and replace it by the theoretical survival ratio of the targeted microbial cells or bacterial spores.

Unlike the F_0 value, the survival ratio is a model-independent measure of the process efficacy. Its calculation does not require that the targeted organism's isothermal survival curves must be all log linear and that the resulting D value's temperature dependence must be log linear too. If this recommendation is ever implemented, the safety factor will be in the form of added time to that required to reach a given survival ratio, deemed to be satisfactory in light of historical experience and incubation tests, and not in terms of hypothetical units of sterility. These, as has been explained, have a clear meaning only for a particular survival pattern, which is the exception rather than the rule, as ample new evidence indicates.

Two MS Excel programs to generate nonisothermal survival curves by the incremental method, based on the Weibullian–log logistic model (Equation 3.1 through Equation 3.7) have now been posted as freeware on the Web. One is set for simulation heat sterilization and the other for pasteurization. They come in two versions: in one, the temperature profile is generated by a formula whose parameters can be adjusted by the user; in the other, the user is allowed to paste his or her temperature profile data. In both, one can also adjust the targeted spore's or organism's survival parameters, namely, the Weibullian power n and the log logistic k and T_c. In addition to the survival curve, the program also produces an "equivalent isothermal time" (in minutes) at the user's chosen reference temperature. It is calculated at each iteration from the expression (Corradini et al, 2006):

$$ t_{\text{equivalent } i} = \left[\frac{-\log_{10} S(t_i)}{b(T_{\text{ref}})} \right]^{\frac{1}{n(T_{\text{ref}})}} \tag{3.17} $$

where T_{ref} is the reference temperature (traditionally 121.1°C [250°F]) for heat sterilization of low-acid foods.

This provides a more realistic measure of the equivalency of isothermal and nonisothermal heat treatments than the traditional F_0 value because the log linearity of the survival curves and of the D value's temperature dependence is not a requirement (Corradini et al., 2006). The Websites from which the programs can be downloaded are listed in the Freeware section of this book. Examples of their outputs are shown in Figure 3.7.

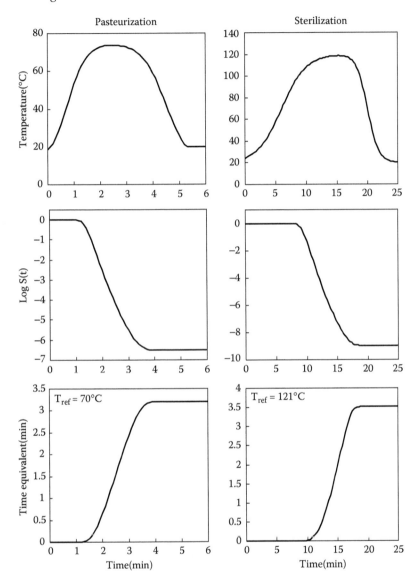

FIGURE 3.7
Two simulated temperature profiles of a pasteurization and sterilization process and corresponding survival and equivalent isothermal time curves produced with the incremental method.

4

Estimation of Microbial Survival Parameters from Nonisothermal Inactivation Data

> We are usually convinced more easily by reasons we have found ourselves than by those which have occurred to others.
>
> **Blaise Pascal**

Traditionally, and regardless of the assumed kinetics (see previous chapters), the database for calculating microbial survival has been a set of isothermal inactivation curves from which the microorganism's or spore's survival parameters have been obtained. This requires that

- A sample of food or other medium that contains a known number of the cells or spores being studied be heated instantaneously to the desired lethal temperature
- The sample be held at this temperature for a specified time
- The sample be cooled instantaneously to a temperature low enough to stop the destruction and also to disallow resumed growth of the survivors before they are counted

An instantaneous isobaric increase or decrease of a real object's temperature is physically impossible, of course, because of heat transfer considerations. (It can be *almost* accomplished by very rapid pressurization or pressure release; see Chapter 6. However, the temperatures reached due to the mere application of ultrahigh pressure are usually still sublethal if the initial temperature is ambient and the high pressure's own effect needs to be considered in such a case.) By external heating and cooling, it is only possible to approximate a "step temperature increase" to be followed by a "step temperature decrease" by shortening the come-up and cooling times of the sample as much as possible relative to the holding time. In contrast, to maintain a small specimen of cells' or spores' suspension at a constant temperature is usually fairly easy. This can be accomplished by placing the small specimen in a large volume of hot oil or water already at the desired temperature. Cooling is done by placing the small specimen in ice water, which is also at a constant but low temperature. Because even a very small volume of liquid (much less than even 1 ml) can still hold a huge number of microbial cells or spores, reducing the treated specimen's size is not considered a problem in most cases. To accomplish the rapid heating and cooling, the sample containing the cells or spores must be held in or passed through a capillary, to assure the shortest heat penetration distance and largest surface area to volume ratio.

Such an experimental procedure, however, cannot be used if the culture medium is too viscous and/or contains suspended solid particulates. This forces the experimenter to choose between two unattractive alternatives: to use a surrogate medium that can be contained in or forced through a narrow tube or to pack the sample in a thin plastic bag or a capsule prior to the treatment. If the first option is chosen, relevant information regarding the medium's effect on the survival pattern will be obviously lost. If the second option is exercised, the temperature distribution within the sample will be largely unknown and thus the overall reliability of the results might be hampered. Regardless of how the samples are heated and cooled, the previously mentioned technical solutions might be totally ineffective whenever one deals with treatments of very high temperatures and very short times. This is because, in such cases, even a few seconds at a temperature only slightly lower than the target temperature can still result in significant sporal inactivation.

A different situation with similar consequences may exist when the organism becomes extremely heat labile within a very narrow temperature range. For example, an organism like *Salmonella, Shigella,* or *E. coli* perishes very fast at temperatures higher than about 60°C. Thus, an error of a single degree in the neighborhood of this temperature may be manifested in significant survival or a reduction of the cells' number to an undetectable level.

Many of the preceding problems could be eliminated, at least in principle, if one could develop a procedure to determine the cells' or spores' survival parameter *from nonisothermal inactivation data* produced under *slow heating regimes* and, if necessary, slow cooling as well. Although the experimental procedures for such analyses are yet to be developed, one can envision that they will be much simpler and possibly more reliable than at least some of the "isothermal" treatments currently in use. The question that remains is whether the development of such procedures is theoretically possible — i.e., whether an organism's or spore's survival parameters can be retrieved from inactivation data obtained under nonisothermal conditions. What follows will demonstrate not only that this is possible, but also that the concept's validity has already been confirmed for *Salmonella* cells and that there is evidence that it might be applicable to bacterial spores as well.

The concept that kinetic parameters can be retrieved from nonisothermal data is not new, of course (see Mizrahi et al., 1978, for example). The proposed departure from the previous approach is the explicit assertion that the momentary rate of change, whether it is inactivation (see Chapter 1 and Chapter 2) or growth, depends not only on the momentary temperature (and/or other conditions that may apply), but also on the system's momentary state. Or in other words, the momentary rate depends on the *process's history* as well as on the momentary conditions. This means that the validity of neither the first order kinetics model or that of the Arrhenius equation can be taken for granted. For nonisothermal microbial inactivation, this means that, in general, the inactivation rate is a function of the momentary temperature and momentary survival ratio. Organisms or spores whose logarithmic inactivation rate is a function of temperature only are those that follow the first-order mortality kinetics. However, as has been demonstrated, they are only a special case and should be considered the exception rather than the rule (see van Boekel, 2002, for example).

The Linear Case

Linear Survival at Constant Rate Heating

Consider an organism whose isothermal survival curves in the lethal temperature range and on the pertinent time scale are log linear for all practical purposes. The thermal inactivation of *Listeria monocytogenes* as reported by Stephens et al. (1994) can serve as an example. The log linear isothermal survival equation, as has been already mentioned, can be viewed as a special case of the Weibullian–power law model with a shape factor power of unity ($n = 1$):

$$\log_{10}S(t) = b(T)t \qquad (4.1)$$

or

$$\left| \frac{d\log_{10} S(t)}{dt} \right|_{T=\text{const}} = b(T) \qquad (4.2)$$

If the temperature dependence of $b(T)$ can be described by the log logistic model (Peleg et al., 2003; see Figure 4.1):

$$b(T) = \log_e\{1 + \exp[k(T - T_c)]\} \qquad (4.3)$$

Then, under nonisothermal temperature profile, the rate equation will be:

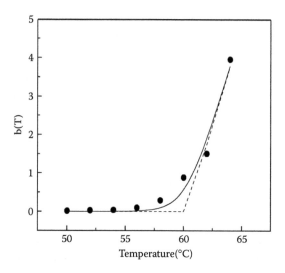

FIGURE 4.1
The $b(T)$ vs. T relationship of *Listeria monocytogenes* fitted with the log logistic model (Equation 4.3) and the discontinuous model (Equation 4.9). The original survival data are from Stephens, P.J. et al., 2004, *J. Appl. Microbiol.*, 77, 702–710. (Adapted from Peleg, M. et al., 2003, *Bull. Math. Biol.*, 65, 219–234. With permission, courtesy of Elsevier Ltd.)

$$\frac{d\log_{10} S(t)}{dt} = -\log_e[1 + \exp\{k[T(t) - T_c]\}] \tag{4.4}$$

In the case in which the temperature increases linearly:

$$T(t) = T_0 + vt \tag{4.5}$$

where T_0 is the initial temperature and the heating rate (in deg.nin^{-1}), in this particular case, the momentary nonisothermal survival rate can be written as:

$$\frac{d\log_{10} S(t)}{dt} = -\log_e[1 + \exp[k(T_0 + vt) - T_c]] \tag{4.6}$$

and the *survival curve* equation will be:

$$\log_{10} S(t) = -\int_0^t \log_e\{1 + \exp[k(T_0 + vt - T_c)]\}dt \tag{4.7}$$

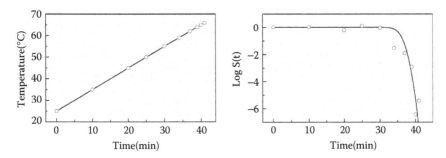

FIGURE 4.2
Simulated linear heating curve (left) and corresponding generated survival data of *Listeria monocytogenes* (right). The open circles are the generated data (with superimposed noise) and solid line the fit of Equation 4.8. For the agreement between the generated and retrieved parameters see text. (From Peleg, M. et al., 2003, *Bull. Math. Biol.*, 65, 219–234. With permission, courtesy of Elsevier Ltd.)

The integral at the right side of the equation has an analytical solution (see the Mathematica® book by Wolfram Research). Thus, the nonisothermal survival curve can be described explicitly by the expression:

$$\log_e S(t) = -\frac{PolyLog\{2, -\exp[k(T_0 + vt - T_c)]\} - PolyLog\{2, -\exp[k(T_0 - T_c)]\}}{kv} \quad (4.8)$$

PolyLog[n,x] (also known as the Jonquire's function) is a standard function in Mathematica® and can be used as an ordinary function in all relevant mathematical operations, *including nonlinear regression*. Thus, if it is known *a priori* that the isothermal survival curves of a given organism at the given medium are all log linear and if a linear heating profile can be programmed — i.e., T_0 and the hating rate, v, are known — then one can calculate that organism's survival parameters, k and T_c, by nonlinear regression using Equation 4.8 as a model. (Notice that only when *all* the isothermal survival curves of the organism in question are log linear will there be only two survival parameters, as in the traditional model. If, however, even some of the isothermal survival curves are curvilinear, then at least three survival parameters will be required to account for the organism's heat resistance — see below.)

The application of the model is demonstrated in Figure 4.2. The figure shows a simulated heating curve where the temperature increases at constant rate (top) and corresponding survival data generated for *Listeria* with Equation 4.8 as a model, to which random noise was added in order to make them realistic (bottom). Also shown in the figure is the fitted survival curve calculated by nonlinear regression again using Equation 4.8 as a model (solid line). To test the procedure to estimate the survival parameters from nonisothermal data, the generation and regression survival parameters k and T_c were compared. They were (Peleg et al., 2003):

Survival parameter	Generation parameter	Regression parameter
n	1.0 (by definition)	1.0 (by definition)
k	$1.01°C^{-1}$	$0.96°C^{-1}$
T_c	60.3°C	60.4°C

Source: Peleg et al., 2003.

As one would expect (see following), the degree of agreement largely depends on the survival data's scatter. More importantly, for the method to work at all, enough data at the lethal temperature region are needed and the more the better. As can be seen from Figure 4.2, T_c can be estimated by visual inspection of the curve — that is, by locating the temperature where the survival curve takes a dip (see figure). In contrast, information regarding the magnitude of k is primarily obtained from the steepness of the survival ratio's drop at temperatures *beyond* T_c, i.e., from data gathered at the lethal regime.

It is already evident from the preceding example that the survival curve of a heat-sensitive organism like *Listeria* beyond its T_c can be *too steep* to allow meaningful determination of k, especially if the experimental survival data have a considerable scatter. Therefore, the *slower* the heating rate is, the more moderate will be the survival curves' dive (more on this issue later). This will make it easier to calculate the magnitude of k and the resulting value will be more reliable. Obviously, the heating rate should not be so slow that it will allow the organism to increase its heat resistance through physiological adaptation; in this case, the method could not be used (see Hassani et al., 2005). This constraint applies to all that follows in this chapter and to the prediction of survival patterns in general as already stated in Chapter 2.

Linear Survival at Varying Heating Rate

In light of heat transfer considerations, creating a linear temperature profile (constant rate heating) requires a programmed and tightly controlled heater. Such a piece of equipment might not be available in many laboratories; thus, it would be advantageous if the survival parameters, k and T_c, could be estimated from survival data obtained under natural temperature profiles of the kind shown in Figure 4.3. Such profiles can be obtained by immersing a well-stirred sample in a hot water or oil bath — an inexpensive standard piece of equipment that exists in almost any microbiology laboratory. (The survival data would be obtained by withdrawing small specimens periodically, cooling them with ice, and then proceeding with the count.)

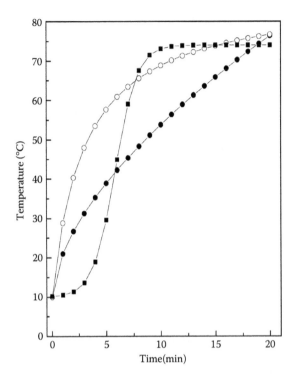

FIGURE 4.3
Simulated temperature profiles using Equation 4.11 through Equation 4.13 as models. Notice that when inserted into Equation 4.9, the survival model's equation (Equation 4.10) will have an analytical solution. (From Peleg, M. et al., 2003, *Bull. Math. Biol.*, 65, 219–234. With permission, courtesy of Elsevier Ltd.)

Unfortunately, when the temperature profile is curvilinear, the rate model's equation (Equation 4.4) cannot be solved analytically. To overcome this problem, one can replace the log logistic component of the model ($b(T)$ in Equation 4.3) by the discontinuous approximation:

$$b(T) = \text{if } [T \le T_c, 0, k\,(T - T_c)] \qquad (4.9)$$

This says that, if $T \le T_c$, $b(T) = 0$; otherwise, it will be $k(T - T_c)$.

The fit of this discrete model to the *Listeria's* survival rate data is shown in Figure 4.1. (Because it does not account for the curvature of $b(T)$ around T_c, its use will provide an *underestimate* of the inactivation rate and thus a more conservative assessment of the process's lethality, at least theoretically.) The expression of the temperature dependence of $b(T)$ as a discontinuous function will render the survival model:

$$\log_{10} S(t) = \mathit{If}[T(t) \leq T_c, 0, -\int_{t_c}^{t} k[T(t) - T_c] dt]$$ (4.10)

where t_c is the time to reach T_c, i.e., $T(t_c) = T_c$.

Notice that if the function that describes the temperature profile, $T(t)$, can be integrated analytically, so would the term $k[T(t) - T_c]$. Consequently, for any such thermal history, the survival curve, $\log_{10} S(t)$ vs. t, can be calculated analytically using Equation 4.10 as a model. Examples of integrable expressions that can fit many experimental heating curves are:

- Monotonic temperature increase with an asymptote $T_0 + 1/a_2$:

$$T(t) = T_0 + \frac{t}{a_1 + a_2 t}$$ (4.11)

- Monotonic temperature increase at a progressively decreasing rate:

$$T(t) = T_0 + a_3 t^m$$ (4.12)

- Logistic increase with an asymptotic temperature T_{target}:

$$T(t) = T_0 + \frac{T_{target} - T_0}{1 + \exp[a_4(t_{ch} - t)]}$$ (4.13)

where
T_0 is the initial temperature
a is a constant
m, T_{target}, and t_{ch} are constants

In fact, the three simulated heating curves shown in Figure 4.3 were produced with Equation 4.9 through Equation 4.11 as models. Most likely, one of the preceding expressions could be used as a regression model for real experimental heating curves. However, if none of the three provides adequate description of the experimental heating curve(s), then one can try an alternative model that can be integrated analytically or use a sum of any number of the same terms (Equation 4.11 through Equation 4.13) with different parameters or any of their combinations (a mixed model). This is because the sum of integrable expressions is also integrable.

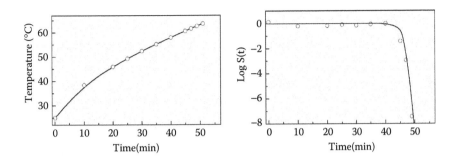

FIGURE 4.4
A simulated temperature profile using Equation 4.12 as a model (left) and corresponding generated survival data of *Listeria monocytogenes* (right). The open circles are the generated data (with superimposed noise) and solid line the fit of Equation 4.10. For the agreement between the generated and retrieved parameters, see text. (From Peleg, M. et al., 2003, *Bull. Math. Biol.*, 65, 219–234. With permission, courtesy of Elsevier Ltd.)

Modern statistical software allows expressions with an 'if statement' to serve as regression models, as already mentioned. Therefore, once $T(t)$ has been defined in the preceding manner and Equation 4.10 is solved, the solution in the form of $\log_{10}S(t) = f(k, T_c, t)$ can be used to estimate the survival parameters k and T_c by nonlinear regression. This is demonstrated with a simulated heating curve using Equation 4.12 as a model (Figure 4.4, left). The corresponding survival data, generated with superimposed noise to make them appear realistic, are also shown in Figure 4.4 (right). The solid line is the fitted survival curve using the solution of Equation 4.10 as the regression model when Equation 4.12 represented the temperature profile, $T(t)$. The regression yielded the following results:

Survival parameter	Generation parameter	Regression parameter
n	1.0 (by definition)	1.0 (by definition)
k	$1.01°C^{-1}$	$0.81°C^{-1}$
T_c	$60.3°C$	$59.8°C$

Source: Peleg et al., 2003

The regression parameters demonstrate that although T_c could be estimated fairly accurately using the discontinuous model, the magnitude of k was about 20% lower than the correct value. Whether such a discrepancy is significant in the practical sense can be determined by comparing the model's predictions with survival ratios obtained experimentally, as will be shown below.

The Nonlinear Case

Weibullian–Power Law Inactivation at Arbitrary Heating Rate History

The previous discussion addressed the special and probably rare case of cells or spores whose isothermal survival curves are all log linear. In reality, one would expect that most isothermal survival curves would not be log linear, but rather follow the Weibullian–power law model. In one of its simplest forms (see Chapter 1 and Chapter 2), the shape factor, or power, n, is fixed and the temperature dependence of the rate parameter, $b(T)$, can be described by the log logistic model (Equation 4.3) or any alternative two parameters secondary model. For any given temperature profile, $T(t)$, the momentary survival rate is the differential equation:

$$\frac{d\log_{10} S(t)}{dt} = -\log_e[1+\exp\{k[T(t)-T_c]\}]\cdot n \cdot \left\{\frac{-\log_{10} S(t)}{\log_e[1+\exp\{k[T(t)-T_c]\}]}\right\}^{\frac{n-1}{n}} \quad (4.14)$$

As previously shown, once the organism's survival parameters, namely, n, k, and T_c, are all known and $T(t)$ can be written as a continuous or discrete algebraic expression, this equation can be solved numerically by Mathematica® to produce the corresponding curve, $\log_{10} S(t)$ vs. t.

It need not concern us here that the model's equation can also be solved by the incremental method (Chapter 3), even when $T(t)$ is not expressed as an algebraic term. It is of concern that Equation 4.14 with $n \neq 1$ can only be solved numerically and therefore it is not a proper model for standard or "canned" nonisothermal regression procedures of the kinds that come with statistical software packages.

To determine the survival parameters from a given set of experimental inactivation data, therefore, requires the programming of a special numerical procedure. One such procedure is based on the simplex or FindMinimum algorithm. When either is employed, the rate equation is *solved numerically* with initial guesses and subsequently newly generated values of the survival parameters. The mean square error is calculated at each iteration, and the process is repeated until the survival parameters calculated in this way yield a solution to the model's equation within the specified accuracy. The result is a set of estimated survival parameters that can be used to test the model's validity against *fresh* data.

Testing the Concept with Simulated Data

To test the preceding procedure, we have used a set of simulated survival data corresponding to three different temperature profiles (Figure 4.5),

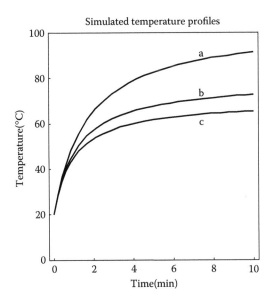

FIGURE 4.5
The simulated temperature profiles generated with the models listed in Table 4.1. (From Peleg, M. and Normand, M.G., 2004, *Crit. Rev. Food Sci. Nutr.*, 44, 409–418. With permission, courtesy of CRC Press.)

with known survival parameters, n, k, and T_c, on which a random noise was superimposed. The temperature profiles' equations and corresponding results are shown in Table 4.1 and Table 4.2, respectively. Examples of the simulated curves with two noise levels are shown in Figure 4.6.

In principle and especially with simulated data when one has control of the number of entries and the random deviations, one can determine all three survival parameters simultaneously. However, this might not be possible when the survival ratios that serve as the database are few and have a considerable scatter. The reason is that letting the power n be an adjustable parameter can result in a totally unrealistic estimate of the value of T_c. Remember that although T_c is a component of a secondary model determined by regression, it still carries a clear physical meaning — i.e., it is the marker of the temperature region in which lethality starts to accelerate. Thus, its calculated estimate must be congruent with the actual temperature where this happens, regardless of whether a different value will result in a better fit as judged by statistical criteria.

To avoid this problem, we have fixed the magnitude of n (at various values) and only determined the values of k and T_c by the previously mentioned literature procedure. That we got the same results with the FindMinimum and simplex algorithms provided mutual verifications (Peleg and Normand, 2004). (As will be seen later, justification of fixing

TABLE 4.1

Empirical Models Used to Characterize Various Temperature Profiles Shown in the Work

Medium	Profile	Equation	Data source
Simulation	a	$T(t) = 20 + \dfrac{t}{0.020 + 0.0200t}$	N/A
	b	$T(t) = 20 + \dfrac{t}{0.020 + 0.017t}$	
	c	$T(t) = 20 + \dfrac{t}{0.020 + 0.012t}$	
Broth[a]	a	$T(t) = 78.1 - \text{Ln}\{1 + \text{Exp}\,[8.7(6.3\text{-}t)]\}$	Mattick et al., 2001
	b	$T(t) = 78.8 - \text{Ln}\{1 + \text{Exp}\,[5.2(10.4\text{-}t)]\}$	
	c	$T(t) = 78.3 - \text{Ln}\{1 + \text{Exp}\,[4.3(12.4\text{-}t)]\}$	
	d	$T(t) = 78.5 - \text{Ln}\{1 + \text{Exp}\,[2.6(20.5\text{-}t)]\}$	
Ground chicken breast[a]	a	$T(t) = \dfrac{70.4(t-0.41)}{[0.11 + (t-0.41)]\{1.0 + Exp[50.0(0.49-t)]\}}$	Murphy et al., 1999
	b	$T(t) = \dfrac{67.6}{1.0 + Exp[0.65(1.8-t)]}$	
	c	$T(t) = \dfrac{66.5}{1.0 + Exp[0.51(1.6-t)]}$	

[a] For the actual fit, see Figure 4.7 and Figure 4.10.

Source: Peleg, M. and Normand, M.D., 2004, *Crit. Rev. Food Sci. Nutr.*, 44, 409–418. With permission, courtesy of CRC Press.

the magnitude of n comes from the model's ability to predict the outcome of different processes correctly, regardless of the chosen value of n, as long as it is not too far away from the "correct" value.) The procedure was tested with simulated nonisothermal survival data of the kind depicted in Figure 4.6. The purpose was to demonstrate that the generation parameters could be retrieved with reasonable accuracy even from scattered survival ratios. Table 4.2 shows that, with only five simulated "replicates," one can get fairly close estimates of the generated parameters k and T_c.

As could be expected, the quality of the estimates declined as the scatter's amplitude increased. Obviously, the replicates' spread is not a parameter that the experimenter can tightly control in microbial inactivation studies. The benefit of simulations of the type shown here is that they allow one to determine the number of replicates that would be required in order to estimate the cells' or spores' survival parameters, k and T_c, reliably with one or more assumed but realistic values of the power n. Again, remember that once the program to estimate the parameters k and T_c has been installed and is running, a large number of simulations under any number of assumed conditions can be generated within a few minutes.

TABLE 4.2

Estimated survival parameters, T_c and k, of a Hypothetical *Salmonella*-Like Microorganism[a]

	Profile										
	a			b				c			
	Scatter's maximum span (log₁₀ units)										
	1.5		2.0		1.5		2.0		1.5		2.0	
Run	T_c	k	T_c	k	T_c	k	T_c	k	T_c	k	T_c	k
---	---	---	---	---	---	---	---	---	---	---	---	---
1	61.0	0.56	59.1	0.48	59.1	0.43	59.8	0.49	59.8	0.49	59.7	0.45
2	60.5	0.52	61.6	0.61	60.3	0.53	58.4	0.40	60.5	0.60	60.3	0.59
3	59.7	0.50	62.4	0.66	59.8	0.51	60.2	0.51	59.1	0.35	57.1	0.23
4	61.0	0.53	59.7	0.48	59.2	0.44	59.7	0.48	59.6	0.41	59.6	0.37
5	59.3	0.43	62.8	0.71	59.6	0.48	59.7	0.46	61.0	0.72	59.7	0.50
Mean	60.3	0.51	61.1	0.59	59.6	0.48	59.6	0.47	60.0	0.51	59.3	0.43

[a] Calculated from simulated survival curves with two levels of superimposed random noise.

Notes: The "smooth" part of the cure was produced with Equation 4.14 as a model with $n = 0.5$, $T_c = 60°C$, and $k = 0.50°C^{-1}$. The temperature profiles, a, b, and c, are shown in Figure 4.5 and their model equations are in Table 4.1.

Source: Peleg, M. and Normand, M.D., 2004, *Crit. Rev. Food Sci. Nutr.*, 44, 409–418. With permission, courtesy of CRC Press.

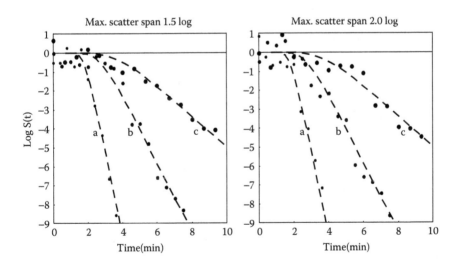

FIGURE 4.6
Examples of generated survival data (with two levels of superimposed noise) of a *Salmonella*-like organism that correspond to the profiles shown in Figure 4.5, fitted with the survival model using the simplex algorithm. The agreement between the generation and retrieved parameters is shown in Table 4.2. (From Peleg, M. and Normand, M.G., 2004, *Crit. Rev. Food Sci. Nutr.*, 44, 409–418. With permission, courtesy of CRC Press.)

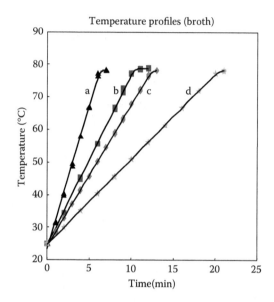

FIGURE 4.7

The four temperature profiles to which the *Salmonella* in broth was subjected, fitted with the models listed in Table 4.3. The original data are from Mattick, K.L. et al., 2001, *J. Food Prot.*, 64, 606–613. (From Peleg, M. and Normand, M.G., 2004, *Crit. Rev. Food Sci. Nutr.*, 44, 409–418. With permission, courtesy of CRC Press.)

('Conditions' here means the number of survival ratios and their spacing, the noise level, and number of replicates.) As to the estimation of n, its approximate value can be deduced from previous results or from literature data reporting isothermal survival curves of the same organism in a different medium or related organisms known to have a Weibullian–log logistic inactivation pattern.

Testing the Method with *Salmonella* Survival Data

Salmonella *in a Growth Medium*

The database of Mattick et al. (2001), which had been used to determine the nonisothermal survival patterns of *Salmonella* from isothermal inactivation data (see Chapter 2), could also be used to test the procedure of determining the survival parameters from nonisothermal inactivation data (Peleg and Normand, 2004). This was done as follows. The temperature profiles of four nonisothermal heat treatments (Figure 4.7) were fitted with the four empirical models listed in Table 4.1. The excellent fit of these models is shown in the figure.

The resulting $T(t)$ expression was then incorporated into Equation 4.14, which served as a model with three preset fixed values of the power, n,

TABLE 4.3

Survival Parameters of *Salmonella* Using Equation 4.14 as a Model[a]

Medium	n	Profile	T_c (°C)	k (°C⁻¹)	Chi square	Data source
Broth	0.6	a	61	0.20	3.36	Mattick et al., 2001
		b	63	0.22	2.53	
		c	61	0.17	5.76	
		d	62	0.15	3.23	
		Mean	62	0.19		
	0.7	a	61	0.18	3.21	
		b	63	0.19	2.10	
		c	62	0.16	5.47	
		d	64	0.14	3.20	
		Mean	63	0.17		
	0.9	a	62	0.22	2.98	
		b	65	0.26	1.49	
		c	65	0.19	5.12	
		d	69	0.17	3.22	
		Mean	65	0.21		
Ground	0.6	a	61	1.06	0.10	Murphy et al., 1999
chicken		b	63	0.92	1.00	
breast		c	61	1.03	1.97	
		Mean	62	1.00		
	0.7	a	61	0.89	0.06	
		b	63	0.73	1.91	
		c	61	0.83	1.76	
		Mean	62	0.82		
	0.9	a	59	0.61	0.03	
		b	62	0.46	0.74	
		c	61	0.56	1.36	
		Mean	61	0.54		

[a] The actual fit of the model is demonstrated in Figure 4.8 and Figure 4.11.

Source: Peleg, M. and Normand, M.D., 2004, *Crit. Rev. Food Sci. Nutr.*, 44, 409–418. With permission, courtesy of CRC Press.

at 0.6, 0.7, and 0.9. (From previous experience [see Chapter 2], we could deduce that, probably, the magnitude of the "true" value of n would be somewhere in that range. The same can be said about similar or related organisms like *Shigella* or *E. coli*.) With $T(t)$ determined and the assumed three values n in place, the procedure was first applied to curve fit the four survival curves and calculated the corresponding k's and T_c's. The results are listed in Table 4.3. The fit of the model (Equation 4.14) with the three n's and the values of k and T_c thus determined is shown as a solid line in Figure 4.8.

Now, the true test of the model is not that it fits a set of given data because a similar fit could be accomplished with much simpler models; rather, it is that it can be used to *predict* the survival curve under conditions not used in its parameters' determination. However, before applying this

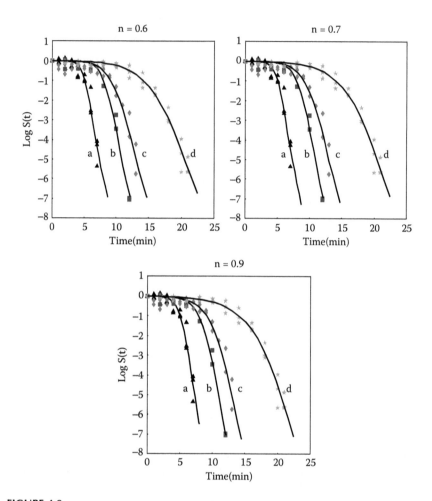

FIGURE 4.8
The experimental survival data of *Salmonella* heat treated in broth that correspond to the four temperature profiles shown in Figure 4.7, fitted with the survival model (Equation 4.14) using the simplex algorithm. The regression parameters are listed in Table 4.3. The original data are from Mattick, K.L. et al., 2001, *J. Food Prot.*, 64, 606–613. (From Peleg, M. and Normand, M.G., 2004, *Crit. Rev. Food Sci. Nutr.*, 44, 409–418. With permission, courtesy of CRC Press.)

test, let us examine the results more carefully. A glance at Table 4.3 shows that the estimates of T_c and k determined using data obtained during the four different treatments were fairly close, but not identical. However, the values of T_c and k showed no discernible trend; that is, the magnitudes of the calculated parameters did not increase or decrease in unison by increasing or decreasing the heating rate. One must therefore conclude that the observed differences in these survival parameters' magnitudes were primarily due to the scatter of the experimental survival ratios from which they were obtained.

Table 4.3 also shows that the chosen value of n had only a very minor effect on T_c, which was not totally unexpected. Because, as already mentioned, T_c is a marker of the temperature range where inactivation accelerates, it is an objective characteristic of the organism, which must not be affected by the selected secondary model whenever the transition from tolerance of the heat to mortality at an accelerated pace is sharp. It so happens that *Salmonella* cells' thermal destruction indeed picks up momentum at a temperature around 60°C, which is manifested in the calculated values of T_c. With T_c hardly changing, the magnitude of k was supposed to decrease as the value of the selected n increased (see below).

However, if such a trend occurred, it seems to have been obscured by the experimental data's scatter. Because the scatter was inevitable, the model's predictive ability was tested using the averaged values of the T_c and k values. This was done as follows:

1. For each chosen n value, the T_c and k of three treatments, starting with a, b, and c, were averaged.
2. The averaged values were then incorporated into the model's rate equation (Equation 4.14) together with the temperature profile equation, $T(t)$, of the fourth treatment, d.
3. The equation was then solved numerically to produce the survival curve that corresponds to that particular treatment, i.e., treatment d.
4. The procedure was repeated with the averaged parameters retrieved from the data of treatments a, b, and d to predict the survival curves of treatment c.

This was done again with the parameters obtained from treatments a, c, and d to predict the survival curve of treatment b and once more using the averaged T_c and k obtained from treatments b, c, and d to predict the survival curves of treatment a. Comparison of the survival curves thus calculated (dashed lines) with the actual experimental survival data are shown in Figure 4.9. They were clearly in general agreement. Support for this statement comes form the simulated data (see "Testing the Concept with Simulated Data"), which had produced similar results. In light of the scatter in the original data, one can confidently claim that the agreement between the predictions and observations has indeed confirmed the model's applicability for *Salmonella* in the particular medium.

Hopefully, similar studies will confirm the model's applicability for other organisms as well. This will only require that an effective procedure to recover and count the survivors be in place and that *a sufficient number*

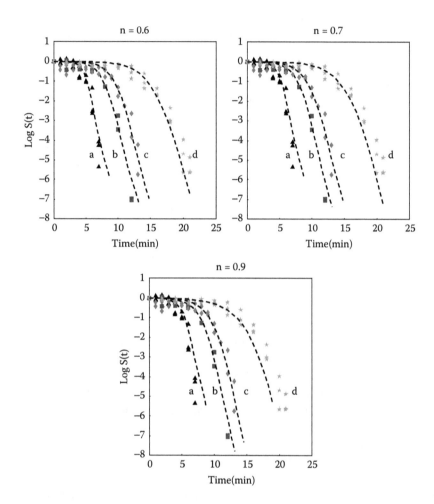

FIGURE 4.9
Predicted survival curves of *Salmonella* in the four heat treatments whose temperature profiles are depicted in Figure 4.7 and the experimentally observed survival ratios (symbols). Notice that the averaged values of the survival parameters k and T_c calculated in three treatments (see Table 4.3) were used to predict the fourth (see text)). The original survival data are from Mattick, K.L. et al., 2001, *J. Food Prot.*, 64, 606–613. (From Peleg, M. and Normand, M.G., 2004, *Crit. Rev. Food Sci. Nutr.*, 44, 409–418. With permission, courtesy of CRC Press.)

of data points in the lethal temperature range be present. This can be accomplished by the application of *low heating rates* that will allow effective sampling of the survival ratios at the region where their magnitude rapidly drops, but not too low so that the organism adaptation is allowed.

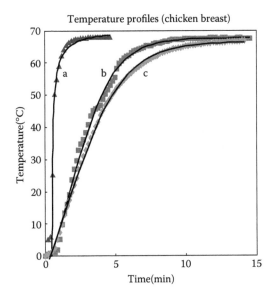

FIGURE 4.10
The three temperature profiles to which the *Salmonella* in ground chicken breast was subjected, fitted with the models listed in Table 4.3. The original data are from Murphy, R.Y. et al., 1999, *J. Food Prot.*, 62, 980–985. (From Peleg, M. and Normand, M.G., 2004, *Crit. Rev. Food Sci. Nutr.*, 44, 409–418. With permission, courtesy of CRC Press.)

Salmonella *in Minced Chicken Meat*

Murphy et al. (1999) reported the survival curves of *Salmonella* in minced chicken meat subjected to different temperature profiles (Figure 4.10). The uniformity of the temperature distribution within the samples was not assured and there were only few scattered data points in the highly lethal temperature range. This is not a criticism of the work, which had not been originally intended to provide data for testing the proposed survival model discussed in this chapter. The results reported by Murphy et al. are probably what one would typically obtain in this type of experiment. Therefore, it would be interesting to see whether the records could still be used as a database for the construction of a predictive survival model.

To examine this possibility, the procedure previously used with Mattick and colleagues' 2001 data was repeated with those of Murphy et al. (1999). The three temperature profiles, $T(t)$, were fitted with the three empirical models listed in Table 4.1. Their almost perfect fit, especially in the lethal region, is shown in Figure 4.10. These expressions of $T(t)$ were incorporated into the survival model's rate equation (Equation 4.14) with the same three preselected values of n (i.e., 0.6, 0.7, and 0.9) used for the *Salmonella* when heated in the broth. With $T(t)$ included in it, the model's equation was solved numerically using the simplex algorithm to yield the estimated values of T_c and k that corresponded to each preset n. The results are given in Table 4.3 and the fitted curves shown in Figure 4.11.

T_c was found to be about 61 to 62°C and, as before, it was practically unaffected by the choice of n. Unlike in the case of *Salmonella* treated in a culture medium, here a clear inverse relationship existed between the calculated magnitude of k and the chosen n value. The calculated magnitudes of k were also two to five times higher than those found for the organism's inactivation in the broth medium. In light of the experimental record's obvious imperfections, the calculated values of T_c and k were still fairly reproducible. Yet, they did not have the same degree of reproducibility achieved in Mattick and colleagues' experiments (2001), in which the specimens were heated while passing through a capillary.

As before, the predictive power of the model was tested by using the averaged k and T_c values calculated from treatments a and b to estimate the survival curve in treatment c, those calculated from treatments a and c to estimate the survival curve of treatment b, and those calculated from treatments b and c to estimate the survival curve of treatment a. Comparison of the estimated predicted curves (dashed lines) and the corresponding reported experimental survival ratios are shown in Figure 4.12. The figure demonstrates that despite the deficiencies of the experimental survival records and the fact that only three treatments' data could be used to test the model, the method still provided reasonable predictions of the survival patterns under nonisothermal conditions. One would expect that, with a database created by treating smaller specimens, heated at lower rates, and with more frequent counts taken at the lethal temperature region, the accuracy of the parameters' estimates would be considerably improved and with it the quality of the method's predictions.

Concluding Remarks

A search of the food microbiology literature will yield an enormous amount of survival data of almost any conceivable organism that has a food relation. Most are in the form of isothermal survival curves determined in different media, from actual foods to model systems with controlled pH, water activity, osmotic pressure, salt concentrations, etc. Surprisingly, however, very few reports, at least in recent literature, can be used to test the proposed method to determine survival parameters by the described procedure. This is because reports on isothermal and nonisothermal survival are quite rare. In those available, the data sets are

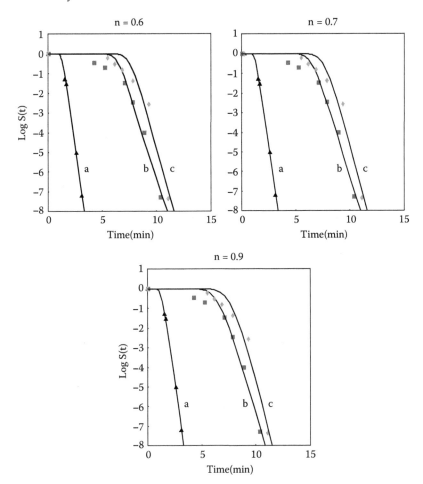

FIGURE 4.11
The experimental survival data of *Salmonella* heat treated in ground chicken breast that correspond to the three temperature profiles shown in Figure 4.10 fitted with the survival model (Equation 4.14) with simplex algorithm. The regression parameters are listed in Table 4.3. The original data are from Murphy, R.Y. et al., 1999, *J. Food Prot.*, 62, 980–985. (From Peleg, M. and Normand, M.G., 2004, *Crit. Rev. Food Sci. Nutr.*, 44, 409–418. With permission, courtesy of CRC Press.)

frequently incomplete or contain too few temperature profiles (isothermal or nonisothermal) to derive a working model.

Upon scrutiny, some published data reveal that the measured temperature could not have been uniformly distributed in the treated specimen, which makes their quantitative interpretation difficult if not impossible. It is to be hoped that adaptation of the proposed method to quantify

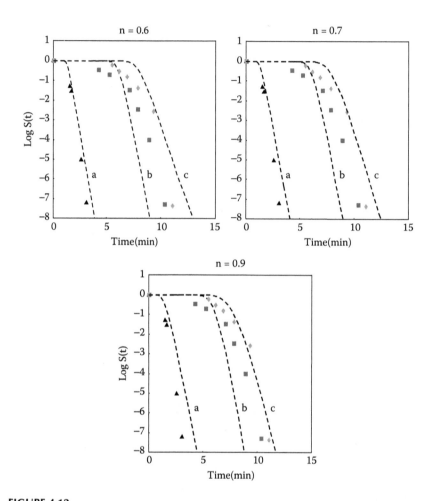

FIGURE 4.12
Predicted survival curves of *Salmonella* in the three heat treatments whose temperature profiles are depicted in Figure 4.10 and the experimentally observed survival ratios (symbols). Notice the averaged values of the survival parameters k and T_c calculated in two treatments (see Table 4.3) were used to predict the third (see text). The original survival data are from Murphy, R.Y. et al., 1999, *J. Food Prot.*, 62, 980–985. (From Peleg, M. and Normand, M.G., 2004, *Crit. Rev. Food Sci. Nutr.*, 44, 409–418. With permission, courtesy of CRC Press.)

survival patterns will also help to produce a more reliable database than that which now exists. Experimental confirmation that nonlinear survival parameters can be used for prediction might encourage microbiologists to determine them in actual foods under nonisothermal conditions. Lists of such parameters could then replace those of D and z values or of rate constants and energies of activation, which are currently an almost exclusive source of information on the heat resistance of microbial cells and spores.

At the time at which this book was written, the evidence to support this view is admittedly very limited indeed. Yet, we have reason to believe that the validity of the model will be confirmed in a growing number of organisms and spores heat treated under a variety of environmental conditions. At the same time, we also hope that the concept could be expanded to non-Weibullian inactivation patterns. There might be more effective ways to extract survival parameters from nonisothermal survival data than the one described in this chapter. No doubt finding them will be a challenge to researchers in the field. What is perhaps most important is the demonstration that, *theoretically*, nonisothermal survival patterns can contain enough information to predict survival patterns under different temperature histories.

5

Isothermal Inactivation with Stable and Dissipating Chemical Agents

Mathematics is the art of giving the same name to different things.

Jules Henry Poincaré

Destruction and inhibition of microorganisms by chemical agents is an option frequently exercised in a variety of fields. Most notable examples are chemical preservation of foods and beverages and the disinfection of drinking water by chlorination or administration of other antimicrobials. In medicine, the use of antibacterial (mostly antibiotics) and antifungal drugs is quite common and the same can be said about drugs that destroy or inhibit protists and worms. Antimicrobial agents of similar activities are also widely used in agriculture. Decontamination of space and equipment and by a gaseous chemical agent, regardless of whether the contaminant is a naturally occurring microorganism or a deliberately introduced bacterium or spore (bioterrorism), is another example of chemical treatment that can have public safety implications.

There are numerous kinds of chemical antimicrobials and they are employed against a large variety of targets. The discussion in this chapter will focus on "traditional" microorganisms, primarily bacteria. However, the principles might be just as applicable to unicellular eukaryotes like yeast, or amoeba, except that the time scale might be quite different.

The chemical structure of antimicrobials and their physical properties also vary enormously, as does the mode of their activity. The way in which antimicrobials are administered and their effective dose also vary dramatically, depending on their chemical species, stability, and, among other things, whether they are in a solid, liquid, or gaseous state. The role of factors such as solubility, diffusivity and chemical affinity in the design of an effective treatment is well recognized and their importance cannot be overemphasized. The same can be said about the biochemical and biophysical interactions that render these chemical agents lethal to the targeted organism. Nevertheless, the focus of the following discussion will be on the inactivation kinetics, as expressed at the population level.

Thus, for the type of modeling addressed in this chapter, the main distinction between the agents is whether their effective concentration, on the pertinent time scale, remains practically unchanged or dissipates appreciably.

The first class includes agents such as benzoates, sorbates, and the like, but also less stable agents like sulfite when administered in a massive dose, in which case an effective lethal concentration is maintained throughout the treatment. The second kind includes chemically unstable or volatile agents whose dissipating rate is high enough so that their effectiveness significantly diminishes during the treatment. The lethality of such agents may disappear altogether and they might need to be replenished so that the treatment will remain effective. For what follows, it will be assumed that in both cases the lethal effect is exclusively produced by the agent's presence and thus that its concentration's profile alone determines the treated microbial population survival pattern. Obviously, changes in temperature or pH, for example, must also affect the survival curve, a fact that has been long recognized.

Some existing models are intended to account for the combined effect of chemicals and heat on the logarithmic inactivation rate of microorganisms. Most of them are based on the incorporation of a temperature and concentration terms into an Arrhenius-type equation. One problem with such models, as has been mentioned in previous chapters, is that the concept of a time-independent logarithmic inactivation rate has a meaning in first-order mortality kinetics only, a kinetics that need not exist in reality (see below). However, even if microbial inactivation by a chemical agent did follow the first-order kinetics model, which is highly doubtful, any assumption regarding the nature of the combined temperature–concentration effect on the logarithmic rate constant, like that implied by the Arrhenius equation, for example, would need to be confirmed experimentally. Again, the confirmation should come through testing the predictive ability of the proposed model and not by demonstrating its fit.

Thus, expressing the logarithmic rate constant as the algebraic sum or product of two or more independent and interactive terms representing the concentration, the temperature and/or pH, etc. is based on the implicit assumption that these factors function independently or in another universally prescribed manner dictated by the chosen model. It is not self-evident that either assumption must always be correct and therefore models constructed in this way ought to be experimentally validated before they can be accepted for use. More on this issue will be discussed in Chapter 13.

For simplicity, we will only address situations in which the chemical inactivation occurs under isothermal or practically isothermal conditions, i.e., in the absence of synergistic or antagonistic effects that might be produced by other factors or agents.

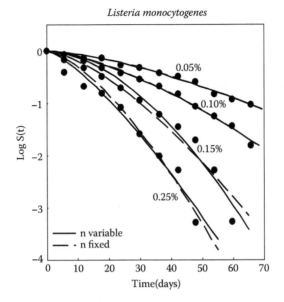

FIGURE 5.1

Survival data of *Listeria monocytogenes* in the presence of potassium sorbate at different concentrations fitted with the Weibullian–power law model, with variable $n(C)$ (Equation 5.1) and with a constant n (Equation 5.2). (The experimental data are from El-Shenawy, M.A. and Marth, E.H., 1988, *J. Food Prot.*, 51, 842–846.)

Chemical Inactivation under "Constant" Agent Concentration

An example of published isothermal survival curves of *Listeria monocytogenes* exposed to potassium sorbate at various concentrations is shown in Figure 5.1. They clearly demonstrate that the inactivation pattern was not log linear, as would have been expected from the first-order mortality kinetics. Moreover, all the semilogarithmic survival curves of the organism when exposed to the sorbate had a noticeable *downward concavity*.

Such a pattern cannot be explained by an unnoticed loss of the sorbate. If this had happened, the *Listeria*'s semilogarithmic survival curves would have to exhibit *upward concavity* (see below). Also, because the reported isothermal semilogarithmic *thermal* inactivation curves of *Listeria* are at most linear but more frequently concave upward, they could not provide, at least in this particular case, any qualitative or quantitative indication as to how the organism might respond to a specific chemical agent. All this suggests that unless there is a compelling reason to assume otherwise, the inactivation pattern of an organism must be determined experimentally; it cannot be assumed *a priori* that it follows the first-order kinetics.

The patterns shown in the figure could be described by the Weibullian–power law model as their thermal counterparts, albeit with a very different shape factor or power, i.e., $n > 1$ instead of the $n \leq 1$ found for the isothermal heat inactivation curves. Expressed mathematically, the Weibullian model has the form:

$$\log_{10} S(t) = -b(C)t^{n(C)} \tag{5.1}$$

where $b(C)$ and $n(C)$ are concentration-dependent coefficients. In the case of a constant power, $n(C) = n$:

$$\log_{10} S(t) = -b(C)t^{n} \tag{5.2}$$

The actual fit of the model with constant and variable $n(C)$ is shown as dashed and solid curves in Figure 5.1. Either way, the two versions of the Weibull model, like their equivalents in thermal activation, are consistent with the notion that the logarithmic inactivation rate depends not only on the agent's concentration, but also on the *exposure's time*. In the particular case of the *Listeria* shown in Figure 5.1, the downward concavity of the semilogarithmic survival curves is an indication that a prolonged exposure to the sorbate sensitizes the survivors, presumably by accumulated damage to the cells. Here, too, a log linear survival curve, which traditionally has been treated as evidence of first-order kinetics, would be just a special case of the Weibullian model with $n = 1$.

Once the Weibullian model's parameters have been determined experimentally, they can be used to determine the survival curve at any given agent concentration by *interpolation*. The same can be said about *any alternative model*, as long as the chemical agent can be assumed to be practically stable. The term 'stable' here requires clarification. For chemical agents to be effective, they need to interact with the organism and this may take very different forms. The interaction can be primarily physical. For example, sugar at a concentration of above about 65% generates enough osmotic pressure to desiccate the cells. This osmotic preservation should not concern us here because, in most cases (notably in jam or marmalade preparation), heat and low pH are also involved and/or in the case of honey, the sugar concentration is so high that it is a major *ingredient* rather than a preservative only. The same can be said about alcoholic beverages, in which the alcohol plays more than a preservative role.

The interaction can be chemical, though, with the agent interfering with a crucial metabolic function of the microbial cell, for example. In such a case, the chemical agent is actually consumed, but the assumption is that the amount so lost does not significantly affect its overall concentration. (For the case in which the concentration does change appreciably, see "Microbial Inactivation with a Dissipating Chemical Agent" below.) Also, chemical preservatives, like sorbic, propionic, and benzoic acid, are known to be effective antimicrobials only in their undissociated form. However, because these acids have very low water solubility, they are usually administered in their potassium, calcium, and sodium salts, respectively. These salts, especially the potassium sorbate and sodium benzoate, are effective preservatives only when the pH is low, which allows for their partial conversion into the respective acids.

Thus, there is a difference between the nominal concentration of the added salt and the effective concentration of the acid that should be kept in mind. In cured meats, there is a difference between the nominal and effective concentration of the nitrite that, among the added mixture of salts (the other two are sodium chloride and nitrate), is the real preservative. As already stated, any upward concavity of a semilogarithmic survival curve observed under a constant chemical preservative concentration can be in fact or at least partly due to the diminishing concentration of its effective form. This cannot be the case when the observed semilogarithmic survival curve has a clear downward concavity. If the concentration of the effective form of the preservative indeed decreases somewhat with time, the true curvature of the semilogarithmic survival curve would be distorted to some extent.

However, whenever this factor plays a decisive role in shaping the inactivation pattern, the curve's concavity direction would need to be inverted as a result. This creates an asymmetry between the interpretation of upward and downward concavities. Although the latter is a direct manifestation of the inactivation's mode, the former can be the result of the diminishing effective concentration of the agent as well as of the early destruction of the sensitive or weak cells. This will become more clearly evident when we deal with chemical agents known to be unstable or volatile.

Microbial Inactivation with a Dissipating Chemical Agent

Unstable or volatile chemicals are extensively used in the disinfection of water, equipment, packages, and foods. The most notable example is the traditional chlorination of drinking water by perchlorates. More modern disinfectants (apart from ultraviolet light) include ozone generated *in situ*, hydrogen peroxide, and peracetic acid. Like the compounds produced in chlorination, these agents are volatile, very active chemically, and thus unstable. Therefore, unless continuously generated or replenished, their concentration will decrease with time and may reach an ineffective level before the required degree of inactivation has been accomplished.

When administered orally or by injection, antibiotics and other antimicrobial drugs bear no chemical similarity to the agents used in water disinfection, of course. However, their effective concentration in the patient's (or animal's) body also diminishes with time and they too need to be replenished periodically or automatically by slow release. Thus, at least in principle, the same kind of mathematical model developed for assessing the inactivation kinetic of dissipating disinfectants can be just as applicable to the efficacy of antibiotics and other drugs *in vivo*. Obviously, the time scale of the inactivation process, the model's mathematical structure, and the magnitude of its parameters would be quite different. However, the general kinetic considerations leading to the inactivation model's derivation might be very similar despite the dissimilarities in these agent's modes of activity.

A problem that arises in any attempt to model the effect of unstable chemical antimicrobials is that it is very difficult, if not utterly impossible, to obtain reliable microbial survival data under sets of constant agent concentrations. (The problem is much more serious than in modeling the kinetics of thermal inactivation. This is because at least under certain circumstances, it is possible to record survival curves under conditions that are sufficiently close to ideal isothermal heat treatment. These are extremely useful in the establishment of the inactivation that are kinetics, despite their imperfections.) Because of the inherent volatility and stability problem, any mention of constant agent concentration or isoconcentration in what follows will refer to *hypothetical* and not actual experimental conditions.

Traditional Models

When a microbial population is exposed to a lethal chemical agent with diminishing intensity, its semilogarithmic survival curve almost invariably has an upward concavity, i.e., it would exhibit notable "tailing." This has been consistently observed in water treated with a variety of chemical disinfectants. The explanation of this finding is straightforward. As the agent's concentration decreases so does its lethal effect. Therefore, the inactivation rate progressively diminishes; theoretically, when the agent's concentration reaches a threshold level determined by the targeted organism's tolerance (which might be zero), the effect of the treatment vanishes altogether. It was recognized very early that the lethal effect of a volatile agent or microorganisms depends not only on its initial concentration, but also on the exposure duration. Consequently, unlike traditional survival curves that depict the number of surviving organisms or the survival ratio as a function of time, survival curves of organisms exposed to a dissipating chemical disinfectant are frequently presented in the form of a plot depicting the number of cells' survival ratio as a function of a combined concentration–time parameter.

By far the most common combined parameter of this kind is the product of the momentary concentration, $C(t)$, multiplied by the time, t, i.e., $C(t) \cdot t$. The purpose of this modification of the time axis has been to compare experimental survival ratios obtained under conditions in which the agent dissipates at different rates. By plotting the survival ratio against a common variable that accounts for the fact that the agent's concentration varies with time, it has been hoped that the transient nature of the process would be captured. The difficulty with this approach is that the effects of time and concentration are assigned, arbitrarily, a universal reciprocal relation that probably does not exist in reality. To demonstrate the point, consider the following two hypothetical scenarios:

- An agent is applied at a constant concentration of 1000 ppm for 1 h, $C(t) \cdot t = 1000$ ppm h, that is.
- The same agent is applied at a constant concentration of 5 ppm for 200 h, i.e., $C(t) \cdot t$ is also 1000 ppm h.

If the concept is correct, then the survival ratio at the end of the two treatments should be exactly the same, which is highly doubtful. If the agent's concentration is not constant, but decreases with time, then according to the $C(t) \cdot t$ concept, the same lethality should be observed at any equivalent concentration–time combination *regardless of the agent's initial concentration and dissipation rate*. Examination of published records of disinfection experiments and computer simulations show that this would be a very unlikely occurrence. The same can be said about any alternative concentration–time combination, unless it can be experimentally demonstrated that it produces unique survival curves for any practical combination of the agent's initial concentration level and dissipation rate history.

Incorporating an Arrhenius type or a similar concentration term in the survival curve's model equation would be equally problematic. This is because it requires the existence of a special preconceived relationship between the organisms' response to different concentration–time combinations. Such a relationship needs to be established *prior* to the application of the model. If the existence of an Arrhenius-type relationship is considered merely as an assumption, then its validity must be confirmed by testing the model's ability to predict experimental survival curves produced in treatments that had not been used in the model's formulation.

Alternative General Model

Consider a targeted organism whose hypothetical isoconcentration inactivation at the pertinent agent concentration range follows the Weibullian–power law model (Equation 5.1) with a constant power, n (Equation 5.2). The concentration independence of the power term, i.e., $n(C) = n$, is not a prerequisite for what follows and the purpose of using Equation 5.2 as the primary inactivation model is only to simplify the discussion. The same can be said about a number of alternative non-Weibullian models.

As in thermal processing, we assume that on the pertinent time scale the treatment is rigorous enough so that growth, adaptation, and recovery from injury do not occur. We also assume that the agent concentration's dissipation history, or concentration profile (the $C(t)$ vs. t relationship), can be expressed algebraically, although it will be shown later that this, too, is not a strict requirement. The reasoning that led to the development of the nonisothermal survival models for heat inactivation is also applicable here. A main difference is that, in contrast with heating where the process's lethality progressively increases, the lethal intensity progressively decreases during a treatment with a dissipating disinfectant. (The analog of heating is a rise in the agent's concentration while cooling is analogous to the agent's dissipation.)

If the momentary logarithmic inactivation rate, $d\log_{10} S(t)/dt$, under a continuously changing agent concentration is the rate that the agents' momentary concentration produces at a time, t^*, that corresponds to the momentary of the survival ratio as shown in Figure 5.2, we can write: (Peleg et al., 2004):

$$\left. \frac{d\log_{10} S(t)}{dt} \right|_{C=\text{const}} = -b(C)nt^{n-1} \tag{5.3}$$

and

$$t^* = \left[\frac{-\log_{10} S(t)}{b(C)} \right]^{\frac{1}{n}} \tag{5.4}$$

Combining the two equations results in the model rate equation:

$$\frac{d\log_{10} S(t)}{dt} = -b[C(t)]n \left\{ \frac{-\log_{10} S(t)}{b[C(t)]} \right\}^{\frac{n-1}{n}} \tag{5.5}$$

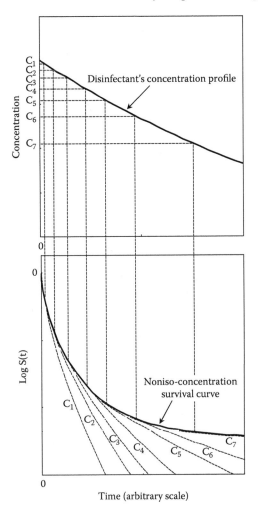

FIGURE 5.2

Schematic view of the survival curve's construction for an organism exposed to a dissipating chemical agent. Notice the assumption that the logarithmic inactivation rate is time dependent even under the (hypothetical) condition of constant agent concentration. (Courtesy of Dr. Maria G. Corradini.)

If the power n is also concentration dependent, then $n[C(t)]$ will replace n in the model equation. Although we do not have sets of experimental isoconcentration survival curves to determine $b(C)$ and n or $b(C)$ and $n(C)$, if $n(C)$ is thought to be concentration dependent too, we can still *assume* that $b(C)$ will follow the log logistic model (Figure 5.3):

$$b(C) = \ln \{1 + \exp[k_c(C - C_c)]\} \qquad (5.6)$$

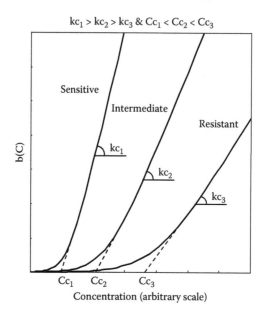

FIGURE 5.3
Schematic view of the log logistic concentration dependence of the rate parameter $b(C)$. Notice that, according to this model (Equation 5.6), C_c can be zero, in which case the agent would be lethal at any concentration.

where C_c marks the concentration level at which lethality accelerates, and k_c the slope of $b(C)$ vs. C at concentrations much higher than C_c. Like before, when $C \ll C_c$, $b(C) \approx 0$ and when $C \gg C_c$, $b(C) \approx k_c(C - C_c)$ (see figure). Again, if the evidence is that the continuation of $b(C)$ beyond C_c is nonlinear, the model can be amended by the addition of a power, m:

$$b(C) = \ln \{1 + \exp [k_c(C - C_c)^m]\} \tag{5.7}$$

Unlike in the application of this model to thermal inactivation, a situation in which the agent has measurable lethality at any measurable concentration cannot be ruled out. In such a case, $C_c = 0$ and Equation 5.6 and Equation 5.7 will be respectively reduced to:

$$b(C) = k_c C \tag{5.8}$$

or

$$b(C) = k_c' C^m \tag{5.9}$$

Recall, though, that because we do not have experimental survival data obtained under constant agent concentration conditions, all the preceding is merely a conjecture. Nevertheless, it would become a *testable hypothesis* if one could retrieve the survival parameters from data obtained under varying agent concentration conditions and use them to predict correctly the outcome of other treatments with different agent dissipation patterns.

More specifically, we can calculate the set of the organism's survival parameters, n, k_c, and C_c, from a treatment under one initial concentration and corresponding dissipation pattern, then insert their values into the model's rate equation and use it to predict the outcomes of other treatments with totally different concentration profiles. Comparison of the predicted survival curves produced against the actual experimental survival data under the new concentration profiles will confirm or refute the model's applicability and the validity of the assumptions on the basis of which it has been derived. The same can be said about any alternative non-Weibullian model and the assumed concentration dependence of its parameters. However, because at the time of this book's writing appropriate data for even the Weibullian model validation have been very scarce (see below), the following will only address the simplified Weibullian inactivation patterns, i.e., survival curves governed by Equation 5.2, with a constant power n as a primary model and Equation 5.6 as a secondary model.

Dissipation and Inactivation

Monotonic Agent Dissipation

Consider a disinfectant introduced into a hypothetical stable environment, where the temperature is constant and there are no physical or chemical disturbances. This is obviously a rather idealized scenario that, in reality, can only be achieved by tight controls. Yet, it might be roughly approximated in certain large water reservoirs and in stored foods, at least for a limited time. Antibiotics and other antimicrobial drugs *in vivo* might also be considered as operating under more or less stable conditions with respect to temperature (unless fever and its suppression are also involved), although the physiological environment might be undergoing cyclical changes. Under such conditions, the small changes in the environment might have only a minor effect on the targeted organism's survival relative to that of the antimicrobial agent's presence. Consequently, the major influence on the inactivation rate will be exerted by the agent's initial concentration and its dissipation pattern. In many instances, the agent's decay would probably be approximately exponential:

$$C(t) = C_0 \exp(-at) \tag{5.10}$$

or

$$C(t) = \Sigma C_i \exp(-a_i t) \quad (\Sigma C_i = C_0) \tag{5.11}$$

where C_0 is the initial concentration and the C_i and a_i are constants. A residual concentration, $C_{residual}$, can be produced by transforming the second model into:

$$C(t) = \Sigma C_i \exp(-a_i t) + C_{residual} \quad (\Sigma C_i = C_0 - C_{residual}) \tag{5.12}$$

Alternatively, one can also use the model:

$$C(t) = C_0 - \frac{(C_0 - C_{residual}) \cdot t}{d + t} \tag{5.13}$$

where d is a constant that represents the diminishing decay rate. In many cases, a model of this kind has a similar fit to that of Equation 5.12 with two or three exponential terms, even though it only has two adjustable parameters: d and $C_{residual}$, which can be zero. Either way, the concentration profile, $C(t)$, however defined can be incorporated into the inactivation model's rate equation, whose solution will be the corresponding survival curve.

Examples of simulated survival curves of a hypothetical microorganism obeying the Weibullian–log logistic inactivation model with $n(C) > 1$ are shown in Figure 5.4. They demonstrate that if the agent dissipates fast enough, the corresponding survival curve can have an upper concavity (right), even though in a treatment with a constant agent concentration (left), they will have clear downward concavity. Thus, whenever, in Equation 5.1, $n(C) \leq 1$ or in Equation 5.2, $n \leq 1$, the organism's semilogarithmic survival curve will always be concave upward whenever the agent's concentration diminishes during the treatment.

However, when $n(T)$ is much larger than one and the agent only slowly dissipates, then a sigmoid survival curve can also be observed, at least in theory, as shown in Figure 5.5. (A similar situation, but with the opposite effect, exists in thermal preservation. Whenever the heating is fast enough, the semilogarithmic survival curve will have a noticeable downward concavity, even for organisms and spores whose isothermal semilogarithmic survival curves have upper concavity in the pertinent temperature range.)

If the agent's concentration falls to a level well below C_c, the treatment's lethality will cease altogether. In such a case, the survival ratio will theoretically remain unchanged. In reality, the agent's loss of potency can result in renewed growth, provided that conditions are favorable and material resources are available. On the other hand, if the targeted organism

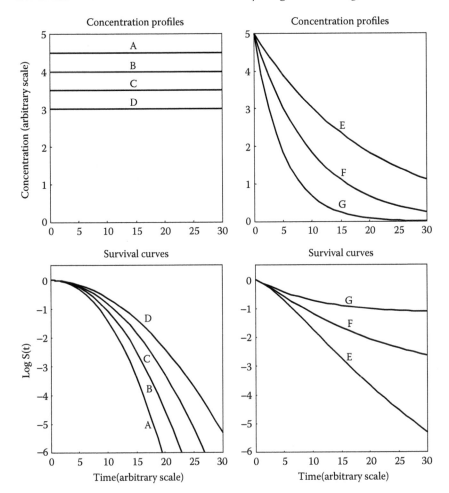

FIGURE 5.4

Simulated constant and diminishing concentration profiles and the corresponding survival curve. Notice that, even if the isoconcentration curves have downward concavity, those under a dissipating agent regime can be concave upward. (From Peleg, M., 2002, *Food Res. Int.*, 35, 327–336. With permission, courtesy of Elsevier Ltd.)

is reduced to below a certain level, it might cease to be viable and considered practically eliminated even if growth conditions are restored. These scenarios are outside the scope of the described model, which deals with the inactivation stage only.

The simulations shown in Figure 5.4 and Figure 5.5 were produced by incorporating the concentration profile into Equation 5.5, together with hypothetical survival parameters, namely, n, k_c, and C_c, and solving the equation numerically with Mathematica®. As has been demonstrated

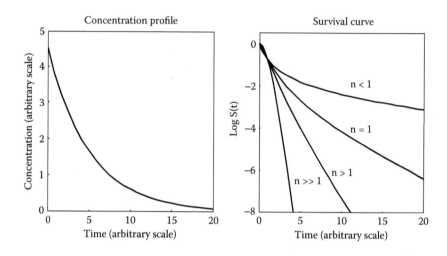

FIGURE 5.5
Demonstration of the effect of the power n on the shape of the survival curve during a chemical agent's dissipation. Notice that when $n \leq 1$, the survival curve will always have upper concavity, but when $n > 1$, it can appear sigmoid, depending on the dissipating pattern. (From Peleg, M., 2002, *Food Res. Int.*, 35, 327–336. With permission, courtesy of Elsevier Ltd.)

earlier (Peleg, 2002c), the equation can also be solved for Weibullian survival patterns, where the concentration dependence of the parameters is described by mathematical models other than the one of constant $n(C)$ and log logistic $b(C)$. It is easy to show that the same calculation procedure also applies to non-Weibullian survival patterns in which the concentration-dependent parameters are described by totally different models.

Agent Dissipation with Regular and Random Oscillations

Maintaining a food or a quantity of water under constant conditions for a considerable time requires control. Obviously, such control is impossible to achieve in water tanks placed outdoors and in open water reservoirs. The same would be true for many commercial food shipments during their storage and, especially, during their transportation, unless the storage space is air conditioned or the truck refrigerated. Diurnal temperature oscillations, which may affect the chemical agent dissipation rate, are unavoidable in every uncontrolled system. However, other fluctuations caused by climatic events of short duration are also quite common. Periods of exposure to the sun, of being under the shadow of trees or buildings or clouds, rain or cold winds can affect exposed systems like an external water tank or commercially transported foods.

We should remember that the temperature fluctuations can affect not only the chemical agent's dissipating rate, but also its potency. Thus, the correct way to deal with fluctuating agent concentration coupled by temperature changes is to incorporate the temperature history and its influence on the agent effectiveness into the inactivation rate equation. We will deal with the yet unsolved problem of how to account for combined lethal effects in Chapter 13. Suffice it to say at this point that data of the kind needed to model and predict the outcome of processes in which the medium's temperature and agent's concentration vary simultaneously are very rare in the literature, if not totally absent. Thus, what follows will relate to grossly oversimplified scenarios in which one can assume that the agent's lethality is solely a function of its concentration, but not of any other factor. In other words, the assumption here is that any simultaneous temperature fluctuations are sufficiently small so that they primarily affect the agent's concentration and, to a much lesser extent, its antimicrobial potency.

Regular Fluctuating Decay Pattern

Traditionally, periodicity or periodicities in physical systems have been described mathematically by incorporating a sinusoidal term or terms into the model's time equation. For example, a fluctuating exponential decay pattern of the kind found in damped oscillating systems can be described by a model such as $y(t) = a_0 \exp(-kt) \cdot (1 - a_1 \sin\omega t)$, where a_0 is $y(0)$, k the exponential decay rate, a_1 the oscillations' amplitudes, and ω their frequency.

Unfortunately, simple models of this kind cannot be used to describe realistic concentration decay patterns because, according to the preceding and similar models, the chemical agent's concentration must rise periodically. In reality, the agent's concentration will never increase, of course, unless it is periodically replenished (see below). For a model to produce a realistic oscillating decay pattern, it must allow for the concentration to decrease or remain constant for a while, but never to increase spontaneously. Mathematically, this translates into the condition that the concentration profile's *derivative*, $dC(t)/dt$, i.e., the momentary dissipation rate, will always be negative or equal to zero:

$$\frac{dC(t)}{dt} \leq 0 \tag{5.14}$$

A decay model that satisfies this requirement can be constructed in more than one way. Perhaps the most convenient models for simulating exponential and linear decay with periodic oscillations are the ones suggested by Professor Claude Penchina (2004, private communication). They are, respectively:

$$C(t) = C_0 \exp\{-[At - B\sin(\omega t)]\} \tag{5.15}$$

and

$$C(t) = C_0 - [At - B\sin(\omega t)] \tag{5.16}$$

where
C_0 is the initial concentration
A is a constant that determines the rate of the underlying exponential or linear decay pattern
B and ω are the oscillations' amplitude and frequently, respectively, bound by the relationship $\omega = A/B - \varepsilon$, where $\varepsilon \geq 0$

Figure 5.6 shows examples of fluctuating decay patterns produced by these two decay models and the corresponding survival curves, when Equation 5.5 serves as the survival rate model.

An alternative model can be constructed from a series of Fermian (logistic) decay terms in the form (Peleg, 2002c):

$$C(t) = C_0 - A\sum_{i=1}^{m} \frac{1}{1 + \exp\left(\dfrac{it_{ci} - t}{a}\right)} \tag{5.17}$$

where
C_0 is the initial concentration
A is an amplitude measure satisfying the condition $m \cdot A \leq C_0$, where m is an integer specifying the chosen number of oscillations during the treatment
t_c and a are constants that determine the duration and time between successive concentration "drops"

Examples of decay patterns generated with Equation 5.17 as a model with the corresponding survival curves are shown in Figure 5.7. Although the dissipation model's equation appears cumbersome, it is very easy to write in the notation of a program like Mathematica®. Once written, it can be used to generate a variety of fluctuating decay patterns, all satisfying the condition of a negative or zero slope (Equation 5.14) with a controlled number of concentration drops. The number of drops is determined by fixing their amplitudes through adjusting the parameter A; the fluctuations' durations are controlled by fixing the t_{ci}. The expression of $C(t)$ in the form of Equation 5.17, i.e., with a summation symbol or notation, is not a hindrance to the numerical solution of the survival model's rate

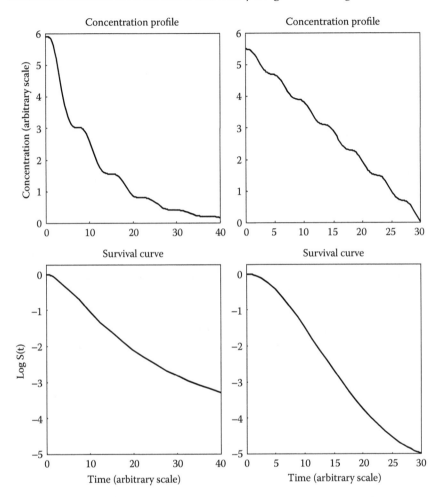

FIGURE 5.6
Simulated exponential oscillatory chemical agent dissipation patterns and corresponding survival curves. The concentration profile was produced by Equation 5.15 and Equation 5.16, suggested by Professor Claude Penchina (2004, private communication) as a model that satisfies the condition set by Equation 5.14.

equation (Equation 5.5) and to plotting of the result, the sought survival curve. In fact, the $\log_{10}S(t)$ vs. t relationship is generated and plotted in about the same time that it takes to produce an "ordinary" survival curve in which the agent concentration decreases monotonically.

The main advantage of Equation 5.17 as a model is that its frame can be easily used to simulate nonuniform and aperiodic concentration fluctuations. Suppose one wants to simulate the diurnal fluctuations in the concentration of an agent added to an open water reservoir or a large

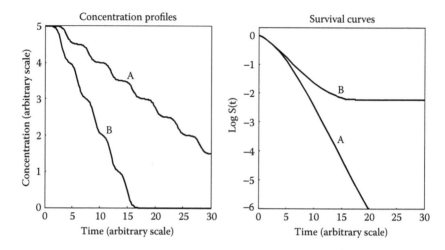

FIGURE 5.7
Simulated dissipation patterns with regular fluctuations produced with Equation 5.17 as a model and corresponding survival curves. (From Peleg, M., 2002, *Food Res. Int.*, 35, 327–336. With permission, courtesy of Elsevier Ltd.)

storage tank exposed to the elements. On the time scale of several days, such a system will experience fluctuations during a more or less constant period of 24 h. Because of weather irregularities, the oscillations' amplitude will not be constant but variable. Such a scenario can be described by the model (Peleg, 2002c):

$$C(t) = C_0 - \sum_{i=1}^{m} \frac{A_{ri}}{1 + \exp\left[\dfrac{it_e - t}{a}\right]} \qquad (5.18)$$

where A_{ri} is random amplitudes satisfying the condition that $\Sigma A_{ri} \leq C_0$.

Examples of concentration profiles generated with this model are shown in Figure 5.8. The figure also shows the corresponding survival curves, calculated with Equation 5.5 as a model. Again, the apparent complexity of the concentration profile, $C(t)$, when expressed in the form of Equation 5.18 has hardly any effect on the speed at which the survival model's rate equation is solved numerically. A dissipation model formulated to produce random concentration fluctuations can be used repeatedly to generate enough survival curves so that these can be subjected to statistical analysis. Thus, if the amplitudes range could be estimated on the basis of previous experience, for example, one could also estimate the *probability* that a certain survival level will be reached after a given time whenever the initial agent concentration is increased or decreased. (For more on forecasting methods, see Chapter 13.)

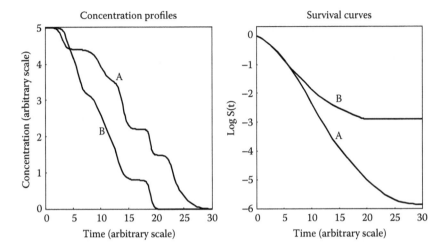

FIGURE 5.8
Simulated dissipation patterns with irregular fluctuations produced with Equation 5.18 as a model and corresponding survival curves. (From Peleg, M., 2002, *Food Res. Int.*, 35, 327–336. With permission, courtesy of Elsevier Ltd.)

Totally Irregular Fluctuating Decay Patterns

The same model frame can also be used to simulate fluctuating concentration decay patterns with random amplitudes *and* frequencies. Such situations probably exist when a food treated with hydrogen oxide or ozone, for instance, is transported over large distances. A closed space disinfected with a gaseous agent may be another example. For these and similar situations, the concentration profile model can be written as (Peleg, 2002c):

$$C(t) = C_0 - \sum_{i=1}^{m} \frac{A_{ri}}{1 + \exp\left[\dfrac{t_{cri} - t}{a_{ri}}\right]} \tag{5.19}$$

with A_{ri}, t_{cri}, and a_{ri} produced by a random number generator. The constraint that $\Sigma A_{ri} \leq C_0$ is still applicable here. (It can be incorporated into the simulation program in the form of a reiterated test of whether the requirement is satisfied whenever a new set of random values is generated.)

Examples of concentration decay patterns generated with Equation 5.19 as a model and the corresponding survival curves are shown in Figure 5.9. As before, the complexity of $C(t)$ as expressed by the preceding dissipation model (with three random entries) is not a hindrance to the

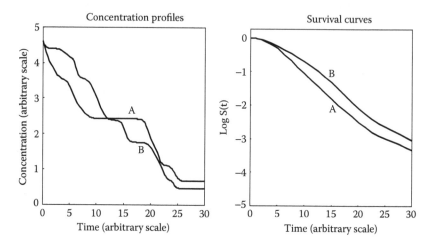

FIGURE 5.9
Simulated dissipation patterns having totally random fluctuations produced with Equation 5.19 as a model and corresponding survival curves. (From Peleg, M., 2002, *Food Res. Int.*, 35, 327–336. With permission, courtesy of Elsevier Ltd.)

numerical solution of the survival model's rate equation (Equation 5.5) into which the expression is inserted. Once the model in the form of Equation 5.19 has been incorporated into the model's rate equation, the model can be used to generate sets of random but realistic survival curves for various contemplated concentration histories. All that is needed is to adjust the dissipation parameters, namely, C_0, A_{ri}, t_{cri}, and a_{ri}, and choose the number of simulations. The results can then be collected in the form of a histogram of the calculated survival ratios after any given time.

As before, the probability of the treatment's success or failure can be estimated from such simulated data. The simulations can also be used to study how changing the agent's initial concentration may affect the disinfection efficacy. Conversely, the repeated simulations can produce the distribution of the *times* at which a chosen survival ratio will be reached given a certain agent initial concentration. The resulting histogram will provide another way to assess the safety of current or planned chemical treatments under realistic uncontrolled situations.

Agent Replenishment

Recall that the survival model (Equation 5.5) sets no limits on the mathematical structure of the concentration profile, $C(t)$. Thus, the agent's replenishment can be modeled by choosing a mathematical expression where $C(t)$ is allowed to increase and decrease alternately. Of course, an unlimited number of functions of different mathematical complexity can be used to account for an occasional or periodic replenishment of the agent. All that is needed is that they produce cycles where a period or periods in which the agent's concentration increases rapidly are followed by periods of substantial dissipation. The dissipation pattern can be "regular" (that is, monotonic or with a constant periodicity) or "irregular" (that is, having concentration fluctuations with a random amplitude and/or frequency; see "Agent Dissipation with Regular and Random Oscillations" above). To demonstrate the model's capabilities, we will show two simulations:

- An administration of the agent and its replenishment in two large doses

- A slow release of the agent in a manner in which its concentration initially increases and then fluctuates for a while until it is eventually exhausted

The first profile can be described by a discontinuous model containing several "if" statements. The chosen example, written in syntax similar to that of Mathematica®, is

$$
\begin{aligned}
C(t) = \ & \text{If}[t \leq t_1, \exp(k_1 t), \text{If}[t \leq t_2, \exp(kt_1)\exp(-k_2(t - t_1)), \\
& \text{If}[t \leq t_3, \exp(k_1)t\exp[-k_2(t_2 - t_1)] \\
& + \exp k_3(t - t_2), \{\exp(k_1 - t_1)\exp[-k_2(t_2 - t_1)] \\
& + \exp[k_3(t_3 - t_2)]\{\exp(-k_4 t - t_3)\}]]]
\end{aligned}
\tag{5.20}
$$

This nested and obviously cumbersome model has three times t_1, t_2, and t_3 that mark the administration of the agent and its complete dispersion as well as four exponential rate constants. Two, rate constants k_1 and k_3, characterize the agent's dispersion and the other two, k_2 and k_4, its exponential dissipation pattern. The model can be simplified by having $k_1 = k_3$ and $k_2 = k_4$ or further complicated by adding one or more replenishment cycles. Either way, it can produce a realistic looking concentration profile (Figure 5.10). Once $C(t)$ is defined, it can be inserted into the model's rate equation to produce the corresponding survival curve. (Again, in Mathematica®, the term $C[t]$, previously defined and expressed in the right syntax, is inserted into the model's rate equation. Once formulated in this way, the function $C[x]$ is recognized by the program as a normal function, on equal footing with the more familiar and canned functions like $\log[x]$, $\exp[x]$, etc.)

The second scenario is described by the empirical model:

$$
C(t) = \frac{C_{\text{target}} \cdot t[1 + a\cos(\omega t)]}{(k_1 + t)(1 + \exp)[k_2(t - t_c)]}
\tag{5.21}
$$

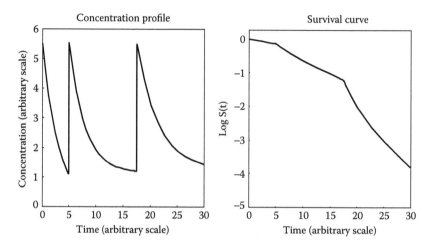

FIGURE 5.10
Survival curves in simulated chemical treatments in which the agent is replenished, using Equation 5.20 as the concentration profile model, and the corresponding survival curves.

where

C_{target} is the targeted concentration
a is the amplitude of the concentration's oscillations
ω is their frequency

The constant k_1 controls the concentration increase rate at the beginning of the treatment and k_2 controls the final dissipation rate. The parameter t_c controls the time at which the final dissipation occurs.

Here, again, the purpose of presenting this six-parameter model is only to demonstrate that the concentration profile complexity is immaterial as far as the solution of the survival model rate equation (Equation 5.5) is concerned. Once $C(t)$ is defined in the appropriate syntax, it is handled by Mathematica® just like any other function already stored in the program.

The survival curves in the two preceding hypothetical scenarios, calculated with Equation 5.5 as the rate model, are shown in Figure 5.10 and Figure 5.11. Once the model is written into the program, it is very easy to change its parameters. One can alter the organism's survival parameters, $(n, k, \text{and } C_c)$ to test how a more sensitive or sturdier organism will survive the treatment, for example. Alternatively, one can examine the effect of changing the chemical agent's concentration profile on the same targeted organism. In the first scenario (see Figure 5.11), this can be done by augmenting the dose (through increasing k_1 and k_2 or extending t_1 and $t_3 - t_2$ in terms of the profile equation), for example, or by shortening or prolonging the time intervals between successive replenishments ($t_2 - t_1$ or $t_4 - t_3$ if a third replenishment is planned).

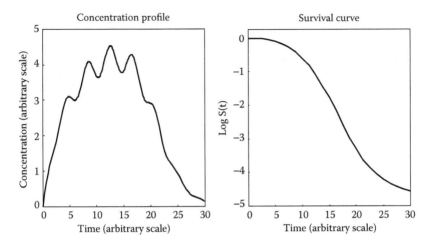

FIGURE 5.11
Survival curves in simulated chemical treatments in which the agent's concentration fluctuates during its administration concentration and dissipation, produced with Equation 5.21 as the concentration profile model, and the corresponding survival curves.

In the second scenario (Figure 5.11), one can increase or decrease the target concentration (C_{target}), prolong or shorten the process (t_c), and change the oscillations' amplitude and frequency (a and t_c, respectively). These are only a few possible modifications of the model; numerous other options are open to the program's user. One can also change *both* the organism's survival parameters and the concentration profile to insert an altogether different concentration profile (e.g., any one of those described in "Traditional Models" and "Alternative General Model") for comparison. All that is needed is to change the pertinent terms in the parameters' definition or replace the concentration profile expression, and run the program to calculate the corresponding survival curve.

Estimation of Survival Parameters from Data Obtained during Treatments with a Dissipating Agent

The real test of a survival model, as has been frequently stressed in the preceding chapters, is its ability correctly to predict survival curves observed under conditions different from those under which the model's parameters have been determined. A most convenient option would be to predict survival patterns under varying agent concentration regimes from data obtained under isoconcentration conditions. Unfortunately, when it comes to microbial inactivation by a volatile and/or unstable chemical agent, survival data obtained under constant agent concentration are very difficult to find. As previously explained, they might not even exist because of the technical difficulty of maintaining constant concentration. Thus, a method to validate the Weibullian–log logistic model similar to the one described in Mattick et al. (2001) or Corradini and Peleg, 2004c for thermal inactivation of *Salmonella* or *E. coli* would not be applicable here, unless a practical and inexpensive way to create isoconcentration conditions is found.

However, if the survival patterns of a given organism exposed to a dissipating lethal agent can be *assumed* to follow the Weibullian–log logistic model, then, at least in principle, the survival parameters could be determined directly from inactivation data obtained under varying agent's concentration, as long as the concentration profile is known (Corradini and Peleg, 2003). For some reason, survival data accompanied by a detailed record of the agent's changing concentration are not easy to find in the literature (see below). Still, the mathematical procedure to determine the survival parameters can be tested by computer simulations.

The test comprises generating survival curves with the model (Equation 5.5) for various concentration profiles, using assumed but realistic survival parameters (n, k_c, and C_c) of a commonly targeted microorganism. A random noise can then be superimposed on the generated survival curve to imitate the experimental scatter that would probably exist if the survival data had been real. The type of noise (with a uniform or normal distribution, for example) can be chosen from the list of random number generators that most modern mathematical softwares offer. The scatter's amplitude can be set at one or more levels that match the spread usually found in experimental survival curves of this kind. The database created can then be used to determine the survival parameters (n, k_c, and C_c) by the simplex or another minimization algorithm (standard options in the latest versions of Mathematica®).

The procedure can be repeated with new sets of random data to simulate experimental replications and the results averaged. The values of the survival parameters calculated — namely, n, k_c, and C_c — can now be compared with those used to generate the original survival data. For the method to work with even a "moderate" noise as in the case of nonisothermal survival (see Chapter 4), fixing the value of n might be needed to avoid unrealistic values of k_c and C_c. In principle at least, simulations of this kind, with the noise level adjusted, can be used to determine the kind of reproducibility required to determine all three parameters simultaneously.

Examples of simulated survival data with two agent dissipation patterns are shown in Figure 5.12. The calculated values of k_c and C_c are shown in Table 5.1. They demonstrate that the simplex algorithm can be used to recover the generation parameters and that fairly reliable estimates of their magnitude can be obtained even with four to five "replicates"

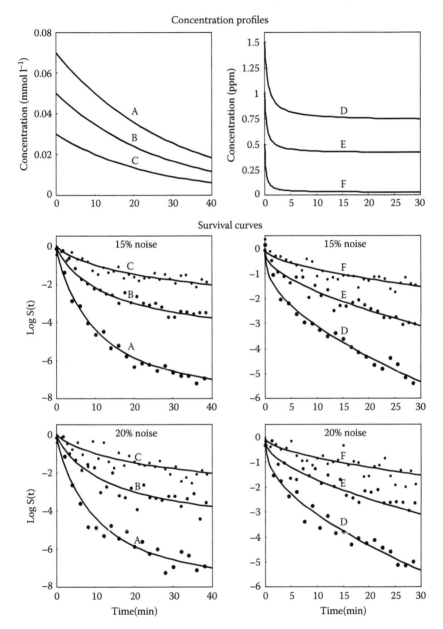

FIGURE 5.12
Simulated exponential chemical agent dissipation profiles generated with Equation 5.10 and Equation 5.13 as models and the corresponding survival data with two levels of superimposed noise. For the models' generation and retrieved parameters, see Table 5.1. (From Corradini, M.G. and Peleg, M., 2003, *J. Appl. Microbiol.*, 95, 1268–1276. With permission, courtesy of Blackwell Publishing.)

TABLE 5.1

Generated and Estimated Microbial Survival Parameters in Disinfected Water

Dissipating agent	Conc. profile	Generation parameters			Noise amplitude			
					15%		20%	
		n	k	C_c	k	C_c	k	C_c
Peracetic acid	Exponential ($C_0 = 0.03$ and $k_{c1} = 0.04$)	0.8	50	0.06	54	0.06	43	0.07
					52	0.06	52	0.06
					52	0.06	42	0.07
					50	0.06	62	0.05
					Mean	Mean	Mean	Mean
					52	0.06	50	0.06
	Exponential ($C_0 = 0.05$ and $k_{c1} = 0.036$)	0.8	50	0.06	51	0.06	63	0.06
					58	0.06	50	0.06
					43	0.06	45	0.06
					43	0.06	56	0.06
					Mean	Mean	Mean	Mean
					49	0.06	54	144
	Exponential ($C_0 = 0.07$ and $k_{c1} = 0.033$)	0.8	50	0.06	48	0.06	50	0.06
					52	0.06	50	0.06
					50	0.06	51	0.06
					50	0.06	48	0.06
					Mean	Mean	Mean	Mean
					50	0.06	50	0.06
Chlorine	$C = C_0 - t/(k_{c2} + k_{c3}*t)$ ($C_0 = 0.5$, $k_{c2} = 0.4$, and $k_{c3} = 2.1$)	0.6	2.0	0.80	2.0	0.83	1.9	0.85
					2.2	0.75	1.9	0.83
					2.0	0.80	1.9	0.85
					1.7	0.93	1.9	0.84
					Mean	Mean	Mean	Mean
					2.0	0.83	1.9	0.84
	$C = C_0 - t/(k_{c2} + k_{c3}*t)$ ($C_0 = 1.0$, $k_{c2} = 0.6$, and $k_{c3} = 1.7$)	0.6	2.0	0.80	2.0	0.81	1.8	0.88
					1.7	0.86	2.2	0.78
					2.2	0.76	2.1	0.79
					1.9	0.82	2.0	0.82
					Mean	Mean	Mean	Mean
					2.0	0.81	2.0	0.82
	$C = C_0 - t/(k_{c2} + k_{c3}*t)$ ($C_0 = 1.5$, $k_{c2} = 0.8$, and $k_{c3} = 1.3$)	0.6	2.0	0.80	2.1	0.80	2.4	0.80
					1.9	0.80	1.9	0.80
					1.8	0.80	1.9	0.77
					2.1	0.80	2.2	0.81
					Mean	Mean	Mean	Mean
					2.0	0.80	2.1	0.80

Source: From Corradini, M.G. and Peleg, M., 2003, *J. Appl. Microbiol.*, 95, 1268–1276. With permission, courtesy of Blackwell Publishing.

(Corradini and Peleg, 2003a). This has not been unexpected, of course, because the same procedure has already been found useful to retrieve the survival parameters from simulated and real data of thermally inactivated bacteria (Peleg and Normand, 2004).

Demonstrations of the Procedure with Published Data

At the time of writing this chapter, we had only one set of published data appropriate for the described analysis, i.e., a set of experimental survival ratios with the corresponding agent's concentration profile. The data shown in Figure 5.13 are of coliforms in water treated with peracetic acid, published by Wagner et al. (2002). The number of data points in the concentration profiles and corresponding survival curves was rather small, four to five to be exact. However, it is probably a typical number that one would expect to find in reports on these kinds of experiments, dictated by technical or logistic considerations.

One set of the published results, which had five complete entries of the dissipated concentration and corresponding survival ratios, was used by Corradini and Peleg (2003a) to estimate the survival parameters of the coliforms when n had three predetermined values. The concentration profile was fitted by an exponential model (Equation 5.10), which was incorporated into the rate model (Equation 5.5). After the application of the simplex algorithm, the following results were obtained:

N (initially fixed)	k_c (l mmol^{-1})	C_c(mmol l^{-1})
0.70	48	0.11
0.75	56	0.11
0.80	65	0.10

As in thermal inactivation, where T_c was hardly affected by the choice of n, here C_c was about the same for all three values of n. As could be expected, increasing the magnitude of the chosen n lowered the corresponding value of k_c, without much effect on the model's fit. The k_c and C_c determined were used to estimate the survival curves obtained under three different patterns of the peracetic acid dissipation. The results (with a fixed $n = 0.75$) and their agreement with the reported experimental values are shown in Figure 5.13.

Admittedly, this is a single demonstration based on a single published source. However, it had data on three different concentration profiles and corresponding survival ratios to test the method. Therefore, the agreement between the predictions and observations, even if far from perfect, is unlikely to be merely a coincidence. Yet, at this stage, the given demonstration should only be considered as a promising sign that the proposed procedure to estimate the survival parameters *might* work. Obviously, much more experimental evidence would be needed to establish that the procedure will be similarly successful for other organisms and for disinfectants that have a different mode of action and dissipation patterns.

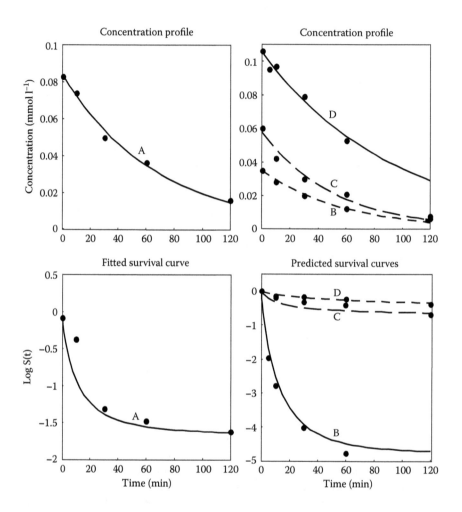

FIGURE 5.13

Published survival curves of fecal coliforms in water treated with peracetic acid at four different initial agent concentration and dissipation rates using Equation 5.10 as a model (top). The dissipation and survival curve at the left was used to predict the three survival curves shown at the bottom right, which correspond to the three concentration profiles shown at the top right. The filled dots are the experimental survival ratios and the smooth lines are the predicted survival curves. The survival parameters used for the predictions were $n = 0.75$, $k = 56 \ l \ mmol^{-1}$, and $C_c = 0.11 \ mmol \ l^{-1}$. The original experimental data are from Wagner, M. et al., 2002, *Water Environ. Res.*, 74, 33–50. (From Corradini, M.G. and Peleg, M., 2003, *J. Appl. Microbiol.*, 95, 1268–1276. With permission, courtesy of Blackwell Publishing.)

Discrete Version of Survival Model

The inactivation rate model (Equation 5.5), like the one proposed for thermal processes, is a differential equation that with few exceptions must be solved numerically to produce the survival curve. If the logarithmic rate $d\log_{10}S(t)/dt$ is approximated by $\Delta\log_{10}S(t)/\Delta t$, where Δt is a small time interval, then the model is transformed into the difference equation:

$$\frac{\log_{10}S(t) - \log_{10}S(t_{i-1})}{t_i - t_{i-1}} = -\frac{b[C(t_i)] + b[C(t_{i-1})]}{2} \cdot n$$

$$\cdot \left\{ \frac{-[\log_{10}S(t_i) + \log_{10}S(t_{i-1})]}{b[C(t_i)] + b[C(t_{i-1})]} \right\}^{\frac{n-1}{n}} \qquad (5.22)$$

where $\Delta t = t_i - t_{i-1}$ and $\Delta\log_{10}S(t) = \log_{10}S(t_i) - \log_{10}S(t_{i-1})$.

Here, again, at $t = 0$, $\log_{10}S(0) = 0$ and $b[C(0)] = b[C_0]$, with C_0 the agent's initial concentration. The first step in the iterative calculation process is to calculate the value of $\log_{10}S(t_1)$ from the equation:

$$\frac{\log_{10}S(t_1)}{t_1} = -\frac{b[C(t_1)] + b[C_0]}{2} \cdot n \cdot \left\{ \frac{-\log_{10}S(t_1)}{b[C(t_1)] + b[C_0]} \right\}^{\frac{n-1}{n}} \qquad (5.23)$$

Because $b[C_0]$ and $b[C(t_i)]$ calculated directly with the concentration profile's equation and t_1 generated or measured, the only unknown in the equation is $\log_{10}S(t_1)$. Its value can be calculated analytically by any general-purpose software (or numerically by the "goal seek" function of Microsoft Excel®). Once $\log_{10}S(t_1)$ is calculated, its value is inserted into Equation 5.22 for the second iteration, with t_2 and the corresponding $b[C(t_2)]$, which is calculated from the concentration profile's equation. With all the other terms known, $\log_{10}S(t_2)$, remains the only unknown in the new equation. Therefore, like $\log_{10}S(t_1)$ before, it too can be extracted analytically or by using the "goal seek" function of MS Excel®. Once $\log_{10}S(t_2)$ is calculated, the iterations can continue to find $\log_{10}S(t_3)$ for t_3, $\log_{10}S(t_4)$ for t_4, etc., until reaching the end of the monitored actual process or the studied simulation.

Examples of survival curves generated with Equation 5.22 as a model are shown in Figure 5.14. These were generated with the program posted as freeware on the Web (see the Freeware section in the back of this book).

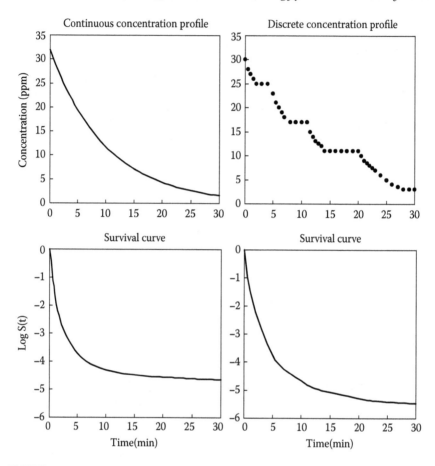

FIGURE 5.14
Concentration profiles and the corresponding survival curve of an *E. coli*-like organism produced by the incremental method using MS Excel with a program posted as freeware on the Web. Left: with the concentration profile generated by a formula; right: with the concentration profile posted by the user. (Courtesy of Dr. Maria G. Corradini.)

Like their counterparts in thermal processing, survival curves generated with the incremental method are practically indistinguishable from those produced by the continuous rate model. Thus, if the survival parameters of the targeted organism, n_c, k_c, and C_c, can be estimated from published data or other sources, one can generate survival patterns under almost any conceivable agent dissipation pattern using any general-purpose software, including the popular MS Excel®.

6

High CO_2 and Ultrahigh Hydrostatic Pressure Preservation

In theory, there is no difference between theory and practice. In practice, however, there is.

Albert Einstein

Pressure technology has been emerging as a potential alternative to heat preservation for some products, notably orange juice and marmalades in Japan and avocado products and oysters in the U.S. The effect of high pressure on microorganisms and on the properties of foods has been a topic of intensive research in recent years and there is a large body of literature on the subject. The kinetics of microbial inactivation under ultrahigh pressure has also received considerable attention in food research and survival parameters of a variety of microorganisms of food safety concerns are listed in Websites or can be found in other publications.

The traditional approach to modeling the inactivation kinetics of microorganisms exposed to ultrahigh pressure has been based on the assumption that it follows the first-order kinetics model and that there is an analogy between the lethal effect of high pressure and that of heat. Thus, most of the published microbial survival parameters posted on the Web are in the form of D_p and z_p values and/or in the form of rate constants at specific temperatures and corresponding energies of activation. The latter have been calculated using a modified Arrhenius equation in which the absolute pressure has replaced the absolute temperature. Alternatively, to account for the combined effect of pressure and temperature, the original Arrhenius equation has been amended by the addition of a pressure term with its own energy of activation.

Suffice it is to restate here that, as in thermal inactivation, there is no compelling reason to assume that the mortality of microorganisms under high pressure is a process that must follow the first-order kinetics. Thus, the meaning of any D_p value would remain uncertain as in the case of thermal and chemical inactivation. Once the D_p value becomes suspect, the usefulness of the z_p value is also seriously undermined. The same applies to its reciprocal, the inactivation rate constant, k_p. Obviously, whenever the semilogarithmic isobaric survival curve is nonlinear, its slope, the momentary logarithmic inactivation rate, would be a function not only of pressure but also of the exposure time. If so, the Arrhenius model cannot be used and the energy of activation loses any meaning that it might have had. Moreover, the rationale for using the Arrhenius equation in the first place has never been explained satisfactorily. The analogy between the effect of high pressure on living cells and the effect of temperature on the reactivity of simple chemical systems is not at all obvious and there is no reason that it should be taken for granted.

As in the case of thermal processing, the Arrhenius equation and the log linear pressure dependence of the D_p value are mutually exclusive models and thus the rationale for listing them side by side in one of the most popular Websites is also unclear. The same can be said about the double logarithmic transformation that the log linear and Arrhenius models require. No evidence indicates that the inactivation rate rises (or falls) by several orders of magnitude in the pertinent applied pressures range.

To complicate matters further, the application of ultrahigh pressure *always* results in adiabatic heating. Therefore, truly isothermal and isobaric conditions are never produced in the laboratory or in any real industrial process. Consequently, for pressure kinetic parameters to be correct, they must be determined from equations that take into account the dynamic and nonisothermal nature of the process. This is because the temperature rise and the exposure time can significantly affect the inactivation rate, at least theoretically. As far as pressure inactivation of microbiology is concerned, at least two very distinct processes exist:

- The decisive lethal or inhibitory effect is of a gas pressure alone, and all other *physical* aspects of the pressure application are negligible. The application of carbon dioxide as in carbonated beverages or other products at pressures of up to about 5 MPa would be a typical example of such a case.

  Admittedly, the *chemical properties* of the gas can play an important role in the inactivation mechanism, as in the case of carbon dioxide and its possible relation to the product's pH. The main temperature effect in this case is on the gas solubility, which may or may not play a major role in the inactivation efficacy. However, at any rate, because the CO_2 pressure levels in carbonated beverages are very low in comparison with those exerted during an ultrahigh hydrostatic pressure treatment, *practically isothermal conditions* can be easily achieved in studies of the inactivation kinetics. This is because the experiment/treatment duration is usually so long that the temperature changes during the pressure applications and release cannot play a decisive role.

- Lethality is produced by the *combined effect* of pressure and temperature. Although a secondary player, the inevitable adiabatic heating by several degrees Celsius or almost adiabatic heating if the vessel is not insulated does affect the inactivation kinetics by enhancing the lethality of the applied ultrahigh pressure, which is on the order of 200 to 800 MPa. The temperature rise effect can be undesirable as far as quality is concerned if the treated product (orange juice or avocado paste, for example) is heat sensitive.

  However, ultrahigh pressure can also be a means to raise the temperature of a product almost instantaneously and, just as importantly, *all over the bulk simultaneously.* This adiabatic heating can be exploited, at least in principle, in thermal sterilization of low acid foods by shortening the come-up time considerably. The result would be a shorter process overall and a product with better nutritional value and organoleptic quality. Either way, though, whenever the pressure is on the order of hundreds of megapascals, the temperature rise is almost always sufficient to affect the efficacy of the treatment and thus should be given an appropriate consideration.

What follows will only address the modeling of the inactivation patterns produced in the two scenarios at the population level. The focus will be on the mathematical implications of the combined pressure–temperature history on the inactivation, without entering the reasons why individual microbial cells are destroyed by the treatments.

Microbial Inactivation under High CO_2 Pressure

As a model system, consider the experimental results of Ballestra et al. (1996), who studied the effect of CO_2 on *E. coli* in a culture medium in which the mildly alkaline pH was shown to remain practically unchanged by the gas's pressure. The survival data of the exposed *E. coli* cells under various pressures at the same temperature are shown in Figure 6.1. Although the reported survival ratios are few and scattered, especially at the lowest CO_2 pressure level, they indicate quite clearly that the organism's inactivation by the gas did not follow the first-order kinetics model. None of the semilogarithmic survival curves were linear and they all had downward concavity that could be described by the Weibullian–power law model (Peleg, 2002d), with $n(P) > 1$:

$$\log_{10} S(t) = -b(P)t^{n(P)} \tag{6.1}$$

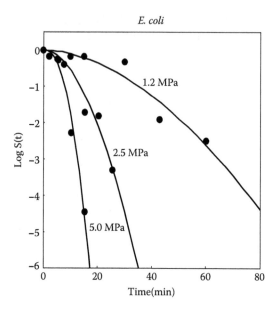

FIGURE 6.1

Experimental survival data of *E. coli* under three levels of CO_2 gas pressure fitted with the Weibull–power law model (Equation 6.1) as the primary model. The original data are from Ballestra, P. et al., 1996, *J. Food Sci.*, 61, 829–831; 836. (From Peleg, M., 2002, *J. Food Sci.*, 67, 896–901. With permission, courtesy of the Institute of Food Technologists.)

where $b(P)$ and $n(P)$ are pressure-dependent coefficients.

The fit of Equation 6.1 is shown in Figure 6.1 and the pressure dependence of $b(P)$ and $n(P)$ is shown in Figure 6.2. With only three pressure levels reported in the original publication, accurate characterization of $b(P)$ and $n(P)$ has been impossible. However, because the emphasis here is on the modeling procedure rather than on the *E. coli*'s response to CO_2 pressure, the mathematical expressions used to describe the $b(P)$ vs. P and $n(P)$ vs. P relationships (both purely empirical; see below) are not really important. Also, no attempt has been made to fix the value of $n(P)$ as has been done in previous applications of the Weibullian–power law model.

As before, the time-dependent isobaric inactivation rate at any given pressure is:

$$\left.\frac{d\log_{10} S(t)}{dt}\right|_{P=\text{const}} = -b(P)n(P)t^{n(P)-1} \tag{6.2}$$

Similarly, the time, t^*, that corresponds to any momentary survival ratio, $\log_{10}S(t)$, at a given pressure is:

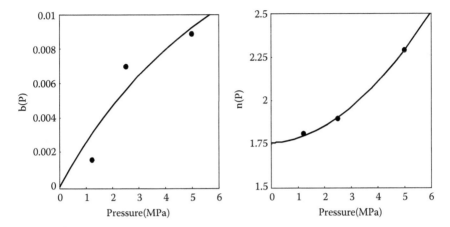

FIGURE 6.2
The Weibullian survival parameters, $b(P)$ and $n(P)$, pressure dependence fitted with ad hoc empirical expressions as secondary models. (From Peleg, M., 2002, *J. Food Sci.*, 67, 896–901. With permission, courtesy of the Institute of Food Technologists.)

$$t^* = \left[\frac{-\log S(t)}{b(P)} \right]^{\frac{1}{n(P)}} \tag{6.3}$$

Thus, for any nonisobaric process in which the pressure varies with time, $b(P)$ and $n(P)$ also vary with time, i.e., $b(t) = b[P(t)]$ and $n(t) = n[P(t)]$, where $P(t)$ is the "pressure profile."

Combining Equation 6.2 and Equation 6.3 yields the rate equation for any isothermal but not nonisobaric process:

$$\frac{d \log_{10} S(t)}{dt} = -b[P(t)]n[P(t)] \left\{ \frac{-\log_{10} S(t)}{b[P(t)]} \right\}^{\frac{n[P(t)]-1}{n[P(t)]}} \tag{6.4}$$

Kinetically, Equation 6.4 is identical to the inactivation model of nonisothermal heat treatments discussed in Chapter 2 through Chapter 4, except that the pressure profile, $P(t)$, replaces the temperature profile, $T(t)$, in the rate equation. Similarly, the theoretical effects of a variety of CO_2 pressure histories could be studied by computer simulations using the rate equation as a model.

In our case, the secondary models $b(P)$ and $n(P)$ vs. P could be described by the ad hoc empirical terms:

$$b(P) = \frac{P}{347.4 + 38.7P} \tag{6.5}$$

and

$$n(P) = 1.76 + 0.0288\, P^{1.808} \tag{6.6}$$

respectively. These in turn could be easily converted into functions of time and subsequently inserted into the model's rate equation as $b[P(t)]$ and $n[P(t)]$ (Equation 6.4).

Effect of Pressure Level and Treatment Duration

A family of realistic pressure profiles of (almost) the same treatment duration can be produced by the superposition of logistic and Fermi terms:

$$P(t) = \frac{P_{target}}{\{1 + \exp[k_1(t_{c1} - t)]\}\{1 + \exp[k_2(t - t_{c2})]\}} \tag{6.7}$$

where
    P_{target} is the target pressure
    t_{c1} and t_{c2} mark the center points of the pressure's come-up and come-down times, respectively
    k_1 and k_2 are constants that account for or control the rate at which the targeted pressure is reached (k_1) and released (k_2)

Thus, a process's duration according to this model is approximately $t_{c2} - t_{c1}$ and the higher the values of k_1 and k_2 are, the closer the profile resembles an instantaneous pressure rise and release. Simulated CO_2 treatments of the kind described by Equation 6.7 are shown in Figure 6.3, together with the corresponding survival curves. The pressure curves were produced by inserting the pressure profile equation (Equation 6.7) with $k_1 = k_2$ and the same t_{c1} and t_{c2}, but with different levels of P_{target}. The survival curves were generated with Equation 6.4 as the rate model into which the organism's survival parameters $b(P)$ and $n(P)$, characterized by Equation 6.5 and Equation 6.6, respectively, were inserted in the nested form of $b[P(t)]$ and $n[P(t)]$. Once formulated, the model's equation was solved numerically by Mathematica® to generate the shown survival curves. These demonstrate how simulations of this kind can be used to determine a pressure level that will produce the desired theoretical reduction in the targeted organism's survival ratio — that of *E. coli* in our particular example.

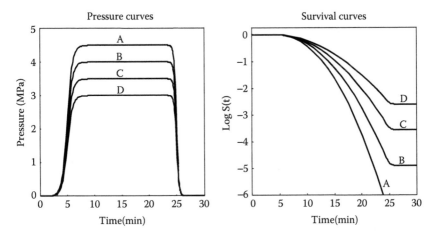

FIGURE 6.3
Simulated CO_2 pressure profile generated with Equation 6.7 as a model with different target pressure levels and the corresponding theoretical survival curves of *E. coli* calculated with Equation 6.14 as a model. (From Peleg, M., 2002, *J. Food Sci.*, 67, 896–901. With permission, courtesy of the Institute of Food Technologists.)

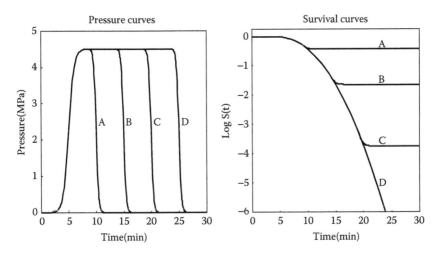

FIGURE 6.4
Simulated isobaric CO_2 pressure profiles of various durations generated with Equation 6.7 as a model and the corresponding theoretical survival curves of *E. coli* calculated with Equation 6.14 as a model. (From Peleg, M., 2002, *J. Food Sci.*, 67, 896–901. With permission, courtesy of the Institute of Food Technologists.)

The same kinds of simulations can be used to determine the *duration* of an effective treatment. This is shown in Figure 6.4. The pressure profiles in this case were also produced by Equation 6.7 as a model, except that

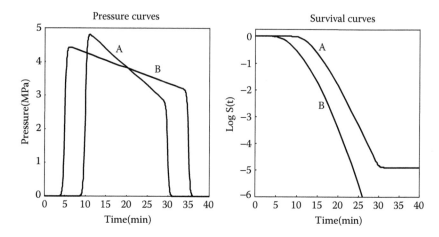

FIGURE 6.5
A simulated CO_2 pressure loss profile generated with Equation 6.8 as a model and corresponding theoretical survival curves of *E. coli* calculated with Equation 6.14 as a model. (From Peleg, M., 2002, *J. Food Sci.*, 67, 896–901. With permission, courtesy of the Institute of Food Technologists.)

the target pressure P_{target}, k_1, and k_2 had been fixed and t_{c2} and t_{c1} were set so that the treatment durations would be approximately 10, 15, 20, and 25 min. The corresponding theoretical survival curves are also shown in the figure. They demonstrate how the process duration needs to be adjusted to achieve a desired level of the targeted organism's destruction, once the pressure level has already been chosen.

The same model can be used to assess the theoretical consequences of simultaneous changes of the target pressure and the treatment duration. Thus, the simulations can be repeated with various pressure profiles until an effective practical combination of the pressure level and process duration can be identified. These, in turn, can be used to select promising treatments prior to actual testing. In a similar manner, the pressurization and pressure release rates can also be varied by adjusting k_1 and k_2, respectively, until the simulated pressure profile matches that of any particular equipment.

Once the model and program are set, they can be used to simulate a variety of scenarios that might be of interest to the process designer or operator. Examples are an error in the pressure gage, premature pressure release, or a leak that would result in a pressure drop. The latter can be simulated by modifying the pressure profile model. Figure 6.5 shows such a hypothetical scenario and its potential consequences in terms of the targeted organism's survival curve. The pressure profile in this case was produced by the model:

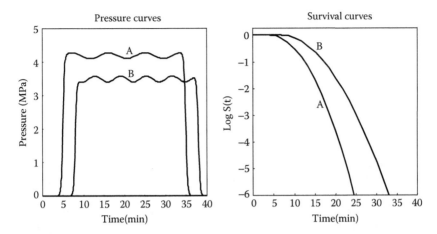

FIGURE 6.6
A simulated oscillating CO$_2$ pressure profile generated with Equation 6.9 and corresponding theoretical survival curves of *E. coli* calculated with Equation 6.14 as a model. (From Peleg, M., 2002, *J. Food Sci.*, 67, 896–901. With permission, courtesy of the Institute of Food Technologists.)

$$P(t) = \frac{P_{\text{target}}[1 - c(t - t_c)]}{\{1 + \exp[k_1(t_{c1} - t)]\}\{1 + \exp[k_2(t - t_{c2})]\}} \qquad (6.8)$$

where c is a constant representing a constant pressure loss rate.

The term '$1 - c(t - t_c)$' in Equation 6.8 can be replaced by an alternative expression if the pressure drops at a varying rather than constant rate or if the pressure loss is accompanied by pressure oscillations. Examples of the latter are shown in Figure 6.6, where the pressure profile was produced by:

$$P(t) = \frac{P_{\text{target}}[1 + a\sin(\omega t)]}{\{1 + \exp[k_1(t_{c1} - t)]\}\{1 + \exp[k_2(t - t_{c2})]\}} \qquad (6.9)$$

where a represents the pressure oscillations' amplitude and ω their frequency.

As has been stated, these are only a few selected examples. Realistic pressure profiles can have discontinuities and regimes of random oscillations. The former can be described mathematically by models that contain 'if statements' and the latter by terms produced by a random number generator. As has been demonstrated in previous chapters, neither is a hindrance to the resulting rate equation's numerical solution with a program like Mathematica®. Therefore, the model's rate equation (Equation

6.4) can be used to generate survival curves under almost any conceivable pressure profile. Once they are produced, these can be used to assess the microbial safety implications of planned and accidental changes in the examined process.

Is the Pressurization Rate a Factor?

One of the outstanding issues in high-pressure technology at the time at which this chapter is written is whether the rates at which the pressure is reached and/or released influence the inhibitory effect of the treatment. The final answer to the question would probably come from the examination of cells that are dead, injured, or surviving intact after processes in which the pressurization and the pressure release rates had been varied. At the population level, the issue can be resolved, at least in principle, by comparing simulated survival curves generated on the basis of the assumption that the pressurization rate plays no role, with experimental survival curves obtained in treatments in which the same target pressure has been reached but at different rates.

A complete agreement between the two groups of survival curves would provide evidence that the pressurization or pressure release rate is *not* a factor, i.e., that the damage to the cells is primarily due to the total duration of the exposure at the treatment's pressure but not to how this pressure has been reached. The conclusion will remain valid if only minor discrepancies are found — discrepancies that could be explained by slight but inevitable differences in the overall exposure to the high-pressure profile and possibly to slight differences in the temperature history. Because the inactivation caused by pressure treatments has a limited reproducibility, the same conclusion would be reached if the differences between the predicted and observed survival ratios fall within the experimental scatter. In such a case, even a statistically significant difference can be judged as practically insignificant and therefore unworthy of further consideration.

However, a large difference that cannot be explained in such manners would constitute evidence that the proposed model (Equation 6.4) is incorrect. This in turn would raise doubt about the validity of the model's underlying assumptions. The most crucial of these is that, in a perfectly isothermal and constant pH treatment, the momentary inactivation rate is only a function of the pressure and momentary survival ratio, but not of the path at which it has been reached and the organism's initial inoculum's size.

An example of a set of simulated pressure profiles in which the rates vary is shown in Figure 6.7. Also shown in the figure are the corresponding survival curves, generated with Equation 6.4 as the rate model with the

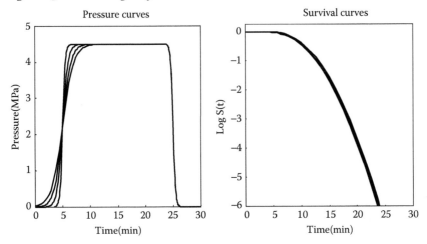

FIGURE 6.7

A set of simulated CO$_2$ pressure profiles of approximately similar duration generated with Equation 6.7 as a model with different pressurization rates. Also shown are the corresponding theoretical survival curves of *E. coli* generated with Equation 6.14 as a model. Notice that comparison of the theoretical survival curves with actual data can reveal whether the pressurization (or pressure release) rates have a synergistic effect, at least in principle (see text). (From Peleg, M., 2002, *J. Food Sci.*, 67, 896–901. With permission, courtesy of the Institute of Food Technologists.)

organism's survival parameters' pressure dependence described by Equation 6.5 and Equation 6.6 as the secondary models. The figure demonstrates that, according to the particular model, when the exposure time is sufficiently long, the come-up time hardly plays a role in the inactivation. Thus, a much larger discrepancy between the experimental and predicted survival ratios would be needed to establish that the pressurization rate is truly an influential factor. The same can be said about the pressure release rate if it too is considered a factor.

Another test of the model is to perform a series of experiments in which the comparison would be between the outcomes of a set of compression–decompression cycles with that of a single cycle of equal (actually almost equal) duration. According to the model and as shown in Figure 6.8, the two kinds of treatments should produce an almost indistinguishable final survival ratio. Thus, if a change in the pressure rate had a lethal effect of its own, the final survival ratio would be significantly lower than that estimated by the model. When ultrahigh hydrostatic pressure is applied, the comparison might not be as straightforward as in the described hypothetical test because alternating the pressure profile will inevitably produce a different product temperature history as well.

Thus, for the pressure rate to be considered a practically significant factor, its effect must be substantially larger than all the other effects that

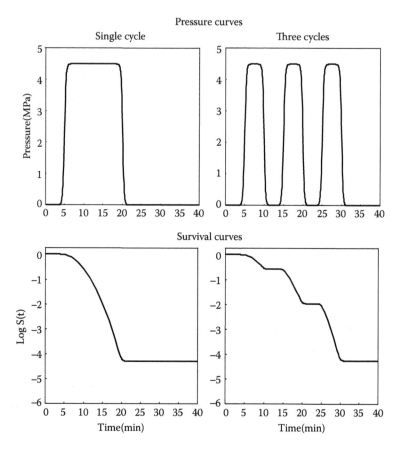

FIGURE 6.8
Two simulated CO_2 pressure profiles of approximately the same total process duration and corresponding theoretical survival curves of *E. coli*. Notice that if Equation 6.14 is a correct inactivation rate model, the two processes will result in a similar final survival ratio. Comparison of the prediction with actual experimental observation can be a sensitive test of whether the pressure application and release rates have a synergistic effect. (From Peleg, M., 2002, *J. Food Sci.*, 67, 896–901. With permission, courtesy of the Institute of Food Technologists.)

altering the pressure profile produces. Admittedly, most of the preceding is based on theoretical considerations only and not on a wide database and personal experience with the process. Yet, the proposed *methodology*, with or without modification, might be applicable not only to research issues, but also to the solution of practical problems. An example is the effect of the variable pressure of CO_2 in beverages after the bottle has been repeatedly opened and closed; with some simplifying assumptions, such scenarios can be modeled in a similar manner to that described in this chapter.

Ultrahigh Pressure

When hydrostatic pressure on the order of 200 to 800 MPa is applied, adiabatic heating is *always* a significant factor and cannot be ignored. In fact, many of the isobaric survival curves reported in the food literature were actually obtained under varying temperature. Because temperature and pressure have a synergistic effect, the lethality of high-pressure treatments is enhanced to at least some extent by the products' temperature rise.

Quantification of the temperature's role in a pressure treatment efficacy is not a simple task. This would be especially true when the pressure vessel is not effectively insulated; in such a case, the temperature drops while the target pressure is maintained. At the time at which this chapter is written, comprehensive data sets from which the pressure and thermal effects could be separated and their interactions quantified are not easy to find. Therefore, what follows is again a theoretical or perhaps a speculative discussion of how models of the combined pressure–temperature effect could be constructed and what kinds of experimental data would be needed to confirm the resulting models' predictive ability.

For the sake of simplicity consider the following assumptions:

- Under constant pressure *and* temperature conditions, hypothetical or real, the survival curve obeys the Weibullian–power law model:

$$\log_{10}S(t) = -b(P,T)\ t^{n(P,T)} \tag{6.10}$$

  Equation 6.10 is an expansion of Equation 6.1, in which the rate parameter, $b(P,T)$, and the curvature measure, $n(P,T)$, are pressure and temperature dependent and not only pressure dependent.

- In a process in which the temperature, pressure, or both vary, the momentary semilogarithmic inactivation rate is uniquely defined as the rate that corresponds to the *momentary combination of pressure and temperature at the time that corresponds to the momentary survival ratio under these conditions.* This assumption is similar to that underlying the modeling of nonisothermal processes under constant pressure and other relevant factors or of varying pressure at a constant temperature and other pertinent factors, except that here the pressure and temperature are allowed to vary simultaneously during the process. (We also assume that recovery from injury is not a factor and that the process is short enough so that the targeted organism has no time to adapt physiologically.)

If and only if these assumptions are valid, then, as before:

$$\left| \frac{d \log_{10} S(t)}{dt} \right|_{\substack{P=\text{const} \\ T=\text{const}}} = -b(P,T)n(P,T)t^{n(P,T)-1} \qquad (6.11)$$

Again, the time that corresponds to the momentary survival ratio, t^*, will be:

$$t^* = \left[\frac{-\log_{10} S(t)}{b(P,T)} \right]^{\frac{1}{n(P,T)}} \qquad (6.12)$$

Combining Equation 6.11 and Equation 6.12 yields the same kind of a rate model as Equation 6.4, except that the time-dependent coefficients are $b(t) = b[P(t),T(t)]$ and $n(t) = n[P(t),T(t)]$. When these are incorporated into the model's equation, it becomes:

$$\frac{d \log_{10} S(t)}{dt} = -b[P(t),T(t)]n[P(t),T(t)] \left\{ \frac{-\log_{10} S(t)}{b[P(t),T(t)]} \right\}^{\frac{n[P(t),T(t)]-1}{n[P(t),T(t)]}} \qquad (6.13)$$

For the sake of simplicity and clarity, we will assume that $n(P,T) \approx$ const. $= n$. The constancy of $n[P(t),T(t)]$ is unlikely to be a realistic assumption in mocosses where the temperature and pressure vary simultaneously and to a large extent. However, as in the previous discussion of nonisothermal heat inactivation, this simplifying assumption is not an essential requirement for what follows. If needed, the pressure and temperature dependence of $n(P,T)$ can always be reincorporated into the model's equation. For $n(P,T) =$ const., the rate model is reduced to:

$$\frac{d \log_{10} S(t)}{dt} = -b[P(t),T(t)]n \left\{ \frac{-\log_{10} S(t)}{b[P(t),T(t)]} \right\}^{\frac{n-1}{n}} \qquad (6.14)$$

- As a first-order approximation, we can assume that the pressure dependence of $b(P)$ under isothermal conditions can be described by the log logistic equation:

$$b(P) = \log_e[1 + \exp\{k_p(T)[P - P_c(T)]\}] \qquad (6.15)$$

where $P_c(T)$ marks the temperature-dependent pressure level at which the inactivation intensifies and $k_p(T)$ the temperature-dependent climb rate of $b(P)$ at pressure levels substantially higher than $P_c(T)$ (as shown schematically in Figure 6.9).

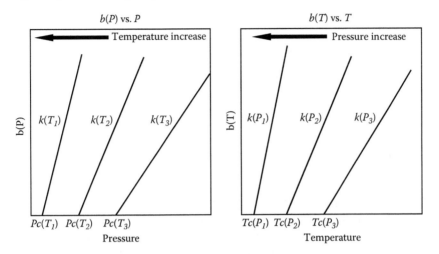

FIGURE 6.9
Schematic view of the pressure and temperature dependence of the Weibullian inactivation rate parameter $b[P,T]$. Notice that a real ultrahigh-pressure treatment is never isothermal because of adiabatic heating. (Courtesy of Dr. Maria G. Corradini.)

- Similarly, we can assume that under isobaric conditions, the temperature dependence of $b(T)$ is also characterized by the log logistic equation (see figure):

$$b(T) = \log_e[1 + \exp \{k_T(P)[T - T_c(P)]\}] \qquad (6.16)$$

In Equation 6.16, $T_c(P)$ marks the pressure-dependent temperature level at which the inactivation intensifies and $k_T(P)$ the pressure-dependent climb rate of $b(T)$ at temperatures substantially higher than $T_c(P)$. Again, this is an assumption made here to demonstrate the suggested methodology rather than to account for the inactivation patterns of any particular organism in any particular medium.

- If all the preceding assumptions are correct, we only need to identify the temperature dependence of $k_P(T)$ and $P_c(T)$ in Equation 6.15 or the pressure dependence of $k_T(P)$ and $T_c(P)$ in Equation 6.16. Again, for the sake of simplicity, let us assume that the temperature dependence of $k_P(T)$ is log logistic:

$$k_P(T) = \log_e\{1 + \exp [a_1(T - T_{c1})]\} \qquad (6.17)$$

and that of $P_c(T)$ is exponential:

$$P_c(T) = P_{Co} \exp (-a_2T) \qquad (6.18)$$

We will also assume that the pressure dependence of $k_T(P)$ is log logistic:

$$k_T(P) = \log_e\{1 + \exp\,[a_3(P - P_{c3})]\} \tag{6.19}$$

and that of $T_c(P)$ is exponential:

$$T_c(P) = T_{Co} \exp\,(-a_4 P) \tag{6.20}$$

where P_{Co}, T_{Co}, T_{c1}, P_{c3}, and the a's are characteristic constants of the organism in question in the particular medium and at the pertinent growth stage.

Equation 6.17 through Equation 6.20 are manifestations of a reasonable but unproven *assumption* that the lethal effect of pressure under isothermal conditions only becomes noticeable beyond a certain pressure level, marked by $P_c(T)$, that decreases (exponentially) as the temperature rises. Similarly, it is also assumed that the lethal effect of temperature under isobaric conditions becomes noticeable only beyond a certain temperature level, marked by $T_c(P)$, that decreases (exponentially) as the pressure is increased. Recall that all the preceding stems from the assumption that, in the region in which the pressure–temperature combination becomes lethal, i.e., where $P \gg P_c(T)$ and/or $T \gg T_c(P)$, the Weibullian inactivation rate constant increases linearly with the pressure or temperature. This assumption, too, is not a prerequisite for the model's development. Alternative expressions of the survival parameters' pressure and temperature dependence can be chosen in light of other considerations and might be just as applicable.

The proportionality constant of the pressure's lethality coefficient, $k_P(T)$, increases with temperature and so does the temperature's lethality coefficient, $k_T(P)$, with pressure, albeit in a manner that becomes noticeable only at a sufficiently high temperature or high pressure, respectively. The assumed temperature and pressure dependence of the survival parameters are shown schematically in Figure 6.10. This elaborate survival model has yet to be confirmed, of course, even as a rough approximation of the characteristics that regulate an organism's mortality pattern during a real ultrahigh-pressure treatment.

Still, if the model is correct, even if only qualitatively, it would imply that, in the general case, an organism's sensitivity or resistance to a combined pressure and heat treatment cannot be fully characterized by less than nine independent survival parameters. These are n, P_{Co}, P_{c3}, T_{Co}, and T_{c1} and the four a's in Equation 6.17 through Equation 6.20. There are conceivable situations in which the number of parameters would be smaller — if some of the preceding constants are related, for example. However, it would not be at all surprising if the survival patterns of certain

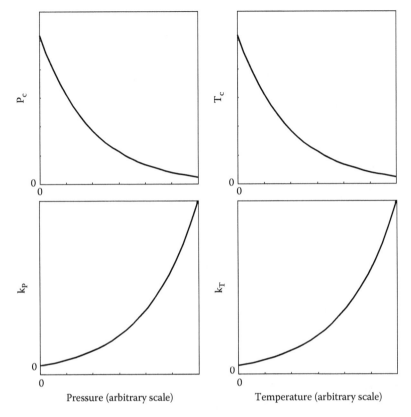

FIGURE 6.10
Schematic view of the temperature and pressure dependence of the log logistic parameters of $b[P,T]$ depicted in Figure 6.9. (Courtesy of Dr. Maria G. Corradini.)

microorganisms would be more rather than less elaborate than the relationships expressed in Equation 6.10 through Equation 6.20 entail. Consequently, the utility of these equations in the derivation of a general quantitative survival model for predicting the outcome of all high-pressure treatments is in serious doubt.

However, if proven reasonable, at least as far as the mathematical form is concerned, the mentioned relationships can serve as *qualitative models* and be used to assess how the targeted organism *might* respond to treatments under different process conditions. The model's qualitative predictions could then be compared with the results of actual experiments to confirm or refute the validity of its underlying assumptions. In other words, although any attempt to determine the model's parameters as outlined in the preceding expressions might be impractical, the conceptual model would be testable through comparison of its qualitative predictions with actually observed trends.

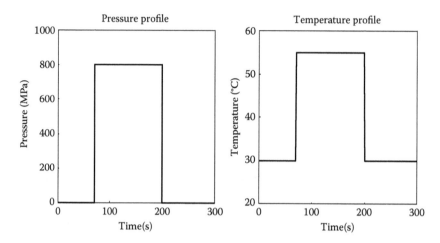

FIGURE 6.11
Simulated ideal isobaric pressure profile in a perfectly insulated pressure vessel and corresponding temperature rise, the result of adiabatic heating. (Courtesy of Dr. Maria G. Corradini.)

Ultrahigh-Pressure Treatment in a Perfectly Insulated Vessel

Consider a high-pressure process in which Equation 6.10 through Equation 6.20 can describe the resistance of the targeted organism to be combined pressure and heat treatments. Again, for the sake of simplicity, let us also assume that the adiabatic heating is expressed by about 3°C rise for every 100 MPa pressure increase, independently of the initial temperature. (In reality, the temperature rise is faster at higher temperatures — an observation that can and should be incorporated in the analysis of actual treatment foods.)

Suppose the heating is perfectly adiabatic, i.e., the vessel in which the pressure is applied is perfectly insulated, and that the treatment has three stages: fast linear pressure increase, holding for a given time, and an instantaneous pressure release (and a corresponding temperature drop). As shown in Figure 6.11, the pressure and temperature profiles will then be (P in megapascals and T in degrees Celsius):

$$P(t) = \text{If } [t \le t_0,\ 0,\ \text{If } [t \le t_1,\ k\ (t - t_0),\ \text{If } [t \le t_2,\ k\ (t_1 - t_0),\ 0]]] \quad (6.21)$$

Equation 6.21 says (in the syntax of Mathematica®) that before the start of the process at time t before t_0, the pressure (gage) is zero. During the pressure application between, t_0 and t_1, the pressure rises linearly and reaches a level $k(t_1 - t_0)$. This pressure is maintained until it is instantaneously released at the time t_2.

Notice that the target pressure, P_{target}, is $k(t_1 - t_0)$ and it is assumed that this pressure is maintained unchanged throughout the holding period. With the assumption of a 3°C rise for every 100 MPa, the corresponding temperature profile will be (Figure 6.11):

$$T(t) = \text{If } [t \leq t_0, T_0, \text{ If } [t \leq t_1, T_0 + \frac{3P_{target}}{100} \cdot \frac{(t - t_0)}{(t_1 - t_0)},$$

$$\text{If } [t \leq t_2, T_0 + \frac{3P_{target}}{100}, T_0] \qquad (6.22)$$

The equation says that before the pressure application at t_0, the food's temperature is T_0. It linearly increases during the vessel pressurization at a rate of 3°C per 100 MPa and remains unchanged at this level (perfectly insulated vessel) until the pressure is released instantaneously. At that time, due to adiabatically cooling the food, the temperature returns to its initial level.

Obviously, the preceding describes an overly simplified scenario. Yet, it can still serve as a demonstration of how simulation of the outcome of ultrahigh-pressure processes can be performed. Because $P(t)$ and $T(t)$ are now expressed as explicit functions of time, they can be used to determine the corresponding varying magnitudes of the survival parameters $k_p[T(t)]$ and $P_c[T(t)]$ (see Equation 6.17 and Equation 6.18). Alternatively, the varying survival parameters $K_T[P(t)]$ and $T_c[P(t)]$ (see Equation 6.19 and Equation 6.20) could also be expressed as a function of time; this should give exactly the same results.

If the power, n, and the coefficients that determine the temperature or pressure dependence of the survival parameters are known or can be estimated with reasonable accuracy from experimental data, one can construct the *time-dependent* coefficient of the process's inactivation rate model (Equation 6.13): namely, $b(t)$. Once $b(t)$ is incorporated into the rate model, the numerical solution of the differential equation will be the theoretical survival curve that the particular combined pressure–temperature profile will produce. In the case in which $n(t)$ must be assumed to be pressure and temperature dependent rather than a constant, one will need to formulate the term $n(t)$ in a manner similar to that used to determine $b(t)$, insert the resulting expression into Equation 6.13, and obtain the theoretical survival curve by solving the new, more elaborate differential equation. Although a time-dependent $n(t)$ will complicate the structure of the model's rate equation considerably, it is doubtful that its numerical solution will take much longer.

Examples of simulations using the preceding procedure with $n = \text{const.}$ are shown in Figure 6.12. Such simulations can be used to examine not

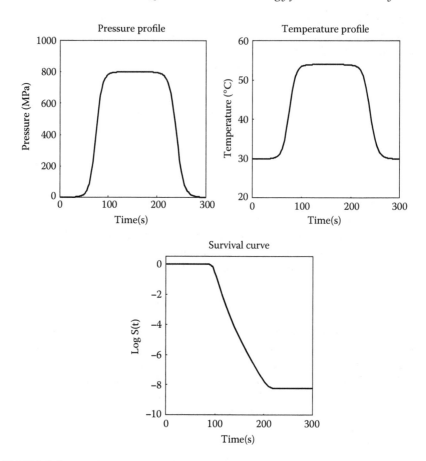

FIGURE 6.12

Simulated ideal pressure and temperature profiles in a perfectly insulated pressure vessel and corresponding theoretical survival curves of hypothetical organisms produced with the Weibullian–log logistic model (Equation 6.14). The temperature and pressure dependence of the organism's survival parameters, $k_P(T)$, $T_c(T)$, $k_T(P)$, and $P_c(T)$ followed Equation 6.17 through Equation 6.20 (see Figure 6.10). (Courtesy of Dr. Maria G. Corradini.)

only the effect of the target pressure and process duration on the process lethality as shown in the figure, but also how preheating, for example, can improve a treatment's efficacy, at least theoretically. Such simulations can also be used to evaluate how the role of an organism's different sensitivities to pressure and temperature might affect lethality of a contemplated combined treatment. Possibly, the difference in the temperature and pressure sensitivities could be exploited and the process's pressure–temperature profile could be adjusted to produce a microbially safe food with minimal damage to its other qualities. A similar type of simulation can be used to assess whether a combined process currently considered safe would still be effective if an emerging pathogen or a resistant mutant had been present in the food.

Treatment in an Uninsulated Vessel

The idealized scenarios discussed in the preceding section can be used to identify trends and qualitatively predict how different processing parameters would affect a treatment's theoretical efficacy. A common industrial problem, however, is how to assess the role of a pressure process's imperfections on its lethality. Perhaps the most common observable deviation from the ideal in ultrahigh-pressure treatments is the temperature drop of the vessel's contents as a result of imperfect insulation. The theoretical effect that such a drop might have on the targeted organism's survival can be studied within the same model's framework, except that the temperature profile would be somewhat different. For example, if the temperature's fall can be described by a single exponential decay term as a first-order approximation, the corresponding temperature profile would be expressed by a term like:

$$T(t) = If[t \le t_0, T_0, If[t \le t_1, T_0 + \frac{3P_{target}}{100} \cdot \frac{t-t_0}{t_1-t_0},$$

$$If[t \le t_2, \left(T_0 + \frac{3P_{target}}{100}\right)\exp[-k(t-t_1)], T_0]]]$$

(6.23)

This says (again in a syntax similar to that of Mathematica®) that the process's temperature rise is the same as previously described; however, after the target pressure has been reached (between t_1 and t_2), the temperature falls exponentially (with a rate constant k) until the pressure is released. At this point, the temperature drops back to its initial value. In other words, the only difference here, when compared with the former insulated scenario, is that after the target pressure has been reached, the temperature does not remain constant. (Again, the temperature drop can be expressed by other models derived from heat transfer considerations, for example.) Examples of such hypothetical scenarios and the corresponding survival curves are shown in Figure 6.13.

Similar survival curves can be generated for pressure drops accompanied by cooling. Also, one can simulate accidents, such as an uncontrolled pressure drop in midprocess, and assess their potential theoretical effect on the process safety. Examples of interrupted pressure profiles and their corresponding survival curves are given in Figure 6.14. They were produced by expanding the pressure and temperature profile equations to include the shown pressure drops. Insertion of the mathematical expressions

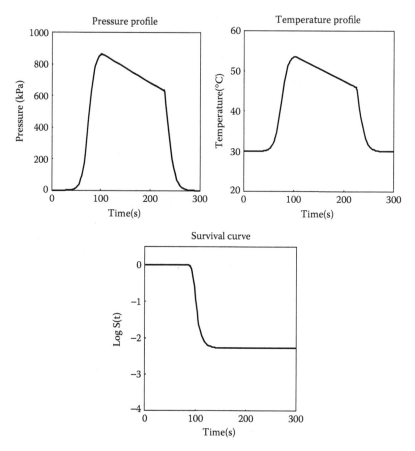

FIGURE 6.13

Simulated pressure and temperature profiles of a process performed in an uninsulated vessel and corresponding survival curves produced with the Weibullian–log logistic model (Equation 6.14). Notice that, at least in principle, simulations of the kind can be used to assess the effect of a process's imperfections on its efficacy. (Courtesy of Dr. Maria G. Corradini.)

that describe the drops into the model's rate equation (Equation 6.14) is all that is needed to calculate the theoretically expected safety consequences of such interruptions and to guide the operator in the choice of a remedy, if needed.

How to Use the Model

All the scenarios depicted in this chapter are hypothetical and no attempt has been made to account for the response of any particular organism to

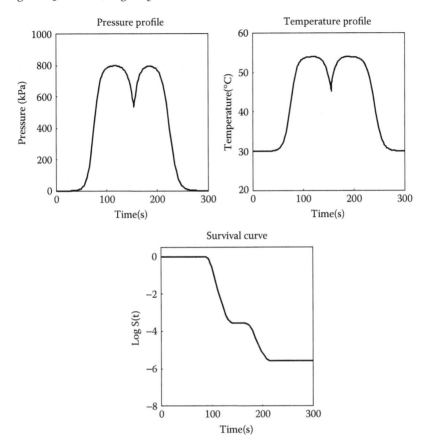

FIGURE 6.14
A simulated interrupted process profile and its effect on the targeted organism's survival. Notice that simulations of this kind can be used to assess the potential safety implications of accidents and suggest remedial measures. (Courtesy of Dr. Maria G. Corradini.)

the combined treatments. Also, none of the primary and secondary models incorporated into the rate equation used to generate the survival curves has been confirmed experimentally. Still, the assumptions from which they have been derived seem to conform to reported observations on the response of pathogens to an applied hydrostatic pressure, at least qualitatively.

Although no claim has been made that models of the kind presented can be used to predict the outcome of actual processes accurately, they can still be useful to those seeking ways to improve or optimize them. Currently, studies aimed at identifying optimal process conditions are based on the response surface methodology experimental design — that is, on analyses in which pressure, temperature, or other factors like pH are treated as independent variables whose interrelationship is determined by regression. This approach might be useful if it is known *a priori* that an optimal combination of conditions within a narrow range can be effectively covered by the planned experimental conditions. However, because the existence of optimal conditions within any particular range of conditions is not guaranteed and, theoretically, the targeted survival ratio can be reached in more than one way, this experimental approach might not always be the most effective.

In contrast, if the temperature is incorporated as a *variable in the rate equation's coefficients,* which is subsequently used to generate a multitude of survival curves under a variety of pressure–temperature–time conditions, identifying conditions that will produce the desired lethality becomes much easier. These conditions could then serve as the starting point for the experimental search and process safety validation. An experimental design based on such a procedure would probably cover a different range of pressure–temperature conditions than one chosen without using computer simulations as a guideline.

As has been demonstrated in this chapter, the number of parameters needed for the construction of a comprehensive pressure–temperature–survival model is so large that their experimental determination might never be a practical option. However, using trial and error methods, one could still derive realistic estimates of at least some of the parameters or establish the limits in which they can vary on the basis of existing data. Running the model repeatedly with different parameter values can easily provide an assessment of the process outcome's sensitivity to potential variations or errors in the model's coefficient estimates. For more on this issue, see Chapter 13. It is to be hoped that future research based on the proposed or similar approach will result in mathematical methods to determine the pertinent survival parameters from experiments performed under realistic conditions, i.e., when the pressure and temperature vary simultaneously.

The development of a mathematical method to determine microbial survival parameters from nonisothermal, nonisobaric data will certainly not be an easy task and it might not even be achievable except for simulated data bases. As in the case of retrieving the heat survival parameters from nonisothermal inactivation data, however, there is no reason to assume that task is theoretically impossible. The same can be said about other factors, such as pH, oxygen solubility, and the like, which also might affect the survival pattern of pathogens and spoilage organisms subjected to lethal combinations of ultrahigh hydrostatic pressure and elevated temperature. One must remember, however, that all the above only refers to situations where the temperature and pressure are related in a predetermined and known way. If not, then the derivation of the survival model will be much more complicated, especially if other factors that may affect the process's lethality vary too.

7

Dose–Response Curves

It is better to have an approximate answer to the right question than an exact answer to the wrong one.

J. Tukey

A dose–response curve is a graphic representation of the number or fraction of survivors in a living population as a function of the dose of a harmful or lethal agent to which the population has been exposed. It can be used as a more elaborate means to represent the lethality of a physical agent or toxicity of a poison than the familiar LD_{50}, the dose that would cause 50% mortality in the affected population. The affected population can be humans, in which case the harmful agent would be not only a directly ingested poison, radiation or a carcinogen, etc. but also a toxin produced by a food-borne or an infectious microbe.

The dose is expressed in the pertinent units, whether mass, e.g., milligrams of a toxin or milligrams per kilogram of body weight, rads in the case of radiation, or the ingested microbial inoculum size that will produce clinical symptoms or death. This type of dose–response curve, especially that associated with the hazard of pathogens to humans, has traditionally been in the domain of medical and food epidemiology. However, although the presence and number of harmful organisms and viruses are of much interest and concern in food microbiology and water microbiology, their effect on human health is clearly an issue outside the scope of our discussion here.

Another kind of dose–response curve consists of plots that describe the relationship between the number or fraction of surviving *microorganisms* and the dose of lethal agent to which they have been exposed. The study of such dose–response curves primarily deals with situations in which the agent is introduced deliberately to reduce or eliminate a targeted microbial population. In this case, the dose is again expressed in units like parts per million, rads, electric field intensity, number of electric pulses, etc. in relation to the microorganisms rather than to humans.

Whenever a lethal agent is applied in microbial inactivation, the affected population is usually reduced by several orders of magnitude. Therefore, the traditional LD_{50} is not a useful measure of the lethality or potency of antimicrobial agents. However, even in cases in which it might be used, it would provide no information on the span of the lethal effect.

In experiments to determine dose–response curves, time is usually not treated as a variable and the affected microbial population's size is determined just before and soon after its exposure to the lethal agent. The mathematical treatment of this kind of dose–response curve is very similar to that of survival curves (Chapter 1 through Chapter 3) except that the exposure time is replaced by the administered dose. It is not inconceivable, though, that although a very high dose of a lethal agent can cause almost instantaneous mortality in the exposed microbial population, a low dose might have a delayed effect.

From a modeling viewpoint, the difference is that, in the first case, time is not a primary factor, but in the second it is. What follows is only pertinent to scenarios in which time is not an issue — that is, the experimenter has already determined the conditions under which assessment of the agent's intensity effects is reproducible and meaningful.

The Fermi (Logistic) Distribution

As in the case of the treatment of survival curves, here, too, we will assume that the treatment allows neither recovery from injury nor enough time for biochemical or genetic adaptations. If correct, then we can also assume that the dose–response curve of an individual microorganism, i, is a step function:

$$\text{If } X \leq X_{ci} \quad S_i(X) = 1 \text{ or } \log S_i(X) = 0$$

$$\text{and} \tag{7.1}$$

$$\text{If } X > X_{ci} \quad S_i(X) = 0 \text{ or } \log S_i(X) \rightarrow -\infty$$

where X is the lethal agent intensity in the pertinent units and X_{ci} is the level at which an individual organism is destroyed.

According to this definition, X_{ci} is a measure of the organism's resistance or sensitivity to the lethal agent. The higher X_{ci} is, the more tolerant the organism is; the lower X_{ci} is, the more susceptible the organism is. The dose–response curve of a discrete microbial population composed of individuals having different resistances, X_{ci}, is therefore:

$$S(X) = \sum S_i(X) \, \Delta\phi_i \tag{7.2}$$

where ϕ_i is the fraction of the microorganisms, or spores, succumbing to the same lethal dose, X_{ci} ($\Sigma\Delta\phi_i$) = 1).

When the microbial population is sufficiently large and the resistance spectrum sufficiently dense, Equation 7.2 can be approximated by a continuous expression based on a continuous parametric distribution of the lethal doses. One way to derive this expression is to approximate the step function described by Equation 7.1 by the previously introduced Fermi function

$$S_i(X) = \frac{1}{1+\exp\left(\dfrac{X-X_{ci}}{a}\right)} \tag{7.3}$$

where a is an arbitrary constant having a very small numerical value.

As can be seen in Figure 7.1, this expression is practically indistinguishable from a true step function when a value of a is sufficiently small — that is, when $a << X - X_c$. (According to Fermi's equation, as has already been shown in the discussion of survival curves, when $X < X_c$ and $a \to$ 0, $(X - X_c)/a \to -\infty$, $\exp[(X - X_{ci})/a] \approx 0$, and $S_i(X) \approx 1$. However, when $X > X_{ci}$ and $a \to 0$, $(X - X_{ci})/a \to \infty$, $\exp[(X - X_{ci})/a] \to \infty$ and $S_i(X) \approx 0$.)

Thus, the dose–response curve of any sufficiently large microbial population can be represented by:

$$S(X) = \int_0^x \frac{d\phi}{1+\exp\left[\dfrac{X-X_c(\phi)}{a}\right]} \tag{7.4}$$

where $X_c(\phi)$ is the lethal dose that corresponds to any fraction ϕ of the population and, as before, a is an arbitrary very small number.

As has been shown, as long as a is sufficiently small, its actual magnitude is unimportant. Although Equation 7.4 can be used as such to describe microbial dose–response curves (Peleg et al., 1997), it is very inconvenient as a mathematical tool. This is primarily because it is difficult to define the term $X_c(\phi)$. An easier way is to treat the dose–response curve, $S(X)$, directly as the cumulative form of a known parametric distribution function that describes the spectrum of the organism's resistances or sensitivities, defined in terms of the agents' dose levels that cause their mortality (Figure 7.1).

One such distribution might be the logistic or Fermi distribution:

$$S(X) = \frac{1}{1+\exp[k(X-X_c)]} \tag{7.5}$$

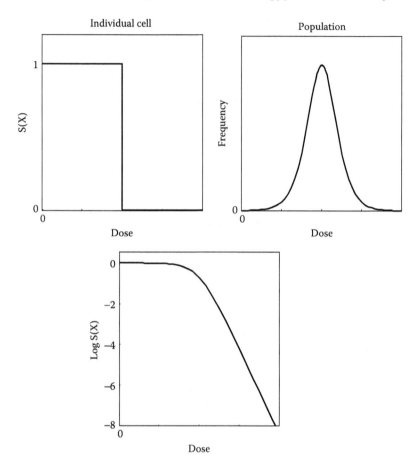

FIGURE 7.1
Schematic view of the dose–response curve of an individual microorganism described by
the Fermi function (left), the distribution of the lethal doses to which individuals within a
population succumb (top right), and the corresponding dose–response curve (bottom).

where k is a constant representing the distribution's spread and X_c its
mode and mean (see Chapter 2). (The model can also be written in the
form of Equation 7.3, in which case $k = 1/a$. To avoid confusion, we will
use the k for the distribution's equation exclusively and leave the former
format to the description of the step function, where a will always have
a small numerical value.)

In Equation 7.5, when k is large relative to $X - X_c$, the distribution is
sharp; when it is small, the distribution is wide. Either way, the Fermi
distribution, like its mirror image the logistic function, is symmetric. If
the survival range covers several orders of magnitude and thus the survival ratio is expressed by its logarithm, then:

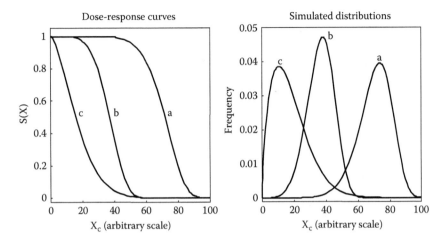

FIGURE 7.2
Simulated dose–response curves of a sensitive and a resistant (a), intermediate (b), and sensitive (c) organism using the Weibull distribution function as a model. (From Peleg, M. et al., 1997, *Bull. Math. Biol.*, 59, 747–761. With permission, courtesy of Elsevier Ltd.)

$$\log S(X) = -\log\{1 + \exp[k(X - X_c)]\} \tag{7.6}$$

As before, this log logistic term implies that, when $X << X_c$, $\log S(X) \approx$ 0 and, when $X >> X_c$, $\log S(X) \approx -k\,(X - X_c)$. Thus, if the lethal dose distribution indeed follows the Fermi (logistic) model, a flat "shoulder" will appear in the dose–response curve after which the continuation will look like an almost perfectly straight line. Also, if the experimental data points are widely spaced and scattered, the smooth transition between the shoulder and semilogarithmic decline might not be detected, regardless of the X_c distribution.

An example of published dose–response curves fitted with Equation 7.5 as a model is shown in Figure 7.3. They are of three types of bacteria cells exposed to ozone and show a very sharp response to the agent that is hardly any measurable effect at concentrations below 0.2 mg.l^{-1} and hardly any measurable survival after exposure to a higher concentration. In other words, the distribution of these organisms' resistances (or sensitivities) to ozone was very narrow. The purpose of showing the figure is to demonstrate the attractiveness of the Fermi (logistic) model to dose–response curves resembling a step function.

The Fermi distribution function is continuous and it has all the derivatives (differentiable everywhere). Thus, it can be used as a model using standard nonlinear regression procedures even for fairly steep dose–response curves of the kind shown in the figure. However, because

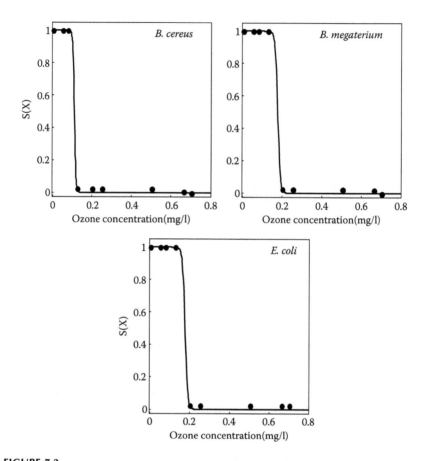

FIGURE 7.3
Sharp dose–response curves of three organisms exposed to ozone, fitted with the Fermi distribution function as a model. The original experimental data are from Clark, D.S. and Takacs, J., 1980, in *Microbial Ecology of Foods*, vol. 1. *Factors Affecting Life and Death of Micro-organisms*. Academic Press, New York, p. 191. (From Peleg, M. et al., 1997, *Bull. Math. Biol.*, 59, 747–761. With permission, courtesy of Elsevier Ltd.)

the magnitude of k is of little importance in such cases (especially when experimental data at and close to X_c are unavailable), it would be prudent to fix the value of k in advance and let X_c be the only adjustable parameter to be determined by the regression. It can be shown that if the preselected value of k is large enough, its actual magnitude will have no noticeable effect on the X_c estimate. This is demonstrated in the example given in Table 7.1, based on Peleg et al. (1997) and the experimental data of Clark and Takacs (1980), where X_c is the concentration of the ozone around which the cells have been destroyed.

TABLE 7.1

Regression Parameters of Three Steep Dose–Response
Cuves Fitted with the Fermi Model (Equation 7.5)
with Fixed and Adjustable k

	Fit with $k = 100$		Fit with adjustable k	
Organism	X_c(mg l⁻¹)	Chi square	X_c(mg l⁻¹)	Chi square
B. cereus	0.10	0.0010	0.11	0.0044
B. megaterium	0.17	0.0020	0.18	0.0066
E. coli	0.17	0.0028	0.18	0.0069

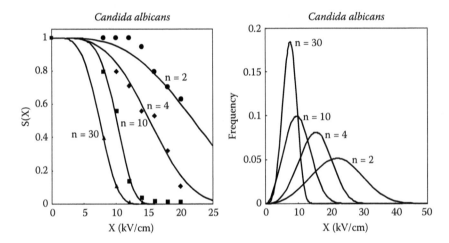

FIGURE 7.4

Response of *Candida albicans* to pulsed electric fields (left) and corresponding frequency
distribution, which can be described by Fermi (logistic) and Weibull distribution functions.
See Peleg (1996b) and Peleg, M. et al., 1997, *Bull. Math. Biol.*, 59, 747–761, respectively. The
original experimental data are from Castro et al., 1993. (From Peleg, M. et al., 1997, *Bull.
Math. Biol.*, 59, 747–761. With permission, courtesy of Elsevier Ltd.)

Examples of dose–response curves that can be described in terms of an
underlying Fermi distribution of resistances with a relatively large span
(Peleg, 1996b) are shown in Figure 7.4 and Figure 7.5. The same curves
can also be described by the Weibull distribution with a small coefficient
of skewness (see following). The advantage of the Fermi model in this
and similar cases is that it enables one to account for shifts in the response
patterns in terms of the distribution's two parameters, k and X_c, whose
significance is intuitively clear and thus easy to compare. Alternatively,
X_c can be treated as the distribution's mean (which in the case of symmetric unimodal distributions is identical to the mode and median) and

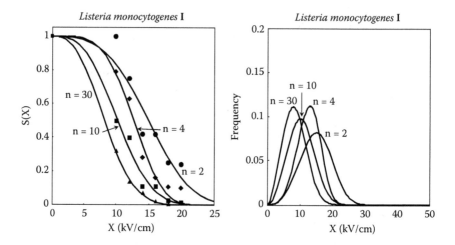

FIGURE 7.5

Response of *Listeria monocytogenes* to pulsed electric fields (left) and corresponding frequency curve (right). Notice the symmetric appearance of the frequency distribution, which can be described by logistic and Weibull distribution functions. See Peleg, M., 1996b, and Peleg, M. et al., 1997, respectively. The original experimental data are from Castro et al., 1993. (From Peleg, M. et al., 1997, *Bull. Math. Biol.*, 59, 747–761. With permission, courtesy of Elsevier Ltd.)

k can be translated directly into the distribution's standard deviation through the formula (Patel et al., 1976):

$$\sigma = \frac{\pi}{\sqrt{3k}} \qquad (7.7)$$

Thus, if a certain factor, such as temperature, pH, or the organism's growth conditions, has a measurable influence on the dose–response curve, its effect can be expressed quantitatively as an increase or decrease in the mean lethal dose and in narrowing or expanding the spread of the lethal doses around it (see below).

The Weibull Distribution

There is no compelling reason why the distribution of sensitivities underlying a microbial dose–response curve will always be symmetric or even approximately symmetric around the mean. Thus, the universality of a model based on the Fermi (logistic) distribution function cannot be assumed. The next best candidate model is the Weibull distribution function, whose mathematical properties have been already discussed in detail in Chapter 2 of this book.

With the appropriate parameters, the Weibull distribution function can account for clearly skewed distributions and almost perfectly symmetric distributions. A dose–response curve reflecting a spectrum of resistances or sensitivities having a Weibull distribution would have the form:

$$S(X) = \exp(-bX^n) \tag{7.8}$$

where, as before, $\beta = 1/b$ is the distribution's location factor and $\alpha = n$ is its shape factor. Because many dose–response curves are presented in the form of $S(X)$ vs. X rather than as $\log_{10} S(t)$ vs. x (as is frequently the case in survival curves — see Chapter 2), the parameters $1/b$ and n in Equation 7.8 correctly represent the distribution's location and shape factors, respectively.

The frequency, or density, form of the Weibull distribution here, $f(X) = dS(X)/dX$, is given by:

$$f(X) = bnX^{n-1} \exp(-bX^n) \tag{7.9}$$

The distribution's mode, X_m, is:

$$X_m = \left(\frac{n-1}{nb} \right)^{\frac{1}{n}} \tag{7.10}$$

Its mean is:

$$\bar{X} = \frac{\Gamma\left(\dfrac{n+1}{n} \right)}{b^{\frac{1}{n}}} \tag{7.11}$$

and its variance is:

$$\sigma_x^2 = \frac{\Gamma\left(\dfrac{n+2}{n} \right) - \left[\Gamma\left(\dfrac{n+1}{n} \right) \right]^2}{b^{\frac{2}{n}}} \tag{7.12}$$

where Γ is the gamma function.

The skewness coefficient, v_1, is given by:

$$v_1 = \frac{\mu_3}{\mu_2^{3/2}}$$ (7.13)

where $\mu_3 = \Gamma(1 + 2/n)/b^{3/2}$ and $\mu_2 = \Gamma(1 + 2/n)/b^{2/n}$.

When actual dose–response curves are fitted with Equation 7.8 as a model, n is expected to assume values on the order of 1.5 to 5 — i.e., $n > 1$, which indicates the existence of a peak frequency (mode) somewhere in the middle of the applied dose range. Consequently, the corresponding b value might have a very small numerical value. For example, if the distribution is not too skewed and has a mean dose on the order of 25 arbitrary units and a shape factor $n = 3$, then b will be on the order of 10^{-5} in the corresponding units.

Such values, especially when combined with a noninteger n, render the model very unattractive if presented in its original form. However, conversion of its parameters b and n into a mean, mode, variance or standard deviation, and a skewness coefficient (using Equation 7.10 through Equation 7.13 as conversion formulas) would greatly facilitate comparison between different organisms. It can also be used to quantify the influence of growth conditions or other factors on the same organism. This is demonstrated in Figure 7.6 and Figure 7.7, in which the excellent fit of Equation 7.8 as a model is also shown.

The first of the two figures depicts the dose–response curves of *E. coli* grown under aerobic and unaerobic conditions and irradiated with x-rays under oxygen and nitrogen. The second figure shows the corresponding frequency (density) distributions of the underlying resistances. These distributions were constructed using Equation 7.9, with the parameters b and n calculated directly from the original dose–response data, i.e., by regression using Equation 7.8 as a model of the data shown in Figure 7.6.

The frequency (density) form of the distribution highlights the effects of the growing conditions and irradiation environment on the organism's survival pattern. Although it is already evident in the original dose–response curve, it becomes much clearer after the conversion because the mode and spread become much easier to discern visually. Quantification of the effects in terms of the Weibull distribution parameters also facilitates their comparison, as shown in the Table 7.2.

Another example is given in the Table 7.3. It shows the effect of the pulsed electric field intensity and number of pulses on the survival of the two organisms whose dose–response curves are shown in Figure 7.4 and Figure 7.5. This is a demonstration that the Weibull model can also be used to fit symmetric distributions and that a model's fit does not imply uniqueness.

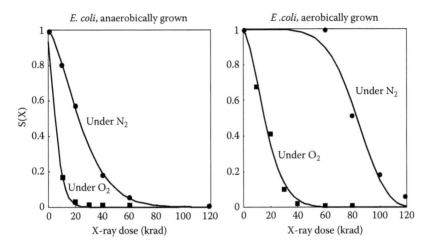

FIGURE 7.6
Dose–response curves of *E. coli* grown and exposed to x-ray irradiation under experimental data are from Ingram, M. and Roberts, T.A., 1980, in *Microbial Ecology of Foods*. vol. 1. *Factors Affecting Life and Death of Microorganisms*. Academic Press, New York, p. 191. (From Peleg, M. et al., 1997, *Bull. Math. Biol.*, 59, 747–761. With permission, courtesy of Elsevier Ltd.)

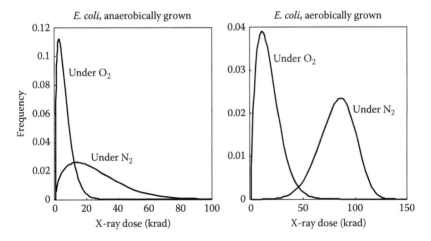

FIGURE 7.7
The frequency distributions that correspond to the dose–response curves of the irradiated *E. coli* shown in Figure 7.6. They were produced by the Weibull model (Equation 7.9) with the parameters b and n calculated from the original dose–response data. (From Peleg, M. et al., 1997, *Bull. Math. Biol.*, 59, 747–761. With permission, courtesy of Elsevier Ltd.)

TABLE 7.2

Dose–Response Distribution Parameters of *E. coli* Grown and Irradiated
by X-Rays under Different Conditions

Growing conditions	Irradiation conditions	b	n	Mode X_{cm} (krad)	Mean $\overline{X}_c$ (krad)	Std. dev. σ_X (krad)	Coeff. skewness v_1 (–)
Anaerobic	N_2	6.6×10^{-3}	1.5	14	2.6	310	1.54
	O_2	8.4×10^{-2}	1.3	2.2	6.0	21	1.68
Aerobic	N_2	1.2×10^{-11}	5.6	86	82	290	1.06
	O_2	8.5×10^{-3}	1.6	11	18	125	1.48

Source: Based on Peleg, M. et al., 1997, *Bull. Math. Biol.*, 59, 747–761. The original experimental data are from Ingram, M. and Roberts, T.A., 1980, in *Microbial Ecology of Foods*, vol. 1. *Factors Affecting Life and Death of Microorganisms*. Academic Press, New York, p. 191.

Frequency (density) distributions and tabulated parameters of the kind shown in Table 7.2 and Table 7.3 can be constructed with alternative distribution functions. Because neither the Fermi nor the Weibull distribution function has been derived from fundamental principles, the possibility that other models will provide a comparable or even better fit cannot be ruled out. However, one should expect that, if the fit of an alternative model to the experimental dose–response data is as good as that of the two mentioned, it will provide very similar estimates of the distribution's mean mode, standard deviation, and coefficient of skewness.

Mixed Populations

When the exposed microbial population consists of two or more subpopulations and each has a different spectrum of sensitivities to the lethal agent, the underlying distribution of the dose–response curve can be uni-, bi-, or multimodal. If the modes of these distributions are sufficiently far apart and their spread sufficiently narrow, the dose–response curve (like the survival curve — see Chapter 2) may have two or more clearly discernible inflection points. When the distributions' modes are close and their spread large and overlapping, the existence of such inflection points will probably be missed, especially if the experimental data are widely spaced and have a considerable scatter.

Either way, the safe way to establish that a given microbial population is a mixture is to isolate the sensitive and resistant subpopulations and determine their dose–response separately. Technically, it will be easier to isolate the more resistant survivors than the sensitive members that perish

TABLE 7.3

Response of *Candida albicans* and *Listeria monocytogenes* to Pulsed Electric Fields

Organism	No. pulses	Fitted parameters		Corresponding distribution parameters			
		b	n	Mode X_m (kV/cm)	Mean $\bar{X}$ (kV/cm)	Std. dev. σ_X (kV/cm)	Coeff. skewness ν (–)
C. albicans	2	2.7×10^{-5}	3.3	22	22	55	1.00
	4	3.8×10^{-5}	3.6	16	15	22	1.12
	10	1.1×10^{-3}	2.8	9.6	10	15	1.18
	30	2.5×10^{-4}	4.0	7.6	7.7	4.4	1.10
L. monocytogenes	2	5.1×10^{-5}	3.5	15	15	22	1.13
	4	2.0×10^{-5}	4.1	13	13	12	1.10
	10	8.2×10^{-4}	2.9	10	10	15	1.18
	30	2.8×10^{-3}	2.6	7.9	8.4	12	1.21

Source: Based on Peleg, M. et al., 1997, *Bull. Math. Biol.*, 59, 747–761, and the original experimental data of Castro et al., 1993.

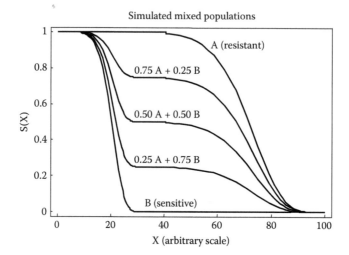

FIGURE 7.8
Schematic view of the dose–response curves of mixed populations with a Weibullian distribution of the lethal doses. Notice that effective destruction of the more resistant subpopulation (A) may require doses outside the experimental range. (From Peleg, M. et al., 1997, *Bull. Math. Biol.*, 59, 747–761. With permission, courtesy of Elsevier Ltd.)

first, but this should not concern us here. In principle at least, one can identify the composition of a mixture by collecting the survivors after treatments of different degrees of lethality and determining their dose–response curves. Although such tests are unlikely to be performed in studies associated with food preservation, they might be quite helpful in other fields.

The emergence of pathogenic bacterial strains resistant to antibiotics might be a case in point. When an experimental dose–response curve ends in a plateau, the presence of a significant number of survivors might be interpreted as a sign that the organism has adjusted to the agent and thus is or has become ineffective. However, an alternative interpretation is the existence of a more resistant subpopulation that could be eliminated only if the dose is substantially increased (Figure 7.8). If the second scenario is plausible, it will be an indication that the effective level of the agent is outside the range tried in the experiment.

Of course, if the experiment has covered the entire practical range of the agent's dose or intensity, one can safely conclude that it should not be used regardless of whether the targeted microbial population is uniform or a mixture.

It is interesting to note that soon after this chapter was written, I was alerted by Dr. Israel Saguy of the Hebrew University in Jerusalem that researchers following the response of individual cells to antibiotics had reported that bacterial cells much smaller than average had survived the exposure and could resume their growth and presumably their division after treatment had been discontinued (Balaban et al., 2004). This demonstrates that the variability within a given microbial population can be the cause of the observed distribution of the cells' responses to the lethal agent.

8

Isothermal and Nonisothermal Bacterial Growth in a Closed Habitat

Microbiology is the only science where multiplication means division.

Anonymous

The Traditional Models

Most growth models have been developed for a monoculture in a closed habitat — that is, for a single microbial "species" or strain in an isolated environment. A closed habitat means that life-sustaining resources, such as nutrients, are finite and no material or energy is exchanged with the outside. Such idealized systems must be quite rare in the context of foods, open water reservoirs, and pathogen multiplication inside or on a human's body, especially if energy and gas exchange is considered. Yet, if material resource availability is the main factor that controls the microbial population growth rate, traditional models developed for a closed system can be quite adequate. Even if total isolation does not exist, the microbial growth pattern in an underprocessed can of ground meat or fish on a supermarket shelf, refrigerated milk during transportation, or standing water in a puddle or a storage tank, for example, is very different from the pattern found in an industrial fermentor or the human gut, where critical resources are continuously replenished and metabolites exuded by the growing organism are removed.

What follows will only deal with the growth of a uniform microbial population residing in a hypothetical, ideal closed habitat. It will address scenarios pertinent to stored and transported foods and to water in a container — that is, to microbial growth patterns that have safety and quality implications rather than to processes designed to maximize harvest and yield. The two kinds of growth might be related, however, as in the case of lactic acid fermentations in yogurt and pickling, where autosterilization might become a factor to consider only at the late stages of the process.

We emphasize 'uniform' here because, although the same factors that regulate the growth rate of a monoculture can also affect that of a mixed microbial population, the two systems can be essentially different. A change in a mixed population's total growth rate can be accompanied by a corresponding change in the population's composition, an aspect that must be taken into account in the growth model formulation. The same can be said about a fresh inoculum introduced into a habitat that already has an indigenous microbial population. Here, too, competition for resources can become a decisive issue and a determinant of the growth pattern and fate of the introduced or extant populations.

The Logistic Equation and the Logistic Function

Many growth curves of an organism introduced into a sterile closed habitat have a typical sigmoid curve of the kind shown in Figure 8.1. Traditionally, it has been treated as representing three growth regimes. After a lag phase or an adjustment period, when the introduced population's size remains practically unchanged (or even declines somewhat), a stage of exponential growth with intensive cell division and cell growth takes place. With finite resources, exponential growth cannot be sustained indefinitely, so the growth rate, after a certain point, starts to decline. It will approach zero asymptotically and the population size, according to most of the currently held growth models, will remain unchanged again.

Nevertheless, it is very unlikely that a habitat depleted of its life-sustaining resources will be able to support a large microbial population for long. Eventually, therefore, cell mortality must follow and the population's absolute size will start to decrease. This has been observed and the growth curve is described in several publications as having *four* regions: lag, exponential growth, stationary, and death or mortality (e.g., Taub et al., 2003; McKellar and Lu, 2004). In foods, the mortality stage may occur when the product is well beyond its designated shelf life and thus has been of a lesser concern to the majority modelers. Also, a food product might become inedible long before its microbial population reaches even a fraction of the size that would exhaust the available nutrients; therefore, only what happens at the early growth stages has practical implications.

However, as shown in Figure 8.2, the onset of mortality can even precede the stationary stage (Peleg, 1997), in which case a true peak growth will be observed. This is a familiar scenario, especially to those working in fermentation. It is a result of the habitat's pollution, competition for space, depletion of an essential nutrient, etc. and any of their combinations. From a purely formalistic viewpoint (as the two figures demonstrate), the

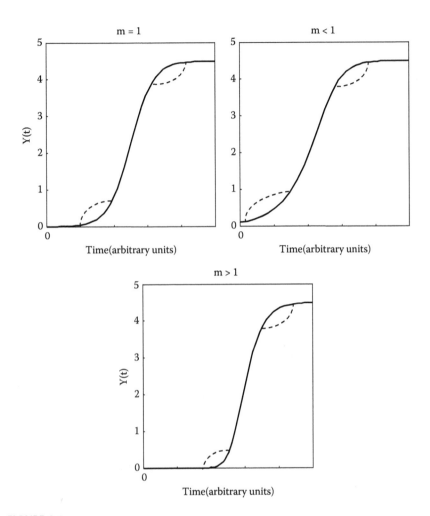

FIGURE 8.1
Schematic view of symmetric and asymmetric sigmoid (logistic) growth curves generated with Equation 8.6 as a model with $m = 1$ and $m \neq 1$, respectively.

growth curve even beyond the inflection point has insufficient information to predict its continuation qualitatively or quantitatively. Therefore, no universal mathematical relationship can unite the lag phase, exponential growth rate, and the population's maximum growth level (see below). Nevertheless, certain ubiquitous growth patterns do emerge and these can be described by kinetic models that share common mathematical features.

When the momentary isothermal growth rate of an isolated population is only proportional to its momentary size *and* to the portion of the yet

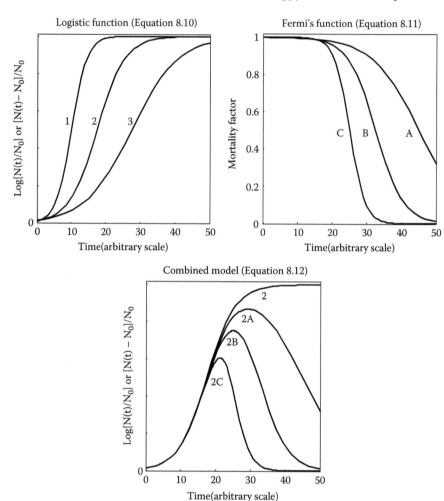

FIGURE 8.2
Schematic view of sigmoid and nonsigmoid growth curves generated with the logistic–Fermi model. Notice that, despite their distinct appearance, all the curves belong to a single family and the differences between them are due to the relative magnitudes of their growth and mortality parameters. Also notice that the classic logistic curve is just a special case of the model where $t_{c\ mortality} \gg t_{c\ growth}$.

unexploited resources in the habitat, the result is a continuous logistic growth pattern. It can be described mathematically by the classic *logistic equation* also known as Verhulst model:

$$\frac{dN(t)}{dt} = rN(t)\left(1 - \frac{N(t)}{N_{asympt}}\right) \tag{8.1}$$

where

$N(t)$ is the momentary number of cells

r is a growth rate constant representing the organism's tendency to grow quickly or slowly in the particular medium at the given temperature

N_{asympt} is the carrying capacity of the habitat, expressed as the number of cells that will exhaust all the available resources

According to this model, the portion of still unutilized resources is represented by the term $1 - N(t)/N_{asympt}$. It is unity when $N(t) = 0$ and approaches zero when $N(t) \rightarrow N_{asympt}$.

A more familiar growth model is the *logistic function* that can describe the growth curve, i.e., the relationship between the actual number of the cells (or biomass) and time. One version of this model is:

$$N(t) = N_0 + \frac{N_{asympt} - N_0}{1 + \exp[k(t_c - t)]} \qquad (8.2)$$

where N_0 is the initial inoculum size and k the (exponential) growth rate at the inflection point.

According to this model, t_c is the time that marks the growth curve's inflection point. This time also corresponds to the time to reach half of the net increase in number of cells that can be theoretically achieved, i.e., $N(t_c) = N_0 + (N_{asymp} - N_0)/2$. According to this model, when k is large and $t \ll t_c$, $N(t) \sim N_0$. Therefore, such a model is particularly convenient in the description of sigmoid growth curves with a long lag. Because the size of microbial populations can increase by several orders of magnitude in a nutrient-rich habitat like a food, it is sometimes more convenient to express it in terms of a normalized *net growth ratio* or a *logarithmic growth ratio* rather than in terms of the absolute number of the individual cells:

$$Y(t) = \frac{N(t) - N_0}{N_0} \qquad (8.3)$$

or

$$Y(t) = \log\left[\frac{N(t)}{N_0}\right] \qquad (8.4)$$

respectively. In either form, at $t = 0$, $N(t) = 0$ and thus $y(0) = 0$.

To simplify the discussion, let us consider a normalized version of the logistic model:

$$Y(t) = \frac{1}{1 + \exp[k(t_c - t)]} \qquad (8.5)$$

where $Y(t)$ is the net or logarithmic growth ratio defined by Equation 8.3 or Equation 8.4, as the case might be.

As a result of the transformation, the domain of the dependent variable is compressed, but the overall sigmoid shape of the growth curve remains basically unchanged. It should be mentioned at this point that N_0, the initial population size, is a measurable quantity and not an adjustable regression parameter. In some cases, N_{asympt}, the asymptotic or stationary population size, can also be determined directly from the experimental growth data; if so, then it, too, should not be treated as an adjustable regression parameter. Also, the logistic equation or similar models (see below) should not be used to determine a population's asymptotic growth level from data in which the monitored microbial population has not yet shown signs of stabilization.

The same applies to the modified Gompertz model (see below), which has been used as an alternative to the logistic equation to describe sigmoid growth curves. This is because the same data can be usually described by several alternative models; each has a very similar degree of fit but a very different asymptotic value. A statistical measure of fit, like the mean square error, cannot serve as a criterion in such a case because the shape of the initial part of the growth curves does not contain enough information to predict the growth pattern at later stages of the population's evolution.

Put differently, at its lag phase and/or during its exponential growth stage, the organism has no way to know what the medium's carrying capacity might be. As will be shown later, microbial cells can only respond to cues from their immediate environment. Therefore, they cannot tell in advance how crowding and habitat pollution might slow down their growth rate or even cause their destruction.

Logistic growth (Equation 8.2 or Equation 8.5) is a symmetric process (see Figure 8.1, left): the regimes of growth acceleration and deceleration mirror one another. However, when a real microbial population is growing in a real habitat, the two regimes are regulated by mechanisms that bear no similarity and, thus, it is unlikely that they will produce such a coherent pattern. Consequently, there is no reason to assume that the symmetry implied by the model will be found ubiquitous, let alone, universal. This has long been recognized and most of the models commonly used in quantitative food microbiology take the potential asymmetry in the growth curve's shape into account.

The logistic function, however, is still a very convenient *conceptual model* that captures, although in a somewhat idealized manner, the important features of a large class of growth patterns. Its convenience stems from its mathematical simplicity and, one might add, elegance, as well as from the clear intuitive meaning of its parameters. The parameter t_c marks the inflection point location, which is also the curve's center and k the steepness of the population rise at the vigorous growth regime. In other words, according to this model, about 90% of the net growth in the population size occurs within the range of $t_c \pm 3/k$. From a purely formal viewpoint, the asymmetry of growth curves can be accounted for by introducing a factor, m, into the model, converting Equation 8.5 into:

$$Y(t) = \frac{1}{1 + \exp[k(t_c - t)^m]} \qquad (8.6)$$

Thus, $m < 1$ will account for milder initial growth acceleration relative to its deceleration at the end of the exponential growth regime, and $m > 1$ for the opposite (Figure 8.1). Although a model based on Equation 8.6 maintains almost all of the original characteristics of the original logistic function (which is a special case of the equation where $m = 1$), the added adjustable parameter, m, makes it a less attractive model for describing experimental microbial growth curves.

Either way, the domain of $Y(t)$ according to Equation 8.5 or Equation 8.6 is from $-\infty$ to $+\infty$ and therefore it has a finite positive value at $t = 0$. This can be corrected by truncation or subtracting the term $1/[1 + \exp(kt_c)^m]$ so that $Y(0)$ will be always zero not only by definition but also according to the growth model (see below).

The Gompertz, Baranyi and Roberts, and Other Growth Models

The most popular growth models for describing microbial growth curves, in foods at least, are the old Gompertz model from the late 19th century and the more recent model of Baranyi and Roberts (1994). Their properties and uses, as well as those of other models, can be found in McKellar and Lu (2004), in several other books' chapters, and in numerous articles and review papers.

When applied to microbial growth curves, Gompertz's model is usually written in the form of (McKellar and Lu, 2004):

$$\log N(t) = A + C \exp\{\exp[-B(t - \mu)]\} \qquad (8.7)$$

where A, B, C, and μ are adjustable constants. The parameter μ of this model has been considered as a measure of the lag time, a concept that deserves special attention and discussion (see below).

Baranyi and Roberts' model is a modified version of the original logistic equation (Equation 8.1) and has the form:

$$\frac{dN(t)}{dt} = \frac{q(t)}{q(t)+1}\mu_{max}\left[1-\left(\frac{N(t)}{N_{max}}\right)^m\right]N(t) \tag{8.8}$$

where

N_{max} is the maximum (actually asymptotic) number of microorganisms that can be reached

μ_{max} is the maximum specific growth rate, defined by the slope of the growth curve at its inflection point

$q(t)$ is a parameter defined by $dq(t)/dt = \mu_{max}q(t)$

m is a constant usually but not necessarily always equal to one

The model was derived from the correct observation that the time dependence of the isothermal growth rate is determined not only by the organism as represented by μ_{max}, which replaces r in the original logistic equation (Equation 8.3) and the habitat's capacity to support the growing population, represented by N_{max}, but also by the physiological state of the introduced population. It has been known that the growth pattern of an organism introduced into a new habitat can strongly depend on its growth stage in the original medium. According to the Baranyi–Roberts model, this physiological state of the organism is accounted for by the term $q(t)/[1 + q(t)]$, which also affects the lag time. According to the model, the expression $q(t)$ is mathematically defined in term of μ_{max}. This entails that the population's growth stage in its original culture medium affect the growth rate in the new habitat in the manner prescribed by the model.

How the function $q(t)$ can be derived independently or directly from the physiological state of the introduced population is not at all obvious and is apparently the topic of new studies. The same can be said about the effect of the physiological state on the parameter m that appears in the more elaborate version of the model, i.e., where N/N_{max} is replaced by $(N/N_{max})^m$ and $m \neq 1$. If the characteristics of $q(t)$ and magnitude of m are determined solely by fitting experimental isothermal growth data with the model, then complementary evidence should be provided that μ_{max} and $q(t)$ are indeed determined by the inoculum's physiological state in the way that the model requires and that the power m can be related to a particular growth mechanism whose existence can be confirmed by an especially designed test.

Until such independent experimental validation is produced, the Baranyi–Roberts model will remain on equal footing with the Gompertz model and all other empirical growth models except that it is formulated as a rate equation, as a true kinetic model should be. Still, its utility, like theirs (see below) should not be judged by its ability to describe isothermal growth — they all can do that — but rather by its ability to predict growth patterns not already used to determine its coefficients.

Sigmoid growth curves with an asymptotic growth level have three discernible parts: the lag phase, exponential growth phase, and stationary phase; thus, they can be described, schematically by three straight lines plotted on semilogarithmic coordinates (Ingraham et al., 1983). On this basis, Buchanan et al. (1997) proposed a discrete three-stage growth model of the general form (using a syntax similar to that of Mathematica®):

$$logN[t] = if [t \leq t_{lag}, logN_0, if [t \leq t_{max}, \mu log(t - t_{lag}), logN_{max}]] \quad (8.9)$$

This model says that during the lag time ($t < t_{lag}$), the initial number, N_0, remains unchanged; in the intermediate region ($t_{lag} < t < t_{max}$), the growth is exponential (i.e., linear on semilogarithmic coordinates); and after t_{max} is passed, the population's size remains constant at a level N_{max}. (Again, N_{asymp} would be a better choice of nomenclature because the model, like the other traditional models mentioned previously, does not really account for a peak growth.)

Modern statistical software can use Equation 8.9 as a regression model. Thus, one can use it to translate experimental growth data directly into three intuitively meaningful parameters, at least in principle: the lag time, the average exponential growth rate, and the asymptotic growth level. Like the original logistic function, however, Buchanan and colleagues' model requires or entails asymmetry between the growth acceleration after the lag time and its deceleration at the end of the exponential growth regime. Such symmetry, however, may or may not be observed in the actual growth curves of microorganisms, as has been mentioned; there is no theoretical reason that it should.

The Lag Time

Much effort has been invested in extracting a lag time, from continuous growth curves. Although the intuitive meaning of a lag time is quite clear — the time it takes a population to reach the exponential growth stage — there is no unique way to determine it. As has been shown by McKellar and Lu (2004), the same growth data of *Listeria monocytogenes* fitted with different models yielded lag times as far apart as 45 and 68 days. Similarly, Arroyo et al. (2005) have found *Listeria's* lag time to vary by a factor of up to two, depending on the growth model chosen and the medium's properties, e.g., 37 to 46 h, 81.5 to 100 h, 4.8 to 11 h, and 8.5 to 17.5 h.

These discrepancies demonstrate that conferring a special status on a point of a monotonic segment of a continuous curve is almost always a risky endeavor. Also, the derivation of models of the Baranyi–Roberts type is based on the implicit assumption that a universal connection exists among the lag time, maximum specific growth rate, and maximum growth level. However, during the fitting of growth data by nonlinear regression, any adjustment for the maximum specific growth rate, for example, will be automatically compensated by a corresponding adjustment of the lag time length and possibly the maximum growth level as well, regardless of whether it is warranted on the basis of the actual state of the growing organism.

The same can be said about the modified Gompertz model's construction (Equation 8.7) in which an adjustment of μ may affect the magnitude of B, which determines the asymptotic growth level. The situation could be corrected if the lag time were redefined as the time needed to double or triple the initial population size, increase it tenfold, or by any other factor deemed meaningful or practical. This was proposed long ago but rarely if ever implemented. Once the lag time is expressed in this way, almost every model with a close fit will produce a similar estimate of its value, regardless of whether the dependent variable is the actual count or a growth ratio.

The Logistic–Fermi Combination Model

All the models previously discussed have been developed to represent the typical sigmoid shape of the *initial* part of the growth curve. Obviously, as already mentioned, a population with a very high density cannot be sustained indefinitely in a closed habitat, especially after most or all of the material resources have been exhausted. Therefore, at some point the population *must* start to decline and, if its evolution is followed for a sufficiently long time, a region of intensive mortality will inevitably follow. Indeed, several textbooks and publications have figures that depict the *four* phases of the population's evolution cycle, with the mortality or death stage commencing after a short or long stationary stage (Taub et al., 2003; McKellar and Lu, 2004).

Must mortality ensue only after a growth plateau has been reached? Should exhaustion of the habitat's resources — the starting point of the classic logistic model derivation and the origin of the term $1 - N(t)/N_{max}$ in its successors' rate equations — always be the main factor that regulates the population's growth rate? The answer to both questions is "no." Although exhaustion of the habitat's resources will always be the ultimate limiting factor, it need not always be the most influential. If pollution of the habitat by released metabolites, which increases simultaneously with the population's destiny, assumes the dominant role early enough, even the first part of the growth curve may not have the sigmoid shape that the traditional models depict. In principle at least, the production of an acid, alcohol, or other metabolites can poison the microbial population (autosterilization) and cause its demise well before it has the opportunity to utilize all the available material resources.

Similar hypothetical scenarios can be contemplated for crowding, although living space can also be viewed as a material resource. The creation of local anaerobic conditions, which can also bring down a population of aerobs before it reaches the expected stationary stage, is another possible scenario. Although the habitat pollution is inevitable in a closed system, the time scale of the toxic or inhibitory metabolite release and those of the cell's growth and division can be quite different in different systems. The same can be true for the diffusion and absorption rates of compounds essential for growth and others that block it. Thus, at least theoretically, the population's decline can start at any stage of the growth cycle, regardless of the habitat's carrying capacity as far as material resources are concerned.

In reality, the cell growth and division rate can be affected simultaneously by all of the preceding items in a manner that might not be known in detail or accurately predicted from the habitat's properties and the organism's growth patterns in other media. These considerations and the fact that, in most cases, we might not even know how many factors actually regulate the growth kinetics should dictate our choice of a growth model. In classic population dynamic models, a mortality term is almost always an integral component of the rate(s) equation(s).

An excellent example can be found in the already mentioned work of Taub et al. (2003), who developed a dynamic population model for non-sigmoidal growth curves. (Their publication also presents a comprehensive coverage of previous attempts to model such growth curves.) The notion that the shape of the growth curve is simultaneously controlled by cell division and mortality can also be implemented, *quantitatively,* in a purely phenomenological growth model construction by combining global growth and decay elements in the same model equation (Peleg, 1996a). The advantage of such models is that the kinetic order of the mortality and growth processes, which is unknown *a priori,* need not be assumed. This relieves the modeler from the burden of verifying the orders experimentally.

For the sake of simplicity, assume that the growth element follows the classic logistic function:

$$Y_{growth}(t) = \frac{Y_{asympt}}{1 + \exp[k_{growth}(t_{c\,growth} - t)]} \tag{8.10}$$

However, superimposed on it is a mortality factor, $R_{mortality}(t)$, which assumes the shape of a Fermi curve, the mirror image of the logistic curve:

$$R_{mortality}(t) = \frac{1}{1 + \exp[k_{mortality}(t - t_{c\,mortality})]} \tag{8.11}$$

The combined growth–mortality curve is therefore described by the product $Y_{growth}(t) \times R_{mortality}(t)$:

$$Y(t) = Y_{growth}(t) \cdot Y_{mortality}(t)$$

$$= \frac{Y_{asympt}}{\{1 + \exp[k_{growth}(t_{c\,growth} - t)]\} \cdot \{1 + \exp[k_{mortality}(t_{c\,mortality} - t)]\}} \qquad (8.12)$$

Generated examples of $Y_{growth}(t)$, $Y_{lethality}(t)$, and their various possible combinations are shown in Figure 8.2, which demonstrates that the familiar plateau with a considerable duration will only appear when $t_{mortality} \gg t_{cgrowth}$. However, when the two are of a comparable magnitude — that is, when $t_{clethality} \approx t_{cgrowth}$ — a true peak growth, followed by a moderate or steep decline, might be observed. In fact, a true growth peak is not at all uncommon; see examples in Peleg (1997) or in the more recent reports of Chorianopoules et al. (2005), Dougherty et al. (2002), and Neysens et al. (2003).

Also, when the two characteristic times, $t_{cgrowth}$ and $t_{cmortality}$, are close, the peak growth can be well *below* the level dictated by the habitat's carrying capacity. In the preceding terminology, $\max[Y(t)] < Y_{asympt}$, where Y_{asympt} is the asymptotic value of $Y_{growth}(t)$ if the growth had been unimpeded. According to the combined model expressed in Equation 8.12, the growth curve, with or without a stationary phase, need not be symmetric. It can be skewed in either direction, depending on whether the growth rate parameter, k_{growth}, is bigger than the decline rate parameter, $k_{mortality}$, or vice versa.

Figure 8.2 also demonstrates that growth patterns that look as if they were of totally different kinds are in fact just special cases of a general growth mode and can be described by the same general mathematical model. In other words, the different shapes of the growth curves are not a manifestation of qualitative differences in the growth and mortality patterns, but rather in the relative magnitude of the time scales of the growth and mortality parameters. Because it is purely phenomenological, the model as expressed by Equation 8.12 does not explain why the growth and mortality parameters assume their absolute or relative magnitudes.

Although the model was found to be consistent with the observed growth patterns of bacteria and cells in a cell culture (as shown in Figure 8.3 through Figure 8.6), it should not be considered predictive. Nevertheless, it can be used to quantify the effect of environmental conditions, such as temperature and pH, on the growth and decline patterns as demonstrated in the figures and in Table 8.1. In the context of this model, the presence of another organism or even of a mixed population in the habitat can be considered as an environmental condition. Like the influence of temperature and pH, its effect can be assessed in terms of the

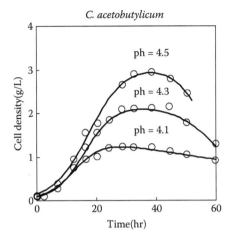

FIGURE 8.3
Growth–mortality curves fitted with the logistic–Fermi model combination. The original experimental data are from Yang, X. and Tsao, G.T., 1994, *Biotechnol. Prog.*, 10, 532–538. (From Peleg, M., 1996, *J. Sci. Food Agric.*, 71, 225–230. With permission, courtesy of Wiley Publishing.)

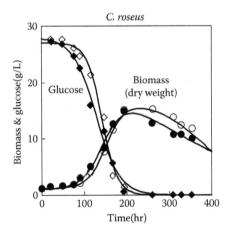

FIGURE 8.4
Growth–mortality curves of fitted with the logistic–Fermi model combination. The original experimental data are from Gulik, W.M. et al., 1994, *Biotechnol. Prog.*, 10, 335–339. (From Peleg, M., 1996, *J. Sci. Food Agric.*, 71, 225–230. With permission, courtesy of Wiley Publishing.)

observed magnitude of the growth and mortality parameters, which can be compared with those determined when the studied organisms are introduced into a sterile habitat.

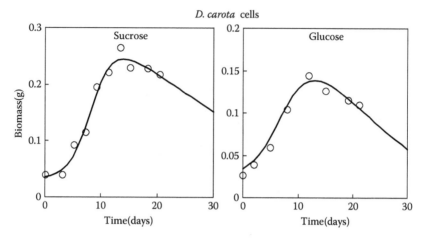

FIGURE 8.5
Growth–mortality curves of fitted with the logistic–Fermi model combination. The original experimental data are from Albiol, J. et al., 1993, *Biotechnol. Prog.*, 9, 174–178. (From Peleg, M., 1996, *J. Sci. Food Agric.*, 71, 225–230. With permission, courtesy of Wiley Publishing.)

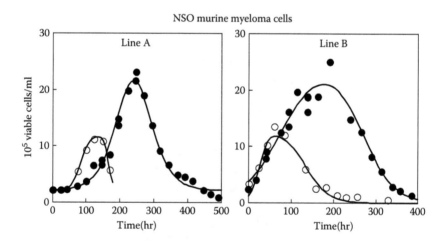

FIGURE 8.6
Growth–mortality curves fitted with the logistic–Fermi model combination. Open circles - batch, filled circles - fed batch. The original experimental data are from Bilbia, T.A. et al., 1994, *Biotechnol. Prog.*, 10, 87–96. (From Peleg, M., 1996, *J. Sci. Food Agric.*, 71, 225–230. With permission, courtesy of Wiley Publishing.)

What might affect the observed growth pattern of an organism in a habitat already occupied by other microbial species is not only the latter initial numbers, but also the relative magnitudes of their growth and mortality parameters. These might be indicative of their vigor or tendency

TABLE 8.1

Regression Parameters of Published Growth Curves Using the Combined Logistic and Fermi Equations as a Model

Organism and conditions	Ns or its equivalent	t_c growth (h)	k_{growth} (h^{-1})	t_c mortality (h)	$k_{mortality}$ (h^{-1})	r^2	Data source
C. acetobutylicum							
pH = 4.5	2.8 g·l^{-1}	16	0.21	80	1.0	0.994	Yang and Tsao (1994)
pH = 4.3	2.2 g·l^{-1}	16	0.20	62	0.13	0.999	
pH = 4.1	2.3 g·l^{-1}	13	0.21	40	0.019	0.999	
C. roseus							
Initial inoculum low	22 g·l^{-1}	180	0.030	248	0.024	0.998	Gulik et al. (1994)
Initial inoculum high	19 g·l^{-1}	103	0.031	187	0.022	0.996	
Shaken flasks	17 g·l^{-1}	152	0.046	389	0.016	0.998	
Stirred fermentor	27 g·l^{-1}	148	0.043	237	0.068	0.997	
D. carota cells							
Sucrose medium	0.38 g	8.2 (days)	0.58 (day^{-1})	20 (days)	0.069 (day^{-1})	0.994	Albiol et al. (1993)
Glucose medium	0.20 g	7.2 (days)	0.55 (day^{-1})	2 (days)	0.11 (day^{-1})	0.998	
Recombinant NSO murine myeloma cells							
Line A batch	12 10^{5}cells.ml^{-1}	84	0.076	166	0.13	1.000	Bilbia et al. (1994)
Line A fed batch	81 10^{5}cells.ml^{-1}	238	0.029	237	0.027	0.991	
Line B batch	13 10^{5}cells.ml^{-1}	25	0.094	134	0.037	0.957	
Line B fed batch	25 10^{5}cells.ml^{-1}	75	0.023	268	0.027	0.985	

Source: Peleg, M., 1996, *J. Sci. Food Agric.*, 71, 225–230. With permission, courtesy of Wiley Publishing.

to perish on their own, which is likely to become a factor when interspecies competition sets in. Remember that when the isothermal growth curve has a true peak, the asymptotic growth level in the absence of mortality cannot be determined directly from the data. Also, experimental relationships with a sharp peak, like those shown in Figure 8.3 through Figure 8.6, are notoriously difficult to curve fit because of a convergence problem. Fitting such curves requires very close initial guessed values for the regression to succeed and even then a unique set of fit parameters is not guaranteed. This is especially the case whenever only a few data points are around the peak.

As mentioned, because even the asymptotic growth level cannot be estimated from the data, the initial guesses problem must be overcome by other means. One way is to add *fictitious data* in order to get estimates of the parameters. Once these are obtained, the fictitious data are removed and the actual data are fitted with the model using the estimates thus obtained as the initial guesses. However, even when the data can be fitted successfully by such methods, as can be seen in the figures, the magnitude of the regression parameters might only be used to estimate the actual conversion of the available resources in the medium into biomass or number of cells. The maximum theoretical population size that could be supported by the habitat, if of interest, *must be determined independently* by monitoring the concentration of essential nutrients (for example, as shown in Figure 8.3) or by following any other factor that might limit the growth level.

Alternatively, the habitat's carrying capacity can be estimated by experiments in which the metabolites that cause the mortality are constantly removed. This is, of course, easier said than done. Such experiments will most likely be very difficult to devise and perform and thus would not be a feasible option in most cases involving microbial growth in foods, for example. (The situation might be different if a continuous production of the organisms is the objective of the study; however, as previously stated, this is a topic outside the scope of the present discussion.) Although a rough estimate of the habitat-carrying capacity can be obtained by mass balance, this will require an estimate of the biomass conversion factor, which might be unknown.

Simulation of Nonisothermal Growth Pattern Using the Logistic–Fermi Model

Consider a hypothetical case in which the isothermal growth curves of an organism in a given habitat at a pertinent temperature range follow the logistic–Fermi model (Equation 8.12). Let us also assume that the initial and asymptotic inoculum sizes (N_0 and N_{asympt}) have been determined experimentally or estimated by other means so that the model can be used in its normalized form by expressing the population size in terms of the growth ratio, $Y(t)$, instead of the actual number of the organism or its logarithm.

In this version of the logistic–Fermi model, the isothermal growth and decay pattern can be characterized by four temperature-dependent parameters only: $k_{growth}(T)$, $t_{c\,growth}(T)$, $k_{mortality}(T)$, and $t_{c\,mortality}(T)$. This is not a strict requirement and it is introduced here for the sake of clarity only. What follows will apply equally to situations in which N_{asympt} is treated as a temperature-dependent variable. It will also apply to all simpler models, including the traditional logistic model (see below) and several of its modifications and substitutes, which do not account for the mortality phase.

Let us start with the oversimplified case in which, in the temperature range of interest, all the isothermal growth curves follow the classic logistic function:

$$Y(t) = \frac{1}{1 + \exp\{k(T)[t_c(T) - t]\}} \tag{8.13}$$

where $k(T)$ and $t_c(T)$ are the temperature-dependent model's parameters.

As in the case of inactivation, we assume that, under nonisothermal conditions, the momentary growth rate is the isothermal growth rate at the momentary temperature at a time that corresponds to the momentary population size. Again, the momentary growth rate under nonisothermal conditions depends not only on the changing temperature (as the traditional Arrhenius model, for example, entails), but also on the growing population's *changing state*. If this assumption is correct, then the momentary growth rate of a population exhibiting logistic growth is:

$$\left|\frac{dY(t)}{dt}\right|_{T=const} = \frac{k_c(T)\exp\{k(T)[t_c(T)-t]\}}{1 + \exp\{k(T)[t_c(T)-t]\}^2} \tag{8.14}$$

as shown schematically at the top of Figure 8.7.

The time, t^*, which corresponds to any momentary level of $Y(t)$ (see figure) is the solution of Equation 8.13 for t:

$$t^* = t_c(T) - \frac{\log_e\left[\dfrac{1}{Y(t)} - 1\right]}{k(T)} \tag{8.15}$$

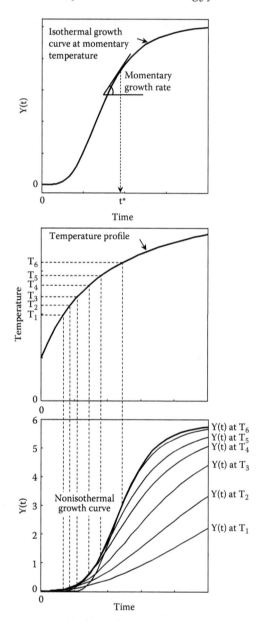

FIGURE 8.7
Schematic view of how the momentary isothermal growth rate is defined (top) and how the nonisothermal growth curve is constructed for a given temperature profile (middle and bottom). Notice that, unlike in the construction of nonisothermal survival curves, the concept that the isothermal growth rate depends on the growth stage is well established and generally accepted. (From Corradini, M.G. and Peleg, M., 2005, *J. Appl. Microbiol.*, 99, 187–200. With permission, courtesy of Blackwell Publishing.)

Combining Equation 8.14 and Equation 8.15 and making the temperature a function of time yield the nonisothermal growth rate:

$$\frac{dY(t)}{dt} = \frac{k[T(t)]\exp[k[T(t)]\{t_c[T(t)] - t*(t)\}]}{[1 + \exp[k[T(t)]\{t_c[T(t)] - t*(t)\}]]^2}$$

(8.16)

where $T(t)$ is the population's temperature history or profile and $t*(t)$, as defined by Equation 8.15, becomes:

$$t*(t) = t_c[T(t)] - \frac{\log_e \frac{1}{Y(t)} - 1}{k[T(t)]}$$

(8.17)

Despite the cumbersome appearance of the model, it is an ordinary differential equation that can be easily solved numerically by commercial software like Mathematica®. Its solution, the sought growth curve $Y(t)$ vs. t, can be calculated for almost any conceivable thermal history, provided that $k(T)$, $t_c(T)$, and $T(t)$ can be expressed algebraically (see below). A schematic graphic view of the model's rate equation solution and the manner in which the nonisothermal growth curve is constructed is given in the middle and bottom of Figure 8.7. In the actual construction of the growth curve, the time intervals are, of course, much shorter (thus the appearance of a smooth curve).

Examples of nonisothermal logistic growth curves generated with Equation 8.16 and Equation 8.17 as a model are shown in Figure 8.8 and Figure 8.9. The model can also be written in the form of a *difference equation* (see Chapter 3); in this case, growth curves in the form of $Y(t)$ vs. t (or $N(t)$ vs. t) can be produced with a program like Microsoft Excel® (see below). As in the case of the survival curves, the results will be practically identical if the incremental time intervals are short.

The situation is more complicated if growth and mortality are considered simultaneously. If the normalized logistic–Fermi model is used to characterize the population's growth and decline history (Equation 8.12), the user must contend with five time-dependent terms: $t_{cgrowth}[T(t)]$, $k_{growth}[T(t)]$, $t_{cmortality}[T(t)]$, $k_{mortality}[T(t)]$, and $T(t)$, the temperature history. However, the complication does not arise from the number of terms and the resulting complexity of the differential equation. These would have hardly any effect on even the speed at which the rate equation is numerically solved.

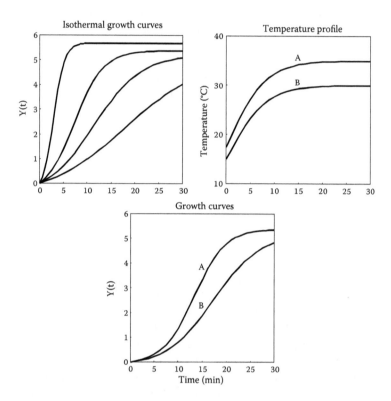

FIGURE 8.8
Examples of simulated nonisothermal logistic growth curves produced with Equation 8.16 as a model for monotonic temperature profiles. (Courtesy of Dr. Maria Corradini.)

The problem here is the same one encountered when dealing with certain inactivation patterns. It stems from the fact that Equation 8.12, which defines the normalized population size, $Y(t)$, as a function of time, has no analytical inverse and thus $t^*[Y(t)]$ cannot be calculated algebraically. However, it can be expressed as the *numerical solution* of the equation and introduced as such into the rate model's equation. Using a syntax similar to that of Mathematica, we define the momentary rate $dY(t)/dt$ as $D[Y[t],t]$ — i.e., the first derivative of $Y[t]$ with respect to t — and the time t^*, *tstar*, as:

$$tstar = t/.x \rightarrow \text{first}[\text{NSolve}[Y(t) == Y[x],x]] \qquad (8.18)$$

That is, t gets the value of the solution of Equation 8.12 for x when $Y(t) = Y(x)$, the momentary value of the normalized growth ratio.

When incorporated into the rate equation, the value of t^* at any real time t would be obtained by solving Equation 8.12 numerically. As was

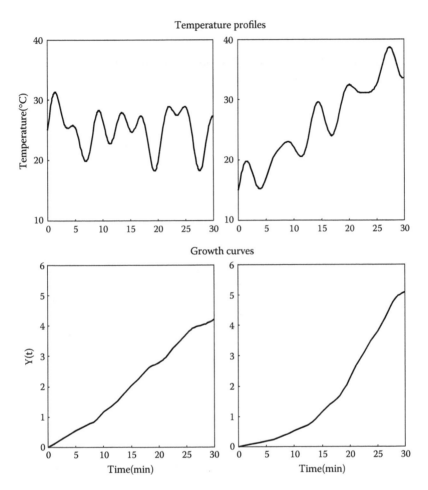

FIGURE 8.9
Examples of simulated nonisothermal logistic growth curves produced with Equation 8.16 as a model for oscillating temperature profiles. (Courtesy of Dr. Maria Corradini.)

shown in the analysis of inactivation patterns, writing the resulting differential equation explicitly is not necessary. We can simply define the growth and mortality parameters' time dependence by nesting, i.e., $k_{growth}[t] = k_{growth}[T(t)]$, $t_{cgrowth}[t] = t_{cgrowth}[T[t]]$, $k_{lethality}[t] = k_{lethality}[T(t)]$, and $t_{clethality}[t] = k_{lethality}[T(t)]$. We can also express the isothermal derivative of Equation 8.10 as a new function, e.g., isothermalDerivative[t], and then replace every T in the expression by $T(t)$ and t by t^* as defined by Equation 8.18. After doing all this, the growth rate equation in the language of Mathematica® becomes:

$$\text{NDSolve}[\{Y'[t] = = \text{isothermalDerivative}[t],$$
$$Y[0] = = 0\}, Y[t], \{t, 0, t\text{Cycle}\}] \qquad (8.19)$$

What Equation 8.19 means for the program is that we want to find and calculate, by solving the differential equation numerically, a function $Y(t)$, our nonisothermal growth curve. We tell the program that the slope of this function with respect to time, $Y'(t)$, at any given time t between zero and the test completion time, tCycle, is the isothermal derivative of $Y[t]$ at the corresponding temperature $T(t)$, at a time t^* as defined by Equation 8.18. The boundary condition here is that at time zero, $N = N_0$ and thus the expression $Y[0] = 0$. (For more explicit models, in which the actual number rather than the normalized ratio is used as the dependent variable [i.e., $N(t)$ vs. t is the sought solution], the boundary condition might be expressed in terms of N_0 or $\log_{10} N_0$ as the case might be.)

As before, the command NDSolve in Mathematica® produces a solution — in our case, the normalized growth curve, $Y[t]$, in the form of an interpolation function. In Mathematica®, an interpolation function is a dense set of numerical values of the function. For almost all practical purposes, as long as its values are within the specified range (i.e., between zero and tCycle, in our case), the interpolation function can be treated and plotted as if it were an ordinary continuous function.

Monotonic Temperature Histories

Examples of simulated growth curves under isothermal conditions and when the temperature increases or decreases are shown in Figure 8.10 and Figure 8.11. They were produced by Mathematica using the preceding procedure, with Equation 8.12 as a model. These figures demonstrate, again, that the model's mathematical complexity is not a hindrance to its differential equation solution by the program. The plots in the figures show how heating and cooling can influence the affected population's growth pattern. Notice that what controls the growth curves according to the described model (Equation 8.12) are two growth and two mortality parameters — namely, k_{growth}, $t_{c\ growth}$, $k_{mortality}$, and $t_{c\ mortality}$, and their temperature dependence. Thus, when these properly chosen scenarios where growth ceases and the population is destroyed by heat (or cold) can be created to simulate the fate of real populations. An example of a hypothetical event of this kind is shown in Figure 8.12.

Regular and Random Temperature Oscillations

Models of the kind expressed by Equation 8.12 can also be used to study the role of temperature fluctuations. These are frequently encountered in

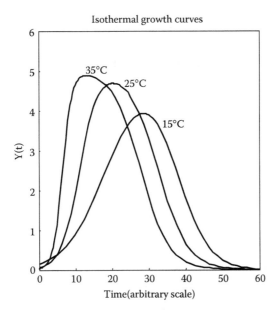

FIGURE 8.10
Example of simulated isothermal growth curves of a hypothetical organism using the logistic–Fermi model combination, produced with Equation 8.12 as a model. Notice that Equation 8.12 has no analytical inverse and thus the time that corresponds to any momentary growth level cannot be expressed algebraically, only numerically as in Equation 8.18. (Courtesy of Dr. Maria G. Corradini.)

transported and stored foods and always in natural water reservoirs like ponds and lakes and in water stored for domestic and industrial use. Thus, the fate of any microbial population originally present or introduced into such a habitat accidentally might be a safety concern. The same can be said, of course, of pathogens accidentally or deliberately released into a food or water source. The main obvious difference between foods and water, though, is that in most cases the food is a rich habitat with respect to material resources; unless water is contaminated with a massive amount of organic matter or polluted with nitrogen containing salts, it is usually a relatively poor growth environment.

On the other hand, unsterilized or unpasteurized foods are usually dried, frozen, refrigerated, or otherwise treated to reduce the microbial load and to halt or slow down any microbial growth. Still, the effect of temperature fluctuations in foods and water can be studied using the same kinds of models. As with microbial inactivation, there are hardly any restrictions on the mathematical structure of the function that describes the temperature history, $T(t)$. Also, the rate equation can be solved numerically whether $T(t)$ is a continuous or discontinuous function. In our case, the model can be used to follow the evolution of microbial

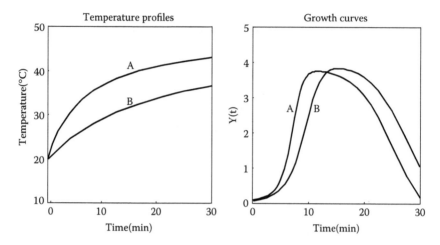

FIGURE 8.11

Simulated nonisothermal growth curves of the hypothetical organism whose isothermal growth curves are shown in Figure 8.10. Notice that the expression of t^* as an iterative numerical solution of the isothermal growth model's equation (Equation 8.18) is not a hindrance to the numerical solution of the differential rate equation into which it is incorporated by Mathematica. (Courtesy of Dr. Maria Corradini.)

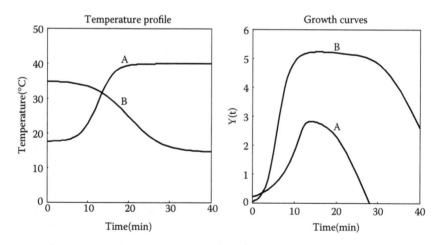

FIGURE 8.12

Example of a simulated scenario in which a particular temperature history (profile) can cause the demise of a growing population. Notice that, in contrast with certain isothermal scenarios shown in Figure 8.2, the fate of the population is determined by the temperature profile and the temperature dependence of the growth and mortality parameters. (Courtesy of Dr. Maria G. Corradini.)

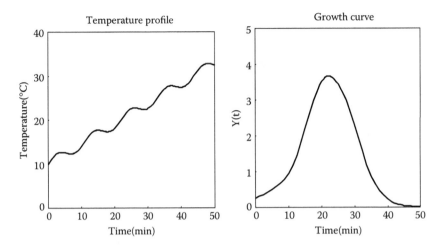

FIGURE 8.13
Examples of simulated growth curves produced by the logistic–Fermi model for a temperature profile characterized by regular temperature oscillations. Notice that the complexity of the regular temperature oscillations' corresponding differential rate equation is not a hindrance to its solution by Mathematica. (Courtesy of Dr. Maria G. Corradini.)

populations exposed to regular temperature fluctuations or abrupt temperature changes.

Examples of simulated growth curves under regular and irregular temperature histories are shown in Figure 8.13 through Figure 8.15. The regular fluctuation pattern (Figure 8.13) can represent, albeit in an oversimplified manner, diurnal temperature oscillations. The random fluctuations (Figure 8.14) can depict the temperature history of an unrefrigerated food during its transportation or poorly controlled storage. Water in a tank exposed to rapidly changing weather conditions may experience similar microbial growth. The discontinuous patterns can represent scenarios of equipment malfunction or breakdown, abuse, accidents, etc.

In principle at least, such simulations can be used to assess the potential microbial consequences of such events. They also allow one to estimate the effect of realistic temperature histories on the microbial quality of foods and water from results obtained under controlled conditions in the laboratory. Forecasting techniques can be employed here, too, to estimate the *probability* of growth to a dangerous level when the food or water temperature varies randomly. Here, again, once the amplitude of the temperature fluctuation can be reasonably estimated, all that is needed is to run the program several times; identify the peak growth or the actual number of organisms after a given time, or several times; and count the number of times the population has reached a dangerous level. Simulation

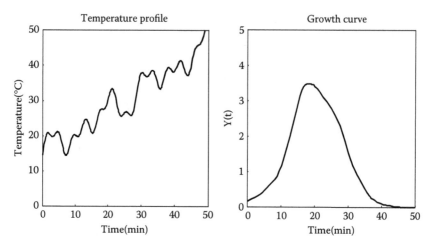

FIGURE 8.14
Examples of simulated growth curves produced by the logistic–Fermi model for a temperature profile characterized by irregular temperature oscillations. Notice that the complexity of the irregular temperature oscillations' corresponding differential rate equation is not a hindrance to its solution by Mathematica. (Courtesy of Dr. Maria G. Corradini.)

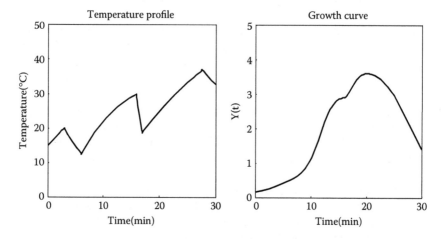

FIGURE 8.15
Examples of simulated growth curves produced by the logistic–Fermi model for a temperature profile characterized by random temperature oscillations. Notice that the complexity of the random temperature oscillations' corresponding differential rate equation is not a hindrance to its solution by Mathematica. (Courtesy of Dr. Maria G. Corradini.)

results of this kind (Figure 8.15) can be arranged in a histogram and used to calculate the mean and standard deviation of the counts after the chosen time or times (Corradini et al., 2005).

These times can be the expected shelf life of a food product or the duration of its transportation and storage, for example. In the case of water, such simulations can be used to determine whether a treatment like chlorination will be necessary. Conversely, the simulations can be used to estimate the *time* at which the probability that microbial growth will reach an unacceptable level would become too high. These kinds of simulations can be used to assess the potential effects of abuse on food products — for example, how long a food product would be safe to eat if stored without refrigeration under conditions in which the temperature can oscillate randomly. For the procedure and simulations of the kind shown in Figure 8.15 to yield realistic probabilities, one must have realistic values of the growth/mortality parameters, based on actual records of the temperature oscillations, amplitudes, and frequencies.

The same procedure can also be employed to evaluate how variations in the model parameters and temperature dependence, i.e., $t_{c\,growth}$ $k_{growth}(T)$, $k_{mortality}(T)$, and $t_{c\,mortality}(T)$, combined with the temperature oscillation amplitude, might affect the expected growth after a given time or times. Alternatively, simulations of the kind shown in the figure can be used to estimate the influence of changes in the growth parameters' magnitude on the time to reach a high probability that the product will become unacceptable or unsafe.

Prediction of Nonisothermal Growth Patterns from Isothermal Growth Data

The ability to predict nonisothermal growth patterns from experimental isothermal growth data has obvious practical benefits. The determination of primary growth parameters using isothermal data is straightforward and can be done using standard nonlinear regression procedures, which are widely available as part of commercial statistical software. As shown in this chapter, one has quite a few mathematical regression models from which to choose. Yet, the advice would be to use Ockham's razor as a guideline and choose the simplest model with the fewest number of adjustable parameters as a model.

Once the primary model is chosen, similar considerations should apply in the selection of the secondary model: it should be of the simplest mathematical structure and with a minimal number of adjustable parameters. Once determined and incorporated into the rate equation, the model can be employed to generate growth curves under almost any conceivable nonisothermal temperature regime, $T(t)$, as long as the predicted curve would be within the time temperature range covered by the experimental isothermal data from which the model's equation has been derived. However, such generated growth curves would be of value only if it could be demonstrated that they would match the actual growth patterns of the organism under the contemplated temperature history. The following demonstrations, created on the basis of published growth data of *Pseudomonads* spp. in refrigerated fish and *E. coli* 1952 in a nutritional broth, show that, indeed, they do.

Recall that, according to the proposed approach, the mathematical character of the chosen primary and secondary models is unimportant if they adequately fit the isothermal growth data and the sought predictions are all within the range covered by the experimental isothermal data from which the models' parameters have been calculated. To illustrate the point, we will use two very different primary models to fit the same two sets of isothermal growth data. Subsequently, we will use their parameters to obtain very different secondary models and incorporate their parameters into the respective models' rate equations. The predictions of the two versions of the rate models will then be compared and matched against the experimental nonisothermal growth data reported in the original publications.

The most common shapes encountered in a growth curve have already been shown in Figure 8.1 and Figure 8.2. However, many such curves (see Figure 8.16 and below) can be described not only by the logistic–Fermi model combination, but also by the totally empirical model (Corradini and Peleg, 2005):

$$Y(t) = \frac{a(T)t^{n(T)}}{b(T) + t^{m(T)}} \tag{8.20}$$

where, again, $Y(t)$ is the normalized growth parameter (i.e., $Y(t) = \log[N(t)/N_0]$ or $[N(t) - N_0]/N_0$) and a, b, m, and n are adjustable parameters.

According to this model, when $t = 0$, $Y(t) = 0$. However, notice that if $n > m$, as $t \to \infty$, i.e., $Y(t)$ will grow indefinitely, which is physically impossible. What $n > m$ therefore means is that an inflection point could be observed within the experiment's duration, but not a stationary stage or peak growth. When $n = m$, $Y(t)$ will have the typical logistic sigmoid shape, i.e., when $t \to \infty$, $Y(t) \to a$. When $n < m$, the growth curve has a true peak at $t = [(m - n)/bn]^{-1/m}$ as can be seen in the figure.

With $n = m$, Equation 8.20 can be a convenient three-parameter model for sigmoid growth curves with a short lag time. The asymptotic growth would be determined by the parameter a; the climb rate primarily, but not exclusively, by b; and the degree of asymmetry, primarily by n. Also, when $n = m$, Equation 8.20 is reduced to:

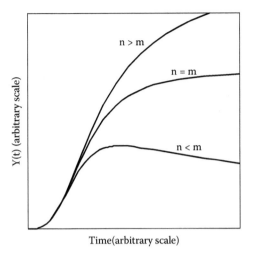

FIGURE 8.16
Simulation of three common isothermal microbial growth patterns using Equation 8.20 as a model. The curves were all produced with the same $a(T)$ and $b(T)$ but different $n(T)$. Notice that the growth curve, even when recorded beyond its inflection point, does not contain enough information to predict its continuation even qualitatively. (From Corradini, M.G. and Peleg, M., 2005, *J. Appl. Microbiol.*, 99, 187–200. With permission, courtesy of Blackwell Publishing.)

$$Y(t) = \frac{a(T)t^{n(T)}}{b(T) + t^{n(T)}} \qquad (8.21)$$

and thus

$$\left. \left| \frac{dY(t)}{dt} \right| \right|_{T=const} = \frac{a(T)b(T)n(T)t^{n(T)-1}}{[h(T) + t^{n(T)}]^2} \qquad (8.22)$$

The time t^* that corresponds to any particular growth level $Y(t)$ is:

$$t^* = \left[\frac{b(T)Y(t)}{a(T) - Y(t)} \right]^{\frac{1}{n}} \qquad (8.23)$$

Combining Equation 8.22 and Equation 8.23 and noticing that, at any time t where the temperature is $T(t)$, the corresponding $a[t] = a[T(t)]$, $b[t] = b[T(t)]$, and $n[t] = n[T(t)]$ yields the growth rate model's equation:

$$\frac{dY(t)}{dt} = \frac{a[T(t)]b[T(t)]n[T(t)]t^{*n[T(t)]-1}}{\{b[T(t)]+t^{*n[T(t)]}\}^2} \tag{8.24}$$

where $T(t)$ is the nonisothermal temperature history, or profile, and t^* is defined by Equation 8.23, i.e., $t^* = (b[T(t)]\,Y(t)/\{a[T(t)]-Y(t)\})^{1/n}$.

Like the previous rate equations, Equation 8.24 is an ordinary differential equation and can be solved numerically with a program like Mathematica. It can also be converted into a *difference equation*, in which case it can be solved with Microsoft Excel® or similar software (Corradini et al., 2005) (see: http://www.unix.oit.umass.edu/~aew2000/MicrobeGrowth-ModelB.html).

An alternative empirical model for sigmoid growth curves is the slightly modified version of the logistic equation:

$$Y(t) = \frac{a'(T)}{1+\exp\{k(T)[t_c(T)-t]\}} - \frac{a'(T)}{1+\exp[k(T)t_c(T)]} \tag{8.25}$$

where the three adjustable parameters are $a'\,t_{c\,\mathrm{mortality}}$, t_c, and k.

The second term in the right side of the equation has been introduced to account for that by definition, $Y(0) = 0$, i.e., that the normalized growth curve starts at the point of origin (0,0). The main advantage of Equation 8.25 over Equation 8.21, despite its cumbersome appearance, is in the description of growth patterns with a long lag time, where the b in Equation 8.21 will be a very big number, and in growth simulations. The parameters of Equation 8.25, like those of the original logistic equation but unlike those of Equation 8.21, have a clear intuitive meaning. The asymptotic growth level can be controlled by adjusting the value of a^1. The whole growth curve can be shifted to the right or left by lowering or raising the value of t_c, respectively, and the steepness of the climb can be regulated by changing the magnitude of k. According to this modified logistic model, the momentary isothermal growth rate is given by:

$$\left.\frac{dY(t)}{dt}\right|_{T=\text{const}} = \frac{k(T)a'(T)\exp\{k(T)[t_c(T)-t]\}}{[1+\exp\{k(T)[t_c(T)-t]\}]^2} \tag{8.26}$$

and the time t^* that corresponds to any given growth ratio $Y(t)$ by:

$$t^* = \frac{1}{k(T)}\log_e\left[\frac{\exp[k(T)t_c(T)]\cdot\{a'(T)+Y(T)\{1+\exp[k(T)t_c(T)]\}\}}{a'(T)\exp[k(T)t_c(T)]-Y(t)\{1+\exp[k(T)[t_c(T)]]\}}\right] \tag{8.27}$$

Combining the two equations and, as before, noticing that, under nonisothermal conditions, $a(t) = a[T(t)]$, $k[t] = k[T(t)]$, and $t_c[t] = t_c[T(t)]$ results in the model's rate equation:

$$\frac{dY(t)}{dt} = \frac{k[T(t)]a'[T(t)]\exp[k[T(t)]\{t_c[T(t)]-t^*\}]}{[1+\exp[k[T(t)]\{t_c[T(t)]-t^*\}]]^2} \quad (8.28)$$

where t^* is defined by Equation 8.27, except that $a'(T)$, $k(T)$, and $t_c(T)$ become $a'[T(t)]$, $k[T(t)]$, and $t_c[T(t)]$, respectively. Again, Equation 8.28 is an ordinary differential equation that can be solved numerically by a program like Mathematica®. It too can be converted into a *difference equation*, in which case it can also be solved with a program like Microsoft Excel or similar software (Corradini et al., 2005) (see http://www-unix.oit.umass.edu/~aew2000/MicrobeGrowthModelA.html).

There is nothing really special about the two models. They were chosen in light of the principle of parsimony (Ockham's razor), i.e., they have three adjustable parameters — the smallest number required to describe asymmetric sigmoid curves with an asymptote parallel to the abscissa. All that these two models can do is to describe isothermal growth data mathematically and no physiological meaning should be attached to any of their parameters. Another demonstration of why one should be extremely careful in the interpretation of adjustable regression parameters is that both models can be used interchangeably to fit the same data, despite their different mathematical construction (see below).

The same can be said about the secondary models, which describe the temperature dependence of these parameters. Here, too, because $a(T)$, $b(T)$, and $n(T)$ are not the same as $a'(T)$, $k(T)$, and $t_c(T)$ and each parameter triplet is interconnected in a different way, none of these parameters can have a universal mathematic interpretation unless it can be supported by independent evidence. In fact, one of the main reasons for using the two models is to show that they are not unique. The same would apply to any other three-parameter model that could fit the experimental growth data and to models that have four or more adjustable parameters.

The Growth of *Pseudomonas* in Refrigerated Fish

Several isothermal and nonisothermal growth data sets of *Pseudomonas* spp. on refrigerated fish stored at 0 to 8°C have been reported by Koutsoumanis (2001). Those among them that started at the same initial inoculum size and had a comparable time span could serve as a proper database to test the proposed concept. Because the issue here is the mathematical method, we will not dwell on the experimental procedure. It is described in great detail in the original publication, which is easily accessible to an interested reader.

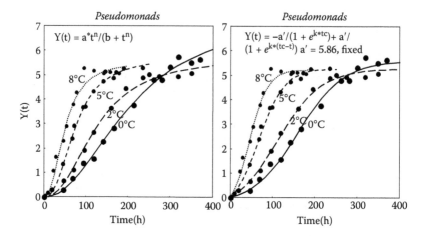

FIGURE 8.17

Isothermal growth curves of *Pseudomonas* spp. in refrigerated fish fitted with Equation 8.21 and Equation 8.25 as primary models, left and right, respectively. Notice the adequate fit of both models. The regression parameters are listed in Table 8.2. The original experimental data are from Koutsoumanis, K., 2001, *Appl. Environ. Microbiol.*, 67, 1821–1829. (From Corradini, M.G. and Peleg, M., 2005, *J. Appl. Microbiol.*, 99, 187–200. With permission, courtesy of Blackwell Publishing.)

In the paper on which the following discussion is based, Corradini and Peleg (2005) describe how the data were retrieved and processed. Suffice it to say here that the growth data, originally in the form of $\log_{10}N(t)$ vs. time relationships, were converted into logarithmic ratios vs. time relationships, i.e., to $\log_{10}[N(t)/N_0]$ vs. time data files. These were subjected to nonlinear regression using Equation 8.21 and Equation 8.25 as models, whose fit is shown in Figure 8.17 and Figure 8.18. The regression results, including those obtained with Equation 8.20 as a model, are given in Table 8.2. The figures and table show that the two models had a similar fit as judged by statistical criteria. They also demonstrate that sigmoid isothermal growth curves of the kind analyzed can be adequately described by at least two alternative three-parameter models and that adding a fourth (Equation 8.20) need not improve the fit dramatically.

The temperature dependence of the primary growth parameters were calculated with the ad hoc empirical secondary models, which are given in the figures. They were chosen solely for their mathematical simplicity and convenience and, needless to say, adequate fit. The nonisothermal temperature profiles reported in Koutsoumanis (2001) were all discontinuous (see below) and thus described by models having 'if' statements of the kind:

$$T(t) = \text{If } [t \leq t_1, T_1, \text{If } [t \leq t_2, T_2, \text{If } [t \leq t_3, T_3, ...]]] \qquad (8.29)$$

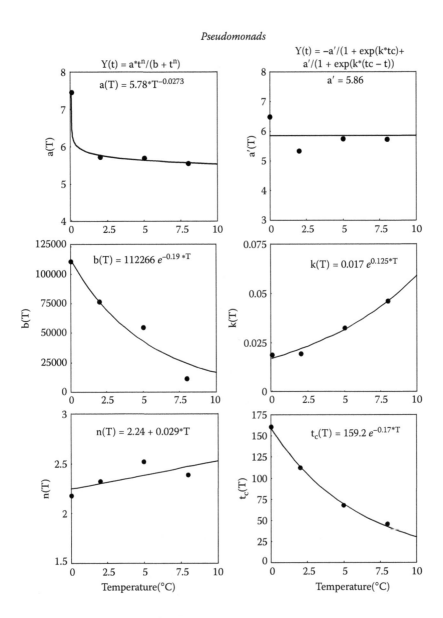

FIGURE 8.18
The temperature dependence of the growth models' parameters of *Pseudomonas spp.* in refrigerated fish (Figure 8.17), Equation 8.21 and Equation 8.25, fitted with ad hoc empirical models. The original experimental data are from Koutsoumanis, K., 2001, *Appl. Environ. Microbiol.*, 67, 1821–1829. (From Corradini, M.G. and Peleg, M., 2005, *J. Appl. Microbiol.*, 99, 187–200. With permission, courtesy of Blackwell Publishing.)

TABLE 8.2
Growth Parameters of *Pseudomonas* spp. in Refrigerated Fish

Temperature (°C)	Equation 8.4 as a model					Equation 8.5 as a model				Equation 8.9 as a model			
	$a(T)$ (h^{m-n})	$b(T)$ (h^m)	$n(T)$ (−)	$m(T)$ (−)	MSE	$a(T)$ (−)	$b(T)$ (h^n)	$n(T)$ (−)	MSE	$a'(T)$ (−)	$k(T)$ (h^{-1})	$t_c(T)$ (h)	MSE
0	8.74	6.7×10^4	2.04	2.06	0.039	7.47	1.1×10^5	2.18	0.035	5.86 (fixed)	0.0188	161	0.045
2	13.3	5.8×10^4	2.06	2.20	0.038	5.73	7.7×10^4	2.32	0.031	"	0.0194	112	0.032
5	24.3	3.7×10^4	2.03	2.28	0.016	5.69	5.5×10^4	2.52	0.018	"	0.0324	68	0.018
8	62.3	1.3×10^4	1.70	2.14	0.061	5.54	1.2×10^4	2.39	0.088	"	0.0460	45	0.045

Sources: The original data are from Koutsoumanis, K., 2001, *Appl. Environ. Microbiol.*, 67, 1821–1829. (From Corradini, M.G. and Peleg, M., 2005, *J. Appl. Microbiol.*, 99, 187–200. With permission, courtesy of Blackwell Publishing.)

The equation says that if $t \le t_1$, the temperature is T_1; otherwise, if $t \le t_2$, it is T_2; otherwise, if $t \le t_3$, it is T_3, etc.

Despite the multitude of 'if' statements that it contains, the function $T(t)$ as defined by Equation 8.29 as a model is treated by Mathematica® as a regular function, which can then be used to define and create the terms $a[T(t)]$, $b[T(t)]$, etc., as previously explained. These, in turn, were incorporated into the two alternative rate equations, Equation 8.24 and Equation 8.28, which were subsequently solved numerically by the program to produce the corresponding growth curves $Y(t)$ vs. t.

Plotted together with the corresponding reported experimental nonisothermal values, these are shown in Figure 8.19. The figure shows that despite the very different mathematical structure of the primary and secondary models used for their derivation, the predictions that Equation 8.24 and Equation 8.28 provided were in agreement with the experimental growth curves. This should not come as a surprise. It can be demonstrated that *any* set of primary and corresponding secondary models that fit the experimental data well will give a very similar estimate (Peleg et al., 2004), as long as they are not used for extrapolation.

It is interesting to note that when the four-parameter model (Equation 8.20) was used as a primary model, the fit was only slightly better, as shown in Table 8.2. However, this minor improvement came with a severe penalty. When we tried to obtain secondary models in this case, coherent relationships between the four parameters a, b, n, and m and temperature did not emerge. The reason is that, for calculating four adjustable parameters instead of three, the number of data points in the experimental isothermal sets must be quite large and their scatter small. If not, then because of the way in which the regression procedure works, an adjustment of one parameter is compensated by a corresponding adjustment in the other three parameters to minimize the mean square error. The result can be totally unrealistic secondary model parameters unless limits to their magnitude are set in advance. This would not only complicate the procedure but also require information that might not be available.

The same problem has already been encountered in the attempt to derive simultaneously the three Weibullian survival parameters from nonisothermal inactivation data (Chapter 4). Nevertheless, it does not arise with simulated data, whose density and spread can be set. This is in contrast to experimental data that might be widely spaced because of practical constraints and whose scatter is not always controllable.

Another point worth mentioning is that Koutsoumanis et al. (2001), who were the authors of the original work on *Pseudomonas*, accomplished similar predictions by a conventional growth model integrated piecemeal. Their method worked successfully in this case because the temperature profile consisted of a set of constant temperature regimes (see Figure 8.19). Thus, a stronger test of the nonisothermal rate model, as expressed in

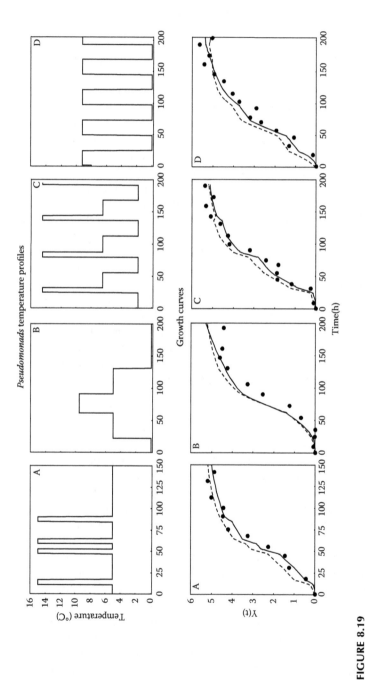

FIGURE 8.19

Comparison of the predictions of the nonisothermal growth rate models, Equation 8.24 and Equation 8.28, solid and dashed lines, respectively, with the experimental nonisothermal growth curves of *Pseudomonas* spp. in refrigerated fish, reported by Koutsoumanis, K., 2001, *Appl. Environ. Microbiol.*, 67, 1821–1829. Notice the similar predictions of the two models despite their different mathematical construction. This is a demonstration that a growth model need not be unique to be predictive. (From Corradini, M.G. and Peleg, M., 2005, *J. Appl. Microbiol*, 99, 187–200. With permission, courtesy of Blackwell Publishing.)

Equation 8.24 and Equation 8.28, and its utility would be its ability to predict growth patterns under totally nonisothermal conditions. Such a test will be described in the next section.

However, before we move to this test, the slight but noticeable discrepancy between the two models' predictions and actual growth data that correspond to the second temperature profile needs a comment. Notice that no growth was initially observed at a temperature and over time that the isothermal data indicated that it should be. Therefore, this extra lag was probably due to the organism's failure to initiate growth in the particular experiment for reasons that are unexplained in the original publication. As could be expected, the original authors' prediction in this case was also somewhat off for the same reason. However, despite the experimental data's (inevitable) imperfections and the two growth models' crudeness, their predictions in almost all the cases were within only about half a log unit of the actual observations. Therefore, considering the ubiquitous variability in microbial growth data, the predictions can be viewed as in agreement with the experimental data.

The Growth of *E. coli*

Fujikawa et al. (2004) have recently reported several isothermal sigmoid growth curves of *E. coli* 1952 in the temperature range of 27.6 to 36°C and three nonisothermal curves in the same growth medium. Their results could be converted into $Y(t)$ vs. time curves where $Y(t)$ is the logarithmic growth ratio, i.e., $Y(t) = \log_{10}[N(t)/N_0]$ (Corradini and Peleg, 2005). The isothermal data could be fitted by the same two three-parameter models, Equation 8.21 and Equation 8.25, used to describe the *Pseudomonas* growth data (see the previous section). The fit of the two models is shown in Figure 8.20 and Figure 8.21 and the corresponding regression parameters are listed in Table 8.3.

As before, these two very different three-parameter models had a very similar fit and could be used interchangeably. Equation 8.25 had a slightly better fit than Equation 8.21, but it was very close to that of Equation 8.20, which has a fourth parameter (see Table 8.3). This is another demonstration that adding adjusted parameters to improve the fit does not necessarily result in a better model. It also strengthens my view that Ockham's razor should serve as a guideline in choosing a growth model.

The temperature dependence of the two primary models — namely, a, b, and n and a', k, and t_c, respectively — is shown in Figure 8.20 and Figure 8.21, together with the fit of the ad hoc empirical models used to describe them mathematically. Again, these empirical models were chosen solely on the basis of their mathematical simplicity and goodness of fit and no specific kinetic meaning has been assigned to their magnitude or interrelationships. The temperature histories to which the *E. coli* has been

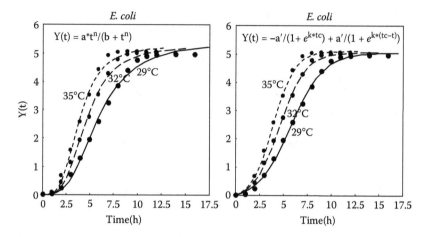

FIGURE 8.20
Isothermal growth curves of *E. coli* in a culture medium fitted with Equation 8.21 and Equation 8.25 as primary models, left and right, respectively. Notice the adequate fit of both models. The regression parameters are listed in Table 8.2. The original experimental data are from Fujikawa, H. et al., 2004, *Food Microbiol.,* 21, 501–509. (From Corradini, M.G. and Peleg, M., 2005, *J. Appl. Microbiol.,* 99, 187–200. With permission, courtesy of Blackwell Publishing.)

exposed were quite complex (see Figure 8.22, top). Nevertheless, they could still be fitted effectively by the model (Corradini and Peleg, 2005):

$$T(t) = \text{if } [t \le t_1, f_0(t), \text{ if } [t \le t_2, f_1(t), \text{ if } [t \le t_3, f_2(t),\ldots]]] \qquad (8.30)$$

where

$$f_0(t) = \frac{c_1}{1+\exp\left[c_2\left(c_3 - t\right)\right]} \qquad (8.31)$$

$$f_1(t) = \frac{c_4}{1+\exp\left[c_5\left(c_6 - t\right)\right]} \qquad (8.32)$$

$$f_2(t) = c_7 + \frac{c_8}{1+\exp\left[c_9\left(t - c_{10}\right)\right]} \qquad (8.33)$$

and so forth. In this series, all the terms with an odd number ($i \ge 3$) have the general format of $f_1(t)$; that is, they have the $c - t$ term in the exponential argument. All that have an even member ($i \ge 4$) (the format of $f_2(t)$, that

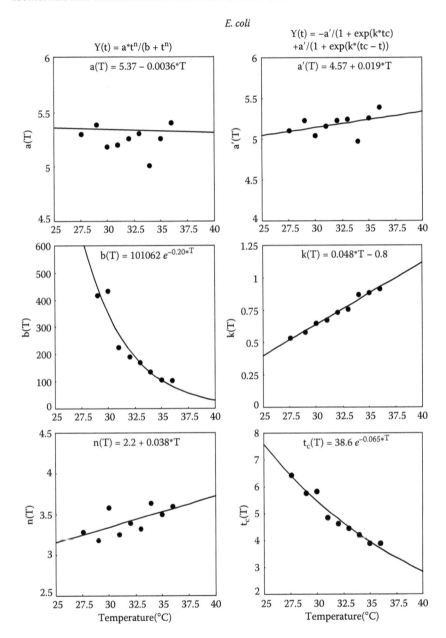

FIGURE 8.21
The temperature dependence of the growth models' parameters of *E. coli* in a culture medium (Figure 8.20), Equation 8.21 and Equation 8.25, fitted with ad hoc empirical models. The original experimental data are from Koutsoumanis, K., 2001, *Appl. Environ. Microbiol.*, 67, 1821–1829. (From Corradini, M.G. and Peleg, M., 2005, *J. Appl. Microbiol.*, 187–200. With permission, courtesy of Blackwell Publishing.)

TABLE 8.3

Growth Parameters of *E. coli* 1952 in Nutrient Broth

Temperature (°C)	Equation 8.4 as a model					Equation 8.5 as a model				Equation 8.9 as a model			
	$a(T)$ (h^{m-n})	$b(T)$ (h^m)	$n(T)$ $(-)$	$m(T)$ $(-)$	MSE	$a(T)$ $(-)$	$b(T)$ (h^n)	$n(T)$ $(-)$	MSE	$a'(T)$ $(-)$	$k(T)$ (hr^{-1})	$t_c(T)$ (h)	MSE
27.6	38.9	1102	2.40	3.08	0.004	5.30	504	3.28	0.017	5.11	0.54	6.44	0.007
29	29.3	556	2.35	2.94	0.005	5.39	288	3.18	0.021	5.23	0.58	5.76	0.008
30	18.3	682	2.76	3.20	0.005	5.32	283	3.17	0.022	5.04	0.65	5.83	0.007
31	21.7	268	2.34	2.85	0.006	5.20	177	3.25	0.024	5.16	0.68	4.85	0.007
32	16.2	215	2.51	2.92	0.008	5.26	184	3.39	0.024	5.23	0.73	4.63	0.007
33	14.2	181	2.56	2.92	0.008	5.31	149	3.32	0.023	5.24	0.76	4.47	0.007
34	12.9	193	2.76	3.12	0.007	5.01	184	3.63	0.020	4.97	0.87	4.21	0.006
35	12.9	127	2.63	2.98	0.009	5.26	117	3.49	0.024	5.26	0.89	3.91	0.007
36	12.4	134	2.74	3.07	0.009	5.40	133	3.60	0.025	5.39	0.91	3.91	0.008

Sources: The original data are from Fujikawa, H. et al., 2004, *Food Microbiol.*, 21, 501–509. (From Corradini, M.G. and Peleg, M., 2005, *J. Appl. Microbiol.*, 99, 187–200. With permission, courtesy of Blackwell Publishing.)

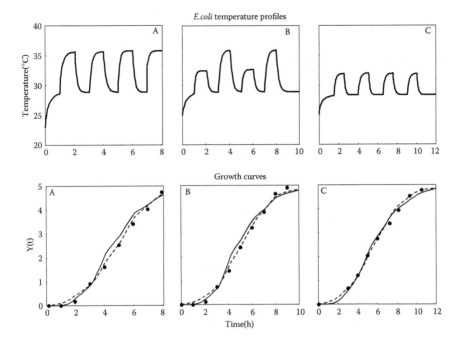

FIGURE 8.22
Comparison of the predictions of the nonisothermal growth rate models, Equation 8.24 and Equation 8.28, solid and dashed lines, respectively with the experimental nonisothermal growth curves of *E. coli* in a culture medium reported by Fujikawa, H. et al., 2004, *Food Microbiol.*, 21, 501–509. Notice the very similar predictions of the two models despite their very different mathematical construction. This is a demonstration that a growth model need not be unique in order to be predictive. (From Corradini, M.G. and Peleg, M., 2005, *J. Appl. Microbiol.*, 99, 187–200. With permission, courtesy of Blackwell Publishing.)

is) have the $t - c$ term in the exponential argument and a constant added to the logistic term.

Once the secondary models and temperature profiles, $T(t)$, could be expressed mathematically, they were combined to create the terms $a[T(t)]$, $b[T(t)]$ and $n[T(t)]$ and $a'[T(t)]$, $k[T(t)]$ and $t_c[T(t)]$, which were incorporated into the corresponding growth rate models' equations (Equation 8.24 and Equation 8.28, respectively). These in turn were used to generate the predicted theoretical growth curves for each temperature history. (Again, the numerical solution of the differential equations was done with Mathematica®, which is a particularly convenient program for such a task.)

Comparison of the predicted curves using the two models and the actual nonisothermal growth data reported by Fujikawa et al. (2004) is shown at the bottom of Figure 8.22. In all three cases, the agreement between the prediction and actual experimental data was very reasonable. This might have been largely due to the dense experimental database, which allowed for the derivation of reliable primary and secondary models. Here, too, the two primary models, despite their very different mathematical construction, gave very similar predictions. Figure 8.22 provides an additional demonstration that a primary growth model need not be mechanistic or unique in order for the rate model based on it to be predictive. Similar predictions would probably be obtained with the Gompertz or Baranyi–Roberts models, if the temperature dependence of each of their parameters were converted into functions of time.

Although empirical rate models like Equation 8.24 and Equation 8.28 do not explain anything, they can still be useful tools in the assessment of the effect of temperature variations on the growth patterns of microorganisms. More importantly, models of this kind are all testable, and their validity, and that of the assumptions from which they have been derived, can be confirmed or refuted by simply comparing their predictions to experimental results obtained under nonisothermal conditions. This is a direct test of validity that does not rely on any preconceived growth kinetics or any mechanistic interpretation of the growth curve shape.

Admittedly, the discussion in this chapter is based on very limited experimental evidence. However, it cannot be merely a coincidence that the described models' predictions almost perfectly matched all the actual nonisothermal growth patterns of *E. coli* 1952 examined and almost all those of the *Pseudomonas* spp. (At the time at which these lines are written, Corradini et al. (2005) have successfully predicted another set of nonisothermal growth curves of *C. perfringens* in chilled ham from the organism's isothermal growth data.) Therefore, the *approach*, at least, is worth further exploration and testing with other microorganisms at different temperatures and in different growth media.

9

Interpretation of Fluctuating Microbial Count Records in Foods and Water

A probability is the desperate attempt of chaos to become stable.

Anonymous

Periodic sampling and determination of microbial counts is a matter of routine in most food processing plants. The samples can be of raw materials or ingredients, intermediate and finished products, and/or the water coming into or leaving the plant. The counts can be of the total microbial population, which is usually indicative of freshness as in meats and fish products, and/or the efficacy of sanitary measures aimed at reducing microbial growth, as in refrigerated raw milk or dried spices. Frequently, the counts are of groups or types of organisms that might have a specific impact on the food's quality or safety. A notable example is the coliform count considered as a measure of a food's or water's contact with fecal material. However, counting other groups, such as total thermophiles or aerobes in baked snacks, yeast in products that contain fruit or fruit juice (including ice creams), and lactobacilli in dairy products, is also a common practice. When a direct safety risk is involved, the focus can be on a particular pathogen such as *Salmonella* in eggs and egg products, *Staphylococci* in refrigerated ground meats, or *listeria* in cured meat products.

The same can be said about water, whether it is fresh from a well or spring, in an open natural or man-made reservoir in a chlorination facility before its delivery, or in a treatment plant upon its arrival and discharge. Again, the counts can be of the total microbial population, coliforms, or specific pathogens, which need not be exclusively bacterial. The same applies to water used for recreation, such as in beaches and swimming pools; again, the total count might be indicative of the overall sanitary conditions at the site, the coliform or *E. coli* count of the degree of the water's fecal contamination, and the counts of specific pathogens (including viruses and protozoa) of the water's safety for the public's use.

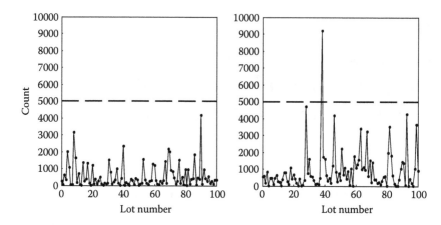

FIGURE 9.1
Simulated industrial microbial count records generated with Equation 9.1 as a model. The dashed horizontal line is the maximum allowed. Notice that the exceptionally high count in the right plot. It had no apparent cause and there were no signs that it was about to occur. Similarly, the return of all the successive counts to the allowed level had no apparent reason too.

The resulting records are of counts taken at very different time intervals from several hours to several days and sometimes even weeks. When plotted against time, the counts often produce a very irregular, apparently random time series (Figure 9.1). In many cases, the record is characterized not only by the counts' irregular fluctuations, but also by the appearance of occasional episodes of explosive population growth or "outbursts." These occur at irregular intervals and their magnitude seems to be random, too.

This chapter will primarily deal with the mathematical properties of such count records and how they can be used to estimate the future frequency of outbursts. The discussion will be limited to scenarios in which the fluctuations and outbursts *have no apparent cause.* One does not need a theoretical model to predict that a failure of the refrigeration system during raw milk transportation will result in intensified microbial growth or that the breakdown in a municipal sewage treatment plant and discharge of untreated water into a lake will affect the coliform count in a nearby recreational beach. Obviously, to predict how raw milk's elevated temperature will affect its microbial population or how an untreated sewage stream will spread bacteria in a lake also require mathematical models, albeit of very different from those to be presented in this chapter. Also, the discussion will focus on the interpretation of count records and thus unculturable organisms will not be considered.

Certain types of microbial growth and decline models have been discussed in other chapters of this book. All of them were developed for situations in which successive counts are close enough so that a continuous pattern of growth, inactivation, or growth followed by inactivation clearly emerges. Such continuous patterns can be described mathematically by continuous models in the forms of an algebraic or a differential equation.

What follows in this and the next three chapters will deal with population histories in which the intervals between successive counts are too large to delineate a continuous history of the system. For this reason, classic population dynamic models (e.g., Brown and Rothery, 1993; Murray, 1989), which are usually used to explain periodic patterns of growth and mortality, offer little help in our case. Chaos theories that have been invoked to explain irregular fluctuating growth and decline pattern are derived from nonlinear differential or difference population dynamic model equations (Ruelle, 1991, 1992). Consequently, they are fully deterministic or, when a random term is added, have a major deterministic component. Unfortunately, all the microbial count records that we have encountered in our studies had no such features and therefore chaos theories were of little use in their analysis and interpretation (Peleg and Horowitz, 2000).

Microbial Quality Control in a Food Plant

Two (simulated) typical microbial control charts of the kind that can be frequently found in a dairy or meat plant, for example, are shown in Figure 9.1. The one on the left shows an acceptable fluctuating pattern in which all the counts are below the level, marked by a horizontal line that if exceeded would render the product unsafe or classified as being of inferior quality. (Standard quality control charts usually also show a lowest permissible level. For obvious reasons, it will be meaningless in the context of microbial counts of most foods and water. The exceptions, such as dry/frozen yeast or a beverage containing probiotic lactobacilli, are too few to warrant a specific discussion in this chapter. Nevertheless, the proposed methodology to estimate the probability of a high count is just as applicable to low counts — see "Extinction and Absence.")

The chart on the right in Figure 9.1 shows a similar pattern of fluctuations except that one count clearly exceeds the permitted limit. This high count is peculiar because there is no prior indication that it was about to happen. The record obviously has no apparent trend and the counts immediately preceding this outburst were all within the expected bounds. The same can be said about the counts that followed the outburst. They too exhibited the same "regular" pattern of oscillations as if this exceptionally high count had never been recorded. Obviously, if a physical reason for the outburst, like a power failure or contamination from an identified source, were known, the matter would have been concluded and no further discussion would be necessary.

However, such outbursts can be observed, not infrequently as a matter of fact, when there is nothing unusual to explain them. As will be shown later, one should not be surprised that such outbursts occur because they may be the natural consequence of the statistics of randomly fluctuating populations. The preceding scenarios are not limited to processed foods. In fact, similar patterns are encountered when the microbial records of water are examined (see below). Moreover, they can also be found in totally different kinds of quality control or 'quality assurance' charts, in which the relevant fluctuations are not in the microbial population's size, but in the concentration of a certain ingredient or nutrient or even in a mechanical or other physical property (Gonzalez–Martinez et al., 2003).

The Origins and Nature of Microbial Count Fluctuations

Imagine the microbial population in shipments of raw milk arriving daily in a dairy plant, for example. The total count varies considerably from day to day (see below), but it fluctuates quite often around a certain characteristic value. Occasionally, at unpredictable time intervals, the count can be quite high (we will return to this point later). The question that immediately arises is, "What causes the fluctuations in the first place?" The disappointing answer is that we really do not know. However, we can surmise that a number of *potential causes* might promote a high count; for example, an accidental contamination of the milk or equipment by fecal material, a slightly warmer day, a less than rigorous washing of the equipment because the person in charge was in a hurry, congested traffic on the way from the farm to the plant or other delays, etc.

However, some factors suppress microbial growth, e.g., a slightly more concentrated chlorination solution used on that day, the tanker being parked in the shadow and thus allowing the milk to be cooled to a temperature slightly lower than the usual, the driver skipping the lunch break for personal reasons, etc. Usually, the growth-promoting and growth-suppressing factors balance one another, although not exactly — thus, the "normal" fluctuations in the microbial population's size. However, there is a certain probability that, on a particular day, an accidental accumulation of growth-promoting effects by far outweighed those of the growth-suppressing factors. Because this was a chance occurrence, the count on the next day will be much lower in most cases or, on a rare occasion, higher still.

Similarly, it is possible that an accidental accumulation of growth-suppressing effects "unbalanced" by those that usually promote growth will bring the microbial population to an unusual low size; again, this may or may not be observed in subsequent counts. Notice that a multitude of factors of both kinds, i.e., growth promoters and inhibitors, can be present and that most of them would be undocumented and some even totally unknown. What characterizes all these factors (see the introduction to this chapter) is that each by itself is insufficient to bring the count to an exceedingly high level that one would call an outburst. Conversely, none of the factors alone can inhibit the population to such an extent that the count would be considered unusually low. It is only the *accumulated effect* of *many such factors* that can sway the count either way. For this explanation to be plausible, successive counts need to be *independent* and the record as a whole should have no trend or long-term periodicity. As will be shown later, whether these conditions are satisfied can be tested by statistical methods.

However, even when a trend is found, in at least a certain kind of records, it can be separated from the random fluctuations and these can be analyzed if they were truly independent. A different scenario is when the outbursts linger and several high counts occur in a row. These will be discussed separately in Chapter 12. For what follows, we will assume that the entries in the available record are indeed random. Thus, if two or slightly more successive counts appear as marking a real peak growth or a lasting depression, they would still be considered independent if the record as a whole meets certain statistical criteria (see below). All the preceding discussion is equally applicable to oscillatory microbial counts in water, except that in most cases the sources of the microbial contamination are quite different. Also, the counts might be very low in comparison with those found in raw foods and long segments of successive zero counts, especially of certain fecal bacteria and pathogens, are not uncommon.

Asymmetry between Life and Death

The concept of a symmetric, bell-shaped distribution is a familiar one. The most recognizable one is the normal or Gaussian distribution, in which the probability or any given deviation from the mean *in either direction* is exactly the same. For many physical phenomena, the normal distribution is a most useful model despite its infinite range. (For physical reasons, the height of any member of group of adult humans, for example, cannot be −10 cm or +10 m. Thus, the probability of observing such a height is not just ridiculously small, as the Gaussian distribution implies, but is actually zero. The same can be said on the size of instant coffee particles in a jar. The probability of finding a particle with a negative diameter or bigger than the jar is exactly zero, not just very small.)

Some systems, however, are characterized by distributions that are necessarily *skewed*; the distributions of microbial count records in foods or water are such systems. This is simply because a microbial population can grow to a huge size when circumstances are favorable, but, even under the most lethal conditions, a population's size cannot be negative. Thus, the distribution of entries within any given microbial count record, whether it is of food or water, is expected to be asymmetric with a positive skew, i.e., tailing to the right. Of course, the distribution of counts in an actual record can *appear* symmetric and even pass the test for normality (see below). This should not come as a surprise or seen as a contradiction. Almost perfectly symmetric frequency (density) distributors can be generated by asymmetric distribution functions, but not vice versa.

Estimating the Frequency of Future Outbursts — the Principle

Suppose a record of independent counts, taken at more or less equal time intervals of the kind shown in Figure 9.1, has no trend or periodicity. If the record is a manifestation of a scenario of the sort described in the first three sections of this chapter — that is, the monitored population size is determined by a multitude of random factors, many or perhaps even most of which are unknown — the magnitude of any count in the series will be random too. However, in light of the very different theoretical limits set for growth and decay, the counts *distribution* within a long enough sequence is expected to be asymmetric. A mathematical model of such a record can be constructed on the basis of the assumption that the counts have a log normal distribution (Peleg et al., 1996):

$$N(n) = 1^{\mu_L + \sigma_L z(n)} \qquad (9.1a)$$

or

$$N(n) = 10^{\mu_L + \sigma_L z(n)} \qquad (9.1b)$$

where
   $N(n)$ is nth count in the series
   μ_L and σ_L are the counts' logarithmic (base e or 10, respectively) mean and standard deviation

$z(n)$ is a random number produced by a standard normal distribution, that is, with $\mu = 0$ and $\sigma = 1$

In fact, the two series shown in Figure 9.1, despite their realistic appearance, are not actual records but simulations generated with Equation 9.1a/b as a model. As will be shown later in this chapter, this same model (Equation 9.1a or Equation 9.1b) can capture the fluctuating patterns of real microbial counts recorded in foods and water.

If this model holds and if we can assume that the general conditions under which the record has been obtained remain unchanged, then the probability that any count N will equal to or exceed a given magnitude, N_c, $P(N \geq N_c)$, will be:

$$P(N \geq N_c) = \frac{1}{\sqrt{2\pi}\sigma_L} \int_{N_c}^{\infty} \exp\left[-\frac{(\log_e N - \mu_L)^2}{2\sigma_L^2}\right]\frac{dN}{N} \tag{9.2}$$

Thus, *if* the record follows the log normal model, then the probability of an outburst's occurrence can be calculated using standard statistical programs. In a case where $N(n)$ has any other density distribution:

$$P(N \geq N_c) = \int_{N_c}^{\infty} f(N)dN \tag{9.3}$$

where $f(N)$ is the particular distribution of the counts. The expected *number* of counts equaling or exceeding any value, N_c, in a series of m successive counts is therefore:

$$\text{Estimated number} = P(N \geq N_c) \cdot m \tag{9.4}$$

The model can be tested and validated by comparing the actual number of counts exceeding one or several chosen values of N_c in a fresh data set with m entries, with the number calculated with Equation 9.4 on the basis of the counts distribution in *another* data set. In other words, the comparison is between the observed number of high counts in a given record, with the number predicted using the μ_L and σ_L, which have been determined in a previous record.

Notice that if the counts are taken at fixed time intervals, Equation 9.4 can be used to estimate the number of outbursts exceeding any given magnitude in any chosen time period. This is, of course, provided that the conditions of independence, absence of trend, and stationarity are satisfied. ('Stationarity' here, or the requirement that the time series' statistical properties are the same, regardless of the starting point). In our case, this would happen only if the general environment has not experienced or is not about to undergo a dramatic change that has affected or will affect the count distribution. If the counts are taken at unequal time intervals, the same reasoning will still apply but only in respect to the probability that an nth count will equal or exceed the chosen N_c. Phrased differently, as long as the general circumstances remain unchanged, Equation 9.4 can be used to estimate the number of outbursts in any future sequence of m entries, regardless of whether they are equally or unequally spaced.

Testing the Counts' Independence

For the described method to be applicable, the magnitude of each count needs to be unaffected by its predecessor and should have no influence on its successor. If a given record has a clear trend, it could probably be detected by even a casual visual inspection. When the data set at hand is very noisy, a plot of a moving average would almost certainly reveal the trend.

A simple visual method to assess the randomness of a time series is the Pachard–Taken plot. It depicts the relationship between the original record and the *same record* shifted by one ('lag one'), for example, or any other chosen lag. Thus, if the original record was 3, 5, 6, 2, 8, 10, 7, the points (x, y) in the Pachard–Taken plot, for lag one in that case, will be (3, 5), (5, 6), (6, 2), (2, 8), (8, 10), (10, 7). For lag two, the (x, y) points will be (3, 6), (5, 2), (6, 8), (2, 10), and (8, 7). For lag three, the points will be (3, 2), (5, 8), (6, 10), (2, 7), and so forth. As demonstrated in Figure 9.2, related data produce a smooth Pachard–Taken plot. It is perhaps more interesting that, whenever the original data series has an underlying periodicity or periodicities, the plot will exhibit geometric (aesthetically appealing features) as also shown in the figure. The same will happen with other lags, except that the resulting pattern will be different.

Similarly, a three-dimensional plot of N_n, N_{n+k}, and N_{n+m} will also produce a geometric pattern in space. Try it yourself and see: If the Pachard–Taken plot is of data produced by deterministic chaos, the result can be an "attractor" with fractal features. However, to visualize such an attractor usually requires a large number of points unlikely to be found in a microbial record. When the data are truly random, the Pachard–Taken plot will also appear random and the points will be scattered within the whole region specified by the entry span as shown in Figure 9.2 and Figure 9.3. (The concept of randomness is a difficult one and a topic of much

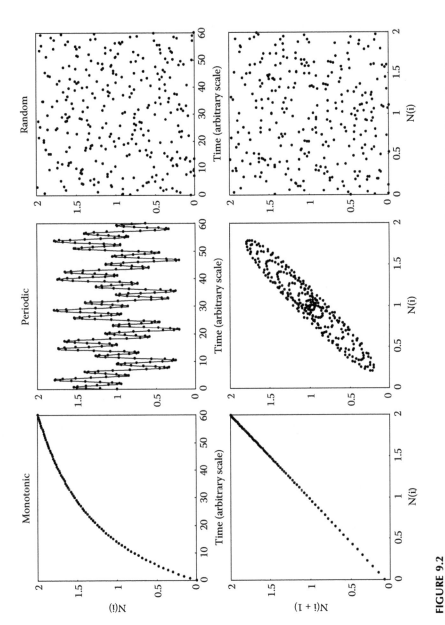

FIGURE 9.2
Simulated examples of different kinds of time series and their corresponding Pachard–Taken plots. Notice the difference between periodicity and randomness. (Plotted by Dr. Maria G. Corradini.)

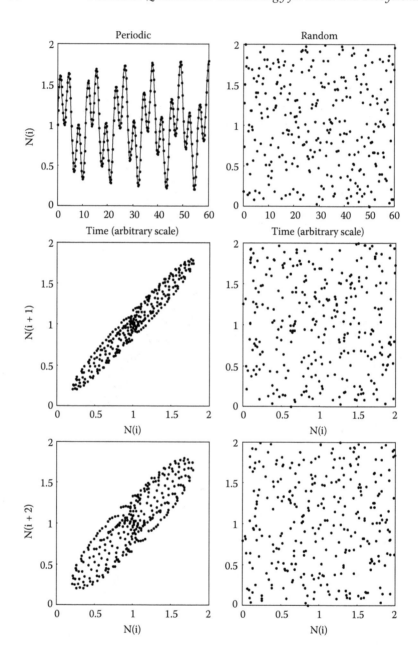

FIGURE 9.3
Simulated examples of periodic and random time series and their corresponding Pachard–Taken plots with two lags. (Plotted by Dr. Maria G. Corradini.)

debate. For what follows, 'random' will refer to passing a statistical test for randomness or to having a pattern in which order could not be found. Admittedly, these are functional definitions but they will suffice for our purpose.)

A more formal test for the existence of a periodicity or periodicities in a given time series is the examination of its autocorrelation function (ACF). It is constructed by successive calculation of the linear regression correlation coefficient, r, of the series after it is "shifted" by successive lags. Thus, the first entry into the ACF is the r for lag one, the second is for lag two, the third is for lag three, etc. (For the lag to be meaningful, it should not exceed about one quarter of the examined series length. Consequently, if the record has 100 entries, the lags should be between 1 and no more than 25).

A series with a clear trend will be highly correlated. For example, the series 1, 2, 3, 4, 5,… will have a correlation coefficient 1.00 for all lags, but a series like 100, 98, 96, 94, 92,… will have –1.00 for all lags. As in the standard linear regression, a positive correlation coefficient, $1 \geq r \geq 0$, signifies that the variables increase or decrease *in unison*; a negative coefficient, $0 \geq r \geq -1$, is an indication of an inverse relation (Figure 9.4).

Periodicity in the examined record will be manifested by a high correlation coefficient for a particular lag (or lags). For example, if, in the record of raw milk tested daily, a particular high microbial count is usually encountered in Sunday's milk as a result of keeping it on the farm during the weekend, there will be a particularly high correlation coefficient for lag seven, assuming that the milk is indeed collected separately every day. Strong seasonal variations can also be manifested in the ACF. If the season is an influential factor, significant correlation at the lag that corresponds to 12 months would probably be found. Sometimes, none of the correlation coefficients is significant but the autocorrelation function is always positive or negative or has a wavy appearance (see below). The former may indicate a mild trend and the latter long-range mild periodicity. (For more on the interpretation of structured ACFs, consult the statistics literature.)

A truly random time series will have no significant correlation coefficient for any lag and, if finite, its ACF will appear random too (Figure 9.4), that is, the probabilities of a positive and negative correlation coefficient, r, will be the same and very small. (A similar, more familiar situation is the distribution of the residuals around a fitted line or curve.) The calculation and plotting of the autocorrelation function is a standard command of statistical packages and can be fairly easily programmed for mathematical and even general-purpose software. Once the ACF is plotted, it can reveal very clearly whether the examined record is *not* random. As will be shown later, the details of the ACF plot can indicate that the

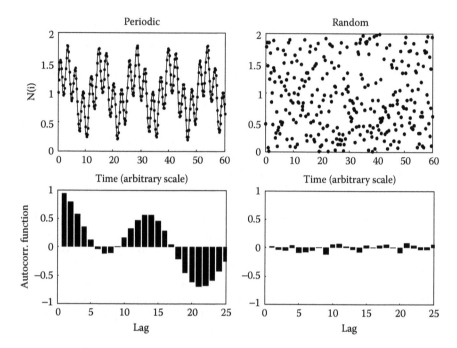

FIGURE 9.4

Simulated time series and their corresponding autocorrelation functions. (Plotted by Dr. Maria G. Corradini.)

sequence might be considered as practically random. Again, the term 'practically random' is rather loose and highly subjective, and its use requires explanation. Clarification of the issue will be provided with the analysis of actual microbial records in Chapter 10 and Chapter 11.

Uneven Rounding and Record Derounding

Microbial counts are frequently rounded. As long as this is done consistently, to the nearest ten, for example, even if the 10 refer to 10s of 1000s, there is no problem. A difficulty arises, for example, when counts under 150 in an industrial plant are rounded to the nearest 10, between 150 to 400 to the nearest 50, and above 400 to the nearest 100 — or (and this is not a hypothetical scenario) the recorded counts are haphazardly rounded. An example would be counts rounded differently, or not at all, by individuals who enter them at different shifts.

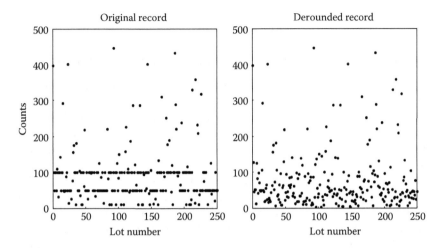

FIGURE 9.5
Simulated "original" and derounded records of *E. coli* in the wash water of an industrial plant. Notice the effect of uneven rounding in the plot at the left. (Courtesy of Dr. Maria G. Corradini.)

Another common situation is that all counts below a certain level, 10 or 20, for example, are reported as such or as 0. In all such cases, examination of the autocorrelation function will reveal that the record is not random and thus that the condition of the count's independence is strongly violated. The situation can be corrected by *derounding* the record. A convenient way to deround a record is to replace its entries by random values generated for the appropriate range. A simulated raw record of *E. coli* counts in the rinse water of a poultry plant is shown in Figure 9.5. The rows of apparently equal counts, especially in the range of 150 to 600, are quite noticeable. Such a record does not pass the test for randomness using the ACF (see "Estimating the Frequency of Future Outbursts — the Principle") and thus it is of very limited value for analysis in it original form. However, subjecting the entries to a *derounding algorithm* "restores" the counts' randomness, thus allowing the record to reveal its statistical properties.

An example of such an algorithm similar to the one reported by Corradini et al. (2001) is:

$$\text{If } 10 < M_n \leq 100, N_n = M_n - 5 + 10\, R_n$$

$$\text{If } 100 < M_n \leq 200, N_n = M_n - 50 + 100\, R_n \qquad (9.5)$$

$$\text{If } 200 < M_n, N_n = M_n - 100 + 200\, R_n$$

where

M_n is the originally reported nth count, i.e., rounded or not

N_n is the corresponding derounded value

R_n is a random number generated with a uniform distribution, $0 \leq R_n \leq 1$

The choice of this particular random number generator eliminates the need to assume the distribution of the true values around the rounded entries. This algorithm resets the values of the counts between 10 and 100, which have been frequently rounded to the nearest 10, to random values in the range of ±5 around the originally entered values. Counts between 100 and 200 are rounded to the nearest 50 and replaced by random values within the range of ±50 around the original entries, and counts above 200 rounded to the nearest 100 are replaced by a random number in the range of ±100 around the original value. (Counts that have not been originally rounded are also replaced, but by a random number of a similar magnitude.) The appearance of the record after the application of the algorithm is also shown in Figure 9.5. All the old ordered rows of entries have disappeared because of the transformation, and large segments of the record could now pass the test of independence (see below).

Another, less striking, example is shown in Figure 9.6. In this case, the derounding was done in order to complement the record for missing values. Thus, entries like $N_n \leq 10$ can be replaced by a generated value produced by the simple algorithm:

$$\text{If } M_n \leq 10,\ N_n = 10\,R_n \tag{9.6}$$

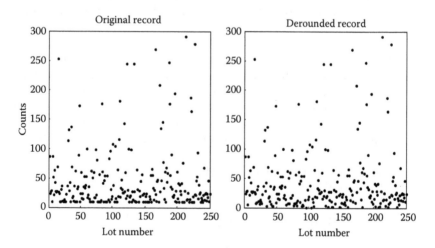

FIGURE 9.6

An example of complementing missing values in a microbial counts record through reporting an upper limit ($N_n \leq 10$ in this case) using Equation 9.6 as an algorithm. Notice that this slight modification of the record may allow it to pass the test of independence. (Courtesy of Dr. Maria G. Corradini.)

This algorithm application has a noticeable effect exclusively on the lowest counts (Figure 9.6). Yet, such a modification can sometimes be sufficient to transform a record that does not pass the test for randomness to one that does making it suitable for estimating the frequencies of future high counts.

Choosing a Distribution Function

The magnitudes of the counts in any given record have a *distribution*. Once determined, this distribution can be used to estimate the frequencies of future counts exceeding any chosen value, if it can be safely assumed that:

- The distribution faithfully describes the counts.
- In the future, or more conservatively in the near future, the conditions that regulate the monitored microorganism's growth and decline pattern will not change dramatically so that the count distribution *will remain practically the same*.

Thus, if one has a long enough record whose counts pass the test of independence, the next stage is to come up with a distribution that describes their fluctuating pattern adequately. To do that, a choice must be made between the use of a *nonparametric* distribution and a *parametric* distribution as the counts' fluctuation model.

Nonparametric Distributions

A nonparametric distribution is constructed by tabulating the existing relative frequencies of entries within specified ranges, as in a histogram. No rule is required and if the general environment remains practically unchanged, as assumed, the relative frequencies of future counts or their probabilities will be the same as that of the counts in the record at hand. This is perhaps the *simplest* way to represent the counts' distribution and estimate the probability of their future occurrence. (A recommended but not generally accepted rule of constructing a histogram is that if the total number of entries in the record is n, then the number of divisions, or "bins," should be Int[$\log_2(n)$] + 1, i.e., the nearest integer to the base-2 logarithm of n, to which one is added.)

The disadvantage of a nonparametric distribution as a model for estimating the future frequencies of counts exceeding any given magnitude is that the probability of events not already observed will be always zero

(Peleg and Horowitz, 2000). This can be a serious limitation if we try to estimate the probabilities of truly rare events from relatively short count series. Records of such exceptional events might not appear in the available data, but their potential occurrence can be of prime importance as far as safety or public health is concerned.

Parametric Distributions

Parametric distributions are continuous distribution functions characterized by a mathematical formula and coefficients, or parameters. The log normal and Weibull distribution functions are two familiar examples but others (see below) might be more suitable in some cases. Because the log normal and Weibull distributions as well as most of their alternatives have a range from zero to infinity, they can be used to calculate the probability of a count exceeding any positive magnitude, regardless of whether it has already been observed or not (Peleg and Horowitz, 2000). Whether such a calculated probability (or corresponding frequency) will be a realistic estimate is another question. However, as has already been stated and will be demonstrated with actual data, the question can be answered by comparing the model's estimates with actual observations.

Calculation of a Distribution's Parameters

Usually there is more than one function that can be used to characterize any given record, especially if the counts' distribution is unimodal. ('Unimodel' is the term assigned to frequency distributions that have a single peak, regardless of whether they are symmetric or skewed.) Apart from the normal and log normal distributions already mentioned, the Laplace (or log Laplace), extreme value, and the beta (or log beta) distribution functions are natural candidates for being a count's fluctuation model. Once a parametric distribution function is considered as a model, its coefficients or parameters can be calculated in three principal ways.

Regression

A set of experimental data can be arranged in the form of a histogram and fitted with the contemplated frequency, or 'density' distribution function, using standard nonlinear regression to calculate its parameters. Recall, though, that standard regression is based on minimizing the mean square error. Thus, when applied to data whose histogram has a bell shape, much weight is given to the central part of the distribution — that is, to the region around the mode (the peak's location) at the expense of the tail's region in which we are really interested. Consequently, *this method is not recommended* for our purpose and should be avoided.

The Method of Moments (MM)

When this method is used, the considered distribution's parameters are assigned in such a way that they would produce the mean (or mode) and variance that the actual record at hand has. (The mean and variance are a distribution's first and second moments, respectively, and thus the method's name.) The calculation is straightforward. In the case of the normal or log normal distribution, the mean (μ) or logarithmic mean (μ_L) and the variance (σ^2) or logarithmic variance (σ^2_L) appear as such in the distribution's formula. Therefore, once they are calculated from the record at hand, their values can be inserted directly into the model distribution's equation.

In other distribution functions, the parameters' calculation requires the use of conversion formulas. Consider the Laplace (or "double exponential") distribution:

$$f(x) = \frac{1}{2b} \exp\left[-\frac{x - \mu}{b}\right]$$

(9.7)

or log Laplace distribution:

$$f(x_L) = \frac{1}{2b_L} \exp\left[-\frac{x_L - \mu_L}{b_L}\right]$$

(9.8)

Here, only the mean, μ, or logarithmic mean, μ_L, appears in the model's equation. Because $b = 2s^2$ or $b_L = 2s_L^2$, where s or s_L is the *record's* linear or logarithmic standard deviation, b or b_L can be easily calculated.

In distribution functions, like the Weibull and beta (or log beta), for example, neither the mean nor the standard deviation, or variance, appears as such in the equation. The parameters of these distribution functions, however, can be estimated by solving simultaneously the equations that relate them to the distribution's mean and standard deviation. The assumption, recall, is that the record's mean and standard deviation are approximately that of the underlying distribution, i.e., that $\mu = \bar{x}$ and $\sigma = s$.

Thus, for a Weibull distribution of a variable x written in the form:

$$f(x) = \frac{a}{b}\left(\frac{x}{a}\right)^{a-1} \exp\left[-\frac{x^a}{b}\right]$$

(9.9)

The distribution's mean, μ, and variance, σ^2, are given respectively by:

$$\mu = b\Gamma\left(1+\frac{1}{a}\right)$$
(9.10)

and

$$\sigma^2 = \frac{b^2}{\Gamma\left(1+\frac{2}{a}\right)-\Gamma\left[\left(1+\frac{1}{a}\right)^2\right]}$$
(9.11)

where Γ is the gamma function. Once a given record's mean and variance have been calculated, the corresponding a and b values can be extracted by solving Equation 9.10 and Equation 9.11 numerically. Similarly, for a beta distribution defined as:

$$f(x) = \frac{\Gamma(p+q)}{\Gamma(p)\Gamma(q)}x^{p-1}(1-x)^{q-1}$$
(9.12)

where p and q are constants and Γ again is the gamma function. (Equation 9.12 requires that the independent variable will have the range from zero to one. It is easy to see that if x is defined by $x = N/N_{max}$ or $x = \log(N/N_{min})/\log(N_{max}/N_{min})$, the beta function can be used for counts in the finite range between zero and N_{max} or N_{min} and N_{max}, respectively.)

The distribution's mean and variance are given by:

$$\mu = \frac{p}{p+q}$$
(9.13)

and

$$\sigma^2 = \frac{pq}{(p+q)^2(p+q+1)}$$
(9.14)

Again, the estimates of p and q can be calculated by solving the two equations (Equation 9.13 and Equation 9.14) simultaneously.

Although only the count or its logarithm has been mentioned, all the preceding is just as applicable to any other transform of the record's entries, such as $x = (N_{n+1} - N_n)/N_n$, $x = \sqrt{N_n}$, $x = \sqrt[3]{N}$, etc.

Maximum Likelihood Estimation (MLE)

When using the MLE method, we seek the contemplated distribution's parameters that would have produced the record at hand with the highest probability. The MLE is considered stronger than the method of moments (MM) and the estimates that it produces are generally more reliable, especially for long records. However, as will be shown in the next chapter, this need not always be the case in the analysis of actual microbial records and the predictions that the MLE method produces may not always be superior to those calculated using the method of moments.

In the case of the normal and log normal distribution, the MLE and MM parameters estimates are the same by definition — that is, $\mu = \bar{x}$ or $\mu_L = \bar{x}_L$ and $\sigma = s$ or $\sigma_L = s_L$ in both methods.

In certain distributions, the calculation of the MLE parameters can be straightforward. A case in point is the Laplace or log Laplace distribution (Equation 9.7 or Equation 9.8, respectively). Here, the MLE estimate of μ or μ_L is the record's *median*, $\tilde{x}$, and the MLE estimate of the spread parameter b or b_L is:

$$b = \frac{1}{n} \sum_{i=1}^{n} |x_{Li} - \tilde{x}| \qquad (9.15)$$

or

$$b_L = \frac{1}{n} \sum_{i=1}^{n} |x_{Li} - \tilde{x}_L| \qquad (9.16)$$

respectively, where n is the number of entries in the series.

In a distribution function like the Weibull, the calculation of the MLE parameters requires a numerical procedure with close initial guesses to start the iterations. When the initial values are not close enough, convergence might not be accomplished and there will be no solution. In such a case, it would be advisable to calculate the MM parameters first — their definition guarantees that there will *always* be a solution — and use the results as initial values or guesses in the calculation of the MLE parameters to assure convergence.

The Q–Q Plot

In most instances in which the histogram of the record at hand indicates an asymmetric count distribution, identification of the parametric distribution function or functions that can describe it is not a trivial task. A helpful tool in screening candidate distribution functions is the quantile–quantile plot, also known as the Q–Q or q–q plot.

In the case of the normal distribution, the Q–Q plot is akin to the data plot on a probability paper. If the data fall on a straight line, this will be an indication that they are normally distributed. In other words, the plot's linearity becomes a test of the distribution's normality. (In the probability paper plot, the line's *slope* can be used to determine the distribution's standard deviation, which is not the case when the Q–Q plot is used.) Basically, the Q–Q plot is a graphical tool to determine whether two data sets come from populations that have the same statistical distribution. However, it can also be used to test the hypothesis that a data set comes from a population that has a given parametric distribution.

If the plot of the expected values according to the contemplated distribution function vs. the actual ones is a straight line with a slope of 45°, one can assume that the data indeed come from a population with this particular distribution. A gross deviation from linearity is a clear indication that the data do not come from such a population. Thus, the test is very effective in the *exclusion* of candidate distribution functions. When testing actual microbial data, the Q–Q plot is rarely a perfect straight line with a 45° slope (see below), although it can be quite close. Determining what constitutes an acceptable agreement between data and model can be a difficult issue. Its resolution may require a more thorough statistical analysis than linear regression and additional tests to confirm that the chosen distribution function is indeed an appropriate one.

Still, the Q–Q plot is a very convenient instrument to *screen* candidate parametric distribution functions. In several commercial statistical packages, the construction of a Q–Q plot is a standard option. The list of distribution functions almost always includes the normal, log normal, and Weibull distributions; it grows as the program is expanded to include additional, less common distribution functions. One can also program a Q–Q plot for a contemplated distribution function if it cannot be found in a handy statistical package. Because mathematical software of the current generation usually does not include *truncated distribution functions* on its menu, the Q–Q plots of these must be programmed by the user whenever the record requires their use (see following).

The Q–Q plot can be used to examine the utility of data *transformations*. A familiar example is that the log normality of a given data set's distribution can be tested by using a normal probability plot in which the entries are not the original values, but rather their logarithmic transforms. Similarly, Q–Q plots can be constructed by using the data not only in their original form, but also after their conversion, whether it is logarithmic, square root or cubic root, etc. Series of tests of this kind with different transformations can reveal the existence of underlying distributions like the log beta or log Laplace. Such distributions are only rarely mentioned in basic statistics textbooks (in contrast with the log normal distribution) and they are not listed as standard distribution functions in all mathematical software. Because no list of distribution functions can cover all possible transformations, the responsibility for the entries' forms (the original values or after being converted) is on the program's user.

Truncated Distributions

When dealing with microbial counts, the physical upper limit usually does not pose a conceptual problem. Obviously, a real microbial population cannot grow indefinitely. Its biomass cannot exceed the total mass of the habitat, for example. However, it is very unlikely that the size of a real microbial population will ever reach a level even near the physical limit, except when resources are really scarce. This is because of biological constraints set by competition for space, the release of toxic metabolites, and the like. Yet, microbial counts, especially in a rich habitat like ground meat, raw milk, or the sludge of water treatment plants, can reach such huge numbers that describing them by distribution functions with a range extending to infinity would be quite appropriate.

The situation is different at the other end, at least in some instances. *Enterococci* in water or *Salmonella* and other pathogens in foods, for example, can be truly absent or present at such low levels that they will not be detected. Consequently, the count records of such organisms in food or water may include several or many zero entries. This means that the parametric distribution one chooses to describe the distribution of such counts must allow for a finite nonzero probability of zero counts. Another familiar situation occurs when there is a practical threshold of detection, for example, 10; in this case, several or many entries in the count record will be in the form of <10. Thus, any distribution function that describes such a record will need to account for a finite nonzero probability of 10 or less. (An alternative is to substitute every such entry by a random number generated in the range of 0 to 10. The modified record so created can then be described by a standard parametric distribution as explained in "Uneven Rounding and Record Derounding" above.)

Truncation at zero of a known distribution function of a random variable Z with a range from $-\infty$ to $+\infty$, $f(Z)$, such as the normal, Laplace, or extreme value, can be done as follows. In a program like Mathematica®, these functions are defined in their frequency (density) and cumulative forms. Because the area under the distribution's frequency (density) curve (that of truncated distributions included) must be one by definition, we can write:

$$\text{For } Z < 0, \quad f_{\text{truncated}}(Z) = 0$$

and $\hspace{11cm}$ (9.17)

$$\text{For } Z \geq 0, \quad f_{\text{truncated}}(Z) = \frac{f(Z)}{\displaystyle\int_0^\infty f(Z)dZ}$$

where Z is the independent random variable — in our case, the original count, N, or any of its contemplated transforms, $Z(N) = N^{1/2}$ or $Z(N) = N^{1/3}$, for example. (Remember that a logarithmic transformation will only apply when the threshold chosen for the truncation is above zero.)

According to Equation 9.17, the probability of a count exceeding any chosen value, N_c, is:

$$P(N \geq N_c) = \frac{\displaystyle\int_{Z(N_c)}^{\infty} f[Z(N)]dZ(N)}{\displaystyle\int_{0}^{\infty} f[Z(N)]dZ(N)} \tag{9.18}$$

Recall that the numerator and denominator of the right side of the equation can be calculated directly with the cumulative form of the original distribution function before its truncation. Consequently, the calculation of $P(N \geq N_c)$ can be done with a program based on functions that are standard in modern mathematical/statistical software.

Similarly, truncation at any other threshold value, $Z_{threshold}$ (like the mentioned ten), will result in the expressions:

$$\text{For } Z < Z_{threshold}, \quad f_{truncated}(Z) = 0$$

and $\tag{9.19}$

$$\text{For } Z \geq Z_{threshold}, \quad f_{truncated}(Z) = \frac{f(Z)}{\displaystyle\int_{Z_{threshold}}^{\infty} f(Z)dZ}$$

In this case, the probability of exceeding any value N_c will be:

$$P(N \geq N_c) = \frac{\displaystyle\int_{Z(N_c)}^{\infty} f[Z(N)]dZ(N)}{\displaystyle\int_{Z_{threshold}}^{\infty} f[Z(N)]dZ(N)} \tag{9.20}$$

Examples of how the outlined truncation method can be implemented in the analysis of microbial counts in water will be shown in Chapter 10.

Extinction and Absence

The phenomenon of species extinction is the hallmark of the evolution of life on our planet. It is claimed that the myriads of species that now exist constitute only about 1/1000 of the number of species that have ever lived (Raup, 1991). No wonder, therefore, that theories and models of extinction have been in active pursuit in quantitative biology for decades (Elliot, 1986). The accelerated extinction of species and of populations within certain geographical regions that is happening today is probably the direct or indirect result of human activity. It occurs primarily by hunting or other forms of direct destruction or the elimination of natural habitats. The intensified interest in the extinction phenomenon has enlarged our knowledge of the conditions that cause the disappearance of living populations.

All this primarily refers to animals and, to a somewhat lesser extent, to plants, but rarely to microorganisms. Without even entering the question of whether the traditional concept of a species really applies to microorganisms, the whole issue of microbial extinction may require a different perspective. In the context of this chapter, let us define extinction or apparent extinction as the spontaneous and sudden disappearance or apparent disappearance of an organism without any traceable cause, or a previous indication that this is about to happen. Obviously, the disappearance of a pathogen is always good news, regardless of the cause, and thus interest in this topic is very limited. Yet, the spontaneous disappearance of a desirable microorganism or even a significant reduction in its population may have practical implications in biotechnology and medicine, so the topic is worth at least a brief discussion.

Formally, the same models that have been used to estimate the probability of a high count exceeding a given magnitude can also be used to estimate the probability of a count below a given level. The only difference is that, if a microbial population needs a critical mass to be viable or a given size to be detected, then once it falls to a level below the critical mass or detection threshold, the count will be zero. The two situations are not identical, of course. In the first case, recovery will be impossible and the organism must be reintroduced in order to be present again; in the second case, it can reappear "spontaneously" after its apparent extinction. Mathematically, the probability that a count will be *below* a certain level can be estimated by the same methods used to estimate the probability of recording counts above a given magnitude. If the fluctuating count's distribution is log normal or log Laplace, for example, and the organism has an extinction threshold of N_E, then, according to our model, the probability of a count reaching a level equal to or below this N_E is:

$$P(N \leq N_E) = \int_0^{N_E} f(N)dN = 1 - \int_{N_E}^{\infty} (N)dN \qquad (9.21)$$

where $f(N)$ is the frequency (density) form of the counts distribution.

We can also use the same distribution functions as models to simulate extinction scenarios. For example, when the fluctuations pattern of a given organism is governed by the log normal distribution, then extinction is introduced into the model by adding the following conditions (Peleg et al., 1997):

$$N_n = 10^{\mu_L + \sigma_L Z(n)}$$

$$\text{If } (N_{n-1} \leq N_E), N(n > n - 1) = 0 \qquad (9.22)$$

This says that, as long as a count is above the extinction level, its successor will be a random number produced by $10^{\mu_L + \sigma_L Z(n)}$, where μ_L and σ_L are the distribution's logarithmic (base 10) mean and standard deviation, respectively, and $z(n)$ a random number with a standard normal probability, i.e., with $\mu = 0$ and $\sigma = 1$. Once a single entry happens to be equal to or smaller than N_E, its immediate and all following successors will be zeros — that is, the population has become extinct.

Examples of simulated scenarios produced by the preceding model are shown in Figure 9.7. They appear to be more realistic than those generated with the traditional "gambler's ruin" model of extinction shown in Figure 9.8, which is based on a random walk algorithm. This is because the new fluctuation model (Equation 9.22) can generate much larger and "spikier" oscillations that do not have the arches and temporary trends that the "gambler's ruin" model generates. (An interested reader can generate his or her gambler's ruin history with freeware posted on the Web. Examples of such sites are http://www.math.ucsd.edu/~anistat/gamblers_ruin.html and http://www.math.utah.edu/~carlson/teaching/java/prob/brownianmotion/4/.)

Special Patterns

Populations with a Detection Threshold Level

The scenario that only numbers below a detection level, N_D, will be recorded as a zero count can be simulated by the similar model:

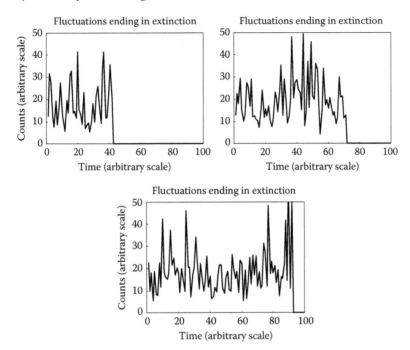

FIGURE 9.7
Simulated life histories that have ended in extinction using Equation 9.22 as a model. Notice the absence of temporary trends of growth and decline and the abrupt extinction. (Courtesy of M.D. Normand.)

$$M(n) = 10^{\mu_L + \sigma_L Z(n)} \qquad (9.23)$$

$$\text{If } M(n) \geq N_D,\ N(n) = M(n)$$

$$M(n) < N_D,\ N(n) = 0$$

The difference here is that the entries that fall below the detection threshold are set to zero but not their successors, unless they too happen to be below the detection level. Examples of records produced by the second model (Equation 9.23) are shown in Figure 9.9. When the number of zero counts is expected to be substantial (20 to 40% of the record, for example), the use of a truncated random distribution function of the kind previously discussed in "Truncated Distributions" can be an alternative fluctuation model.

If only the presence of the organism is recorded but not its population size, then the simplified model discussed in "Records of Positive/Negative Entries" can be used instead.

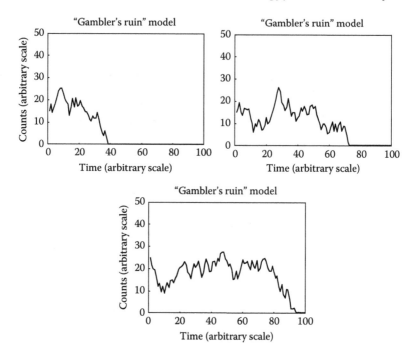

FIGURE 9.8

Simulated life histories that have ended in extinction using the "gambler's ruin" algorithm. Notice the temporary trends (and fractal characteristics) that are absent in the patterns produced by the fully probabilistic model (Figure 9.7). (Courtesy of M.D. Normand.)

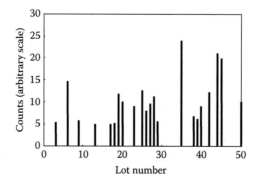

FIGURE 9.9

Simulated industrial microbial records of an organism with a detection level generated with Equation 9.23 as a model. (After Horowitz, J. et al., 1999, *Crit. Rev. Food Sci. Nutr.*, 39, 503–517. With permission, CRC Press.)

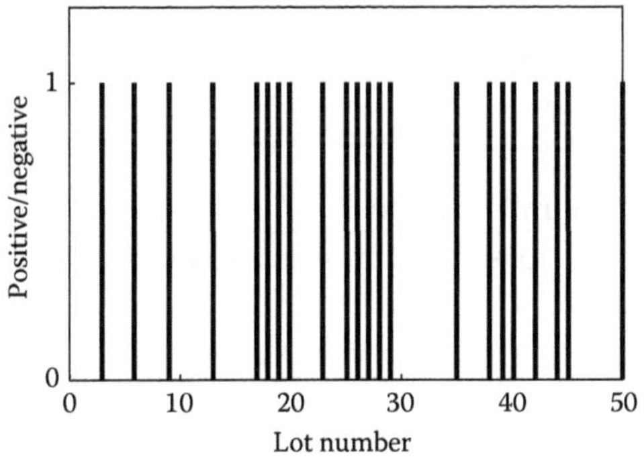

FIGURE 9.10
A simulated microbial record in which the entries are positive (1) and negative (0) using Equation 2.24 as a model. (After Horowitz, J. et al., 1999, *Crit. Rev. Food Sci. Nutr.*, 39, 503–517. With permission, CRC Press.)

Records of Positive/Negative Entries

Perhaps the simplest kind of irregular record is a random sequence of positive/negative entries that refer to the presence or absence of the organism in question. Absence might also be an indication of failure to detect the organism; in this case, we can assign a threshold level, $N_D > 0$, to the detection. If we assign one to the positive entries and zero to the negative entries, the sequence would be defined by:

$$M(n) = 10^{\mu_L + \sigma_L Z(n)}$$

$$\text{If } M(n) \geq N_D, \ N(n) = 1$$

$$\text{and if } M(n) < N_D, \ N(n) = 0 \qquad (9.24)$$

where
  $M(n)$ is the "true" number, which may or may not be known
  $N(n)$ is the corresponding entry to the record, which is zero or one
  N_D is the detection level ($N_D \geq 0$)

A series of this kind is shown in Figure 9.10. If such a series had been an experimental record, it would not tell us the distribution of the microbial loads in the examined food or water. Because the threshold level might be unknown, such data cannot be used to construct a parametric model of the counts' distribution. All that one can do in such a case is to assume that if circumstances are not about to change dramatically, the future *frequency* of positive and negative entries will remain about the same or the estimated probability of a positive entry, $P[N(n) > N_0]$ will be (Horowitz et al., 1999):

$$P(1) \approx \frac{\text{number of "positive" entries in the record}}{\text{total number of entries in the record}} \qquad (9.25)$$

and that of a negative entry, $P(0) = P[N(n) \leq N_D]$, will be:

$$P(0) = 1 - P(1) \approx \frac{\text{number of "negative" entries in the record}}{\text{total number of entries in the record}} \qquad (9.26)$$

The randomness of the distribution can be tested by examining the distribution of the *intervals* between positive (or negative) entries. This would be advisable whenever the existence of an underlying periodicity might be suspected.

Records with a True or Suspected Trend or Periodicity

Records with a clear trend or long-term periodicity or periodicities, especially seasonal, can be identified by superficial examination of their plot or by statistical methods. These may include regression, examination of the autocorrelation function (ACF), and inspection of the Pachard–Taken plot. In many cases, as will be shown later, the trend or amplitude of the long-term oscillations can be so small relative to the amplitude of the counts' random fluctuations that they can be considered negligible. However, totally ignoring long-range trends and periodicities would be justified only if the purpose of the analysis is to obtain rough estimates of the probabilities of major outbursts.

In other cases, the trend or long-term periodicity might or might not be taken into account, at the user's discretion, as long as the counts pass (or almost pass) the statistical test of independence. When a trend is too noticeable to be ignored, one way to predict the frequency of future outbursts is to "normalize" the existing record and to estimate the probability that a count will exceed its predecessor by a *certain ratio* rather than the probability of a count of a certain absolute magnitude.

Consider a count series, $N_1, N_2, N_3,\ldots$ that can be transformed into a sequence of ratios (Corradini et al., 2002):

$$Y_n = \frac{N_n}{N_{n-1}} \qquad (9.27a)$$

that is,

$$Y_1, Y_2, Y_3 \ldots = \frac{N_1}{N_0}, \frac{N_2}{N_1}, \frac{N_3}{N_2} \ldots \qquad (9.27b)$$

We can now examine the randomness and distribution of the Y_n series and estimate the probability that any Y_n will exceed the previous entry Y_{n-1} by any chosen factor — 2, 3, 10, or 100, for instance. This can be done by the same procedures already discussed in this chapter, except that the entries would be the ratios Y_n and not the original counts N_n. Such a method can be useful if and only if the original count record does not have zero entries or very small values followed by large ones, as in the case of raw milk, for example (see Chapter 10). In the first case, the method could not work at all (the prohibited division by zero); in the second, the resulting "spikes" can be so high that the analysis might become meaningless. The same applies to alternative transformations such as the one in which Y_n is defined by:

$$Y_n = \frac{N_n - N_{n-1}}{N_{n-1}} \qquad (9.28)$$

When Y_n is defined in this way, the sought probability will be of an event in which the *relative difference* will exceed a chosen value — 50, 100, or 500%, for example. (Notice that when the normalized transform, Y_n, is defined by Equation 9.28, the series will have negative entries). Here, too, the probability estimates of a future outburst, expressed in terms of a relative difference between successive counts, might be useful only for clearly trendy records in which the counts are relatively high and fluctuations relatively small.

Examples of how the methods described in this chapter can be used to analyze and interpret actual records of microbial counts and of mortality caused by endemic diseases will be given in Chapter 10 and Chapter 11.

10

Estimating Frequencies of Future Microbial High Counts or Outbursts in Foods and Water — Case Studies

It is very difficult to predict, especially the future!

Niels Bohr

The concept that the irregular fluctuations in microbial count records can be translated into probabilities of future high (or low) counts was described in the previous chapter. What follows in this chapter is a set of demonstrations of how the principle and resulting mathematical procedures can be implemented in the analysis of actual industrial microbial records in raw and processed foods and in water of a famous historic lake. The demonstrations will also include a comparison between the predicted frequencies of high counts, calculated with various models, and the frequencies actually observed in fresh data.

The organisms monitored were quite diverse and so were the magnitudes of their counts and the time intervals between them. What is common to all the analyzed records is that they were collected as part of a standard industrial quality control or a routine monitoring operation by a statal water authority. None of the records was originally intended to serve as a database for a statistical study, let alone for the development and testing of a mathematical population model. Thus, in contrast with a well-designed scientific experiment, the counts examined were only rarely determined at fixed time intervals. Although most contain segments of equally spaced counts, the series as a whole does not satisfy this condition. Also, the records available for the analysis were not accompanied by complementary information that could have explained the counts' fluctuations, even partly.

Data, such as the food products' temperature, history, and composition or the weather conditions around the monitored lake, although probably recorded and kept somewhere, were not used in the counts' series analysis. Consequently, the only information that could be extracted from records was that contained by the counts themselves. However, the records' imperfections and the absence of complementary information to explain them, although obviously a setback when it comes to the records' interpretation, had a bright side. It is very unlikely that monitoring the microbial quality of manufactured foods under industrial conditions or of a lake's water as a routine task assigned to a public health or water authority will ever conform to what one would call an appropriate experimental design — the bedrock of experimental research. Therefore, a demonstration that it is possible to extract useful information even from grossly imperfect time series may give new life to past microbial records deemed worthless because of their apparent lack of pattern.

The demonstrations may also save future records from being discarded before analysis. Microbial testing is costly and time consuming. Long count records are therefore records of a considerable expenditure. Thus, it will be frugal to at least attempt to obtain any possible useful information from a record once the investment in its data collection has been made.

Microbial Counts in a Cheese-Based Snack

Analysis of Raw Records

Two examples of microbial industrial record sets are shown in Figure 10.1. They are of the total count (standard plate count, or SPC), anaerobes, and thermophiles found in two different cheese-based snacks, marked A and B, produced over a period of approximately 2 years in a commercial food plant in the U.S. (Peleg et al., 2000). The products were not manufactured continuously, but rather on different dates dictated by logistic considerations at the plant. Consequently, the counts are plotted in their successive order. The actual time gap between successive entries could be from 1 day to about 2 weeks, depending on the particular production schedule. All the records exhibit an irregular pattern of count fluctuations, regardless of the specific product and the type of organism monitored. The corresponding autocorrelation functions of the records are shown in Figure 10.2.

Even upon superficial examination of the autocorrelation functions, most indicate that the count series were not truly random. This is evident in that the positive and negative correlation coefficients were not randomly distributed around the horizontal zero line. Still, a *significant correlation coefficient* was mostly observed for the first lag or few lags only; see Table 10.1. Even when the correlation coefficients were significant,

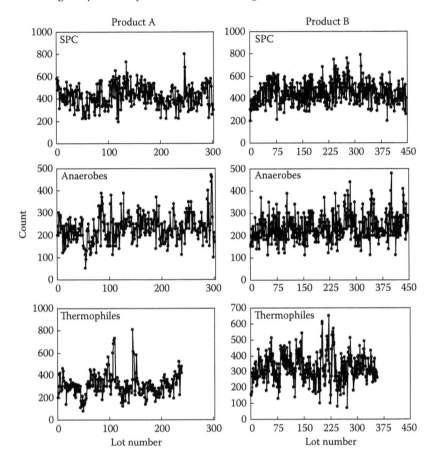

FIGURE 10.1
Industrial microbial records of two cheese snacks; A and B. Note the irregular fluctuation pattern of all three groups of bacteria. (From Peleg, M. et al., 2000, *J. Food Sci.*, 65, 740–747. With permission, courtesy of the Institute of Food Technologists.)

they were still fairly low, mostly on the order of 0.2 to 0.4. This meant that lots or batches produced in succession were more likely to have a similar microbial load than those manufactured several days or weeks apart. Yet, even when the counts of batches produced in succession were correlated, the correlation between them was rather weak. In other words, even when the correlation was found to be statistically significant, it could not be used to estimate the magnitude of counts from that of their predecessors with any acceptable degree of accuracy.

That the listed counts were not determined at fixed intervals only complicates the correlation functions' interpretation and may undermine the notion that the record had a trend or long-range periodicity. As will be

TABLE 10.1

Distribution Parameters of Microbial Counts in Two Snacks (A and B) with Estimated and Observed Numbers of Counts Exceeding Selected Levels

Snack	Organism type	Total cases	Significant autocorrelation for lag(s)[a]	μ_L (First half)	s_L (First half)	N_c	$P(N \geq N_c)$	Number in second half	
								Estimated	Observed
A	SPC	303	1(0.61), 2(0.42), 3(0.32), 4(0.22)	2.61	0.11	500	0.214	32	35
						600	0.0643	10	4
						700	0.00167	2-3	1
						800	0.00393	0-1	1
	Anaerobes	302	1(0.21), 2(0.30), 4(0.18)	2.32	0.13	300	0.119	18	30
						350	0.045	7	8
						400	0.016	2-3	4
						450	0.0057	1	2
	Thermophiles	240	1(0.39), 2(0.24)	2.48	0.16	400	0.217	26	14
						450	0.136	16	9
						500	0.083	10	5
						600	0.030	6	4
						700	0.011	1-2	2
						800	0.004	0-1	1

B		N							
SPC		448	1(0.27), 2(0.15), 3(0.21), 4(0.19), 5(0.25), 7(0.14), 8(0.14), 9(0.13), 10(0.15)	2.63	0.09	500	0.235	53	76
						600	0.059	13	22
						700	0.011	2–3	4
						790	0.0022	0–1	1
Anaerobes		448	1(0.35), 2(0.13), 4(0.15), 5(0.15), 8(0.14), 9(0.16)	2.34	0.10	300	0.087	19	46
						350	0.0206	5	16
						400	0.0043	1	6
						450	0.00082	0–1	1
Thermophiles		356	1(0.58), 2(0.42), 3(0.34), 4(0.28), 5(0.19)	2.49	0.11	400	0.179	32	38
						500	0.038	7	12
						600	0.0067	1–2	3
						650	0.0027	0–1	1

[a] Numbers between parentheses are the corresponding correlation coefficients.

(From Peleg, M. et al., 2000, *J. Food Sci*, 65, 740–747. With permission, courtesy of the Institute of Food Technologists.)

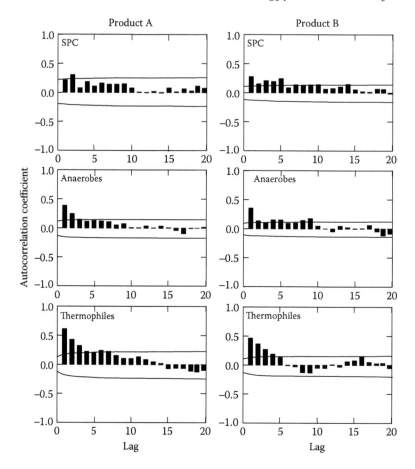

FIGURE 10.2
The autocorrelation functions of the count series of the two cheese snacks, A and B, shown in Figure 10.1. (From Peleg, M. et al., 2000, *J. Food Sci.*, 65, 740–747. With permission, courtesy of the Institute of Food Technologists.)

shown later in this chapter, the impact of the suspected trend, if it had existed, could be reduced by normalizing the data. However, even without such a corrective procedure, the generally low autocorrelation coefficients that were found allowed treating the records as *practically independent and random*, although they were not, as judged by strict statistical criteria (see following). The records' histograms, shown in Figure 10.3, strongly suggested that the counts were log normally distributed. If so, then according to the concept explained in Chapter 9, the probability of encountering a count exceeding a magnitude N_c could be estimated from the formula:

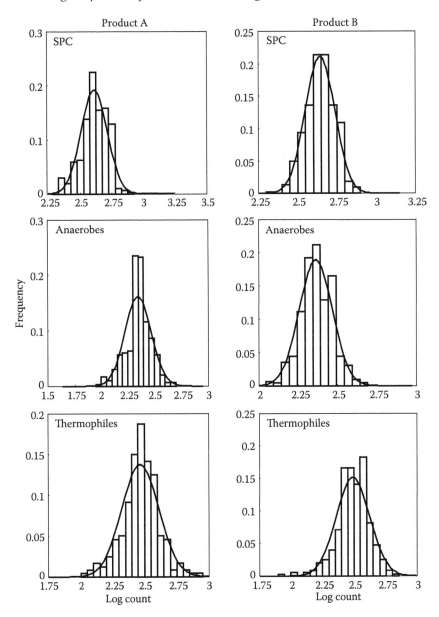

FIGURE 10.3
The histograms of the count records of two cheese snacks, A and B, shown in Figure 10.1. Notice the log normality of the counts' distribution. (From Peleg, M. et al., 2000, *J. Food Sci.*, 65, 740–747. With permission, courtesy of the Institute of Food Technologists.)

$$P(N \geq N_c) = \log_e 10 \frac{1}{\sigma_L \sqrt{2\pi}} \int\limits_{\log_{10} N_c}^{\infty} 10^{-\frac{(\log_{10} N - \mu_L)^2}{2\sigma_L^2}} \frac{dN}{N} \qquad (10.1)$$

where μ_L and σ_L are the logarithmic mean and standard deviation (base 10) respectively.

Once μ_L and σ_L are determined directly from the record, the actual probability of $N \geq N_c$ can be calculated with various general-purpose commercial statistical/mathematical packages, as well as with statistical freeware posted on the Web. (A free program written specifically to calculate the probabilities of microbial counts having a normal or log normal distribution in Microsoft Excel® can be found on the Web site http:// www-unix. oit.umass.edu/~aew2000/microbecount.html; it was developed and posted by Mark D. Normand and the author.)

All the user needs to do is paste the actual counts series that he or she wants to examine in the appropriate spreadsheet column. The program plots the original count record before and after logarithmic transformation and the corresponding two histograms. The program also calculates the linear and logarithmic mean, standard deviation, and coefficient of skewness of the entered record. The probabilities of a set of a user's chosen N_c is automatically calculated, using the normal and log normal distributions as the fluctuations models. The accompanying two Q–Q plots, also generated by the program, provide a statistical test of whether the counts' distribution can be considered normal or log normal.

Whenever the counts' distribution is log normal, the validity of the procedure to estimate the probabilities of high or excessive counts can be tested as follows. The available record is divided into two equal parts. Subsequently, estimates of the number of high counts exceeding chosen values are calculated with the formula (Equation 10.1) using the logarithmic mean ($\overline{N_L}$) and standard deviation (s_L) of each half of the record. These estimates based on the assumption that $\overline{N_L} = \mu_L$ and $s_L = \sigma_L$ (see the method of moments in Chapter 9) are then compared with those actually observed in the other half. Recall that the *number* of counts exceeding or equaling any chosen magnitude, N_c, is given by:

Number of counts exceeding $N_c = P(N \geq N_c)$
$\times$ total number of counts in the sequence   (10.2)

In our case, the latter is the number of entries in each half of the record and therefore the comparison is straightforward: the number predicted by Equation 10.1 and Equation 10.2 vs. the actual number of counts observed in the other half of the record. Six such comparisons are shown in Table 10.1. They and additional such comparisons (see Peleg et al., 2000) demonstrate that, despite the records' deficiencies and the fact that their independence and randomness could not be established, the estimates and actual frequencies were in general agreement. They also indicate that the chosen model, based on the log normal distribution, was quite robust and that its predictions were only slightly affected by the experimental records' imperfections.

That the log normal distribution was found to be an appropriate model for all the data sets without exception (Figure 10.3), regardless of the snack formulation and kind of organisms examined, suggests that log normality might be a rather common feature of microbial records, at least of certain types of commercial foods. Support for this statement was provided by the analysis of totally unrelated microbial records, some of which will be presented in this chapter (Nussinovitch and Peleg, 2000; Nussinovitch et al., 2000).

Analysis of Normalized Data

Because the autocorrelation functions of the snacks' raw microbial records did not show the expected absolute independence and randomness, it was worthwhile to test the possibility that normalizing the data might eliminate any underlying trend or long-term periodicity even if it did not really exist. Normalization has been possible in this case because none of the records had a zero entry. Thus, it made sense to ask about the probability that the next count would be twice or five or ten times as high as its predecessor, for example. The normalized count, Y_n, in this case can be simply the ratio:

$$Y_n = \frac{N_n}{N_{n-1}} \tag{10.3}$$

where N_n and N_{n-1} are any two successive counts.

Now, consider the series of total counts (SPC) of snack A, presented in Figure 10.4 as a normalized record; it was created by the application of Equation 10.3 to the original data shown in Figure 10.1. The transformed series, too, shows an irregular fluctuation pattern with aperiodic outbursts of varying magnitude. Notice, though, that the fluctuations this time are around the horizontal line $Y_n = 1$ and that a clear asymmetry exists between the frequencies of the high and low entries, i.e., where $Y_n > 1$ and $Y_n < 1$, respectively. Even a superficial examination of the normalized record would reveal that it had neither a trend nor any discernible periodicity. This observation is supported by the features of the autocorrelation function (ACF) presented in Figure 10.5. It shows that the waviness clearly evident in the autocorrelation function of the original record has

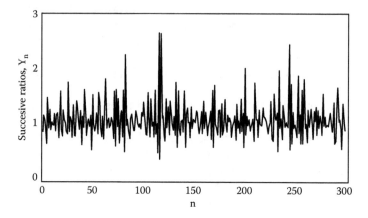

FIGURE 10.4

The total microbial count records of cheese snack A after conversion into a sequence of ratios using Equation 10.3. Notice that the fluctuations of the transformed record are around the line $Y_n = 1$ but that their asymmetry has remained. (From Corradini, M.G. et al., 2002, J. *Food Sci.*, 67, 1278–1285. With permission, courtesy of the Institute of Food Technologists.)

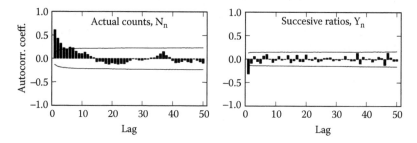

FIGURE 10.5

The autocorrelation functions of the thermophiles' record before and after normalization. Notice the almost complete disappearance of the "waviness" that has resulted from the transformation of the counts to a sequence of ratios. (From Corradini, M.G. et al., 2002, J. *Food Sci.*, 67, 1278–1285. With permission, courtesy of the Institute of Food Technologists.)

almost completely disappeared by conversion of the original counts' record into a sequence of successive ratios.

The only significant autocorrelation encountered in the converted series is for lag 1, which, as before, indicates that lots produced in succession are more likely to have a similar microbial load than lots produced after a long pause. The absolute magnitude of the correlation coefficients, for all the lags, was again very low and about half that of the corresponding correlation coefficients of the raw data. This meant that a successful prediction of a count ratio on the basis of its immediate predecessor would

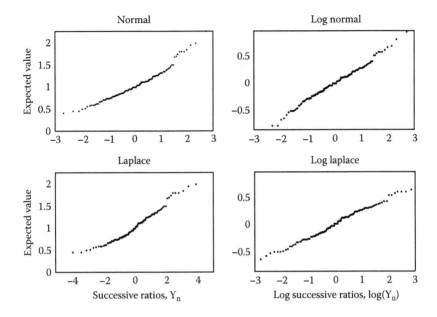

FIGURE 10.6
Examples of Q–Q plots of the normalized records of snack A used to test the applicability of the normal, Laplace, log normal, and log Laplace distributions. Notice the superior fit of the two asymmetric distributions. (From Corradini, M.G. et al., 2002, *J. Food Sci.,* 67, 1278–1285. With permission, courtesy of the Institute of Food Technologists.)

have even a lower probability than that of estimating the actual count. The normalized data's Q–Q plot using the normal, log normal, Laplace, and log Laplace distribution functions as models is shown in Figure 10.6. The actual distribution's asymmetry is manifested in that the corresponding Q–Q plots of the symmetric distributions, namely, the original normal and the Laplace, were much less linear than those of the log normal and log Laplace. This was especially noticeable at the plot's edges, which represent these models' account of the distributions' tails.

The actual distribution's asymmetry is also manifested in the Q–Q plots' regression coefficients, listed in Table 10.2. They indicate that at least as far as these plots' linearity is concerned, the log Laplace was the best representative of the ratios' distribution ($r^2 = 0.994$), followed by the log normal distribution ($r^2 = 0.967$). Very similar regression coefficients were found for the anaerobes and thermophiles, $r^2 = 0.990$ vs. 0.971 to 0.980, respectively (see the table), thus indicating that their count ratios' distribution was asymmetric too. The table also shows the predicted frequencies of high counts' ratios, i.e., those exceeding four chosen levels of Y_c. They were estimated with the distributions' parameters calculated by the

TABLE 10.2

Estimated and Observed Frequencies of Count Ratios in a Commercial Dairy-Based Product Exceeding Specified Values[a]

Microbial population	Y_C	Normal MM and MLE	Log normal MM and MLE	Laplace MM	Laplace MLE	Log Laplace MM	Log Laplace MLE	Observed in second half
SPC	1.8	2–3	4–5	3–4	2–3	5–6	6–7	6
	1.9	1–2	3–4	2–3	1–2	4–5	4–5	5
	2.0	0–1	2–3	1–2	1–2	3–4	3–4	5
	2.1	0–1	1–2	1–2	0–1	2–3	3–4	4
	2.2	0–1	0–1	0–1	0–1	2–3	2–3	1
	r^2	0.959	0.967	0.990		0.994		
Anaerobes	1.7	4–5	7–8	5–6	4–5	7–8	8–9	7
	1.9	1–2	3–4	2–3	2–3	4–5	5–6	5
	2.1	0–1	1–2	1–2	1–2	2–3	3–4	4
	2.3	0–1	0–1	0–1	0–1	1–2	2–3	1
	r^2	0.976	0.971	0.996		0.990		
Thermophiles	1.6	2–3	5–6	3–4	5	5–6	8–9	9
	1.8	0–1	2–3	1–2	2–3	3–4	4–5	7
	2.0	0–1	1–2	0–1	0–1	2–3	3–4	3
	2.2	0	0–1	0–1	0–1	1–2	2–3	1
	r^2	0.986	0.980	0.992		0.990		

[a] The r^2 values are of the Q–Q plots of the corresponding distribution functions.

Source: From Corradini, M.G. et al., 2002, *J. Food Sci.*, 67, 1278–1285. With permission, courtesy of the Institute of Food Technologists.

method of moments (MM) and the maximum likelihood estimation (MLE) from the first half of each record.

Comparison of the predicted frequencies with those actually recorded in the second half of the record (Table 10.2) shows a very good agreement wherever the log Laplace or log normal distribution had been used. A lesser agreement was found with the predictions based on the symmetric normal and Laplace distributions; this provides another demonstration that whenever the ratios' distribution (or that of the original counts) is clearly skewed, the asymmetry must be taken into account in order to improve estimates' reliability. Even a superficial visual comparison of the predictions' quality vis a vis that of those obtained using the counts rather than their ratios (see Table 10.1) shows that the transformation resulted in a substantial improvement in the predictions' quality. However, this would not necessarily be the case if the autocorrelation function of the original count record had shown the absence of any trend or periodicity, true or apparent.

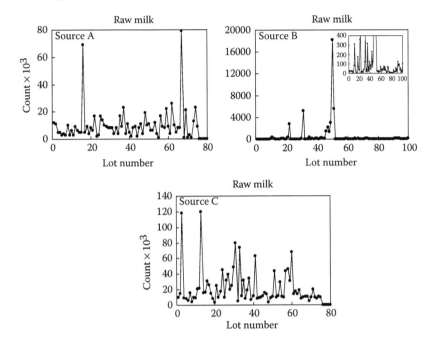

FIGURE 10.7
Examples of total count records (SPC) of raw bovine milk from three farms. Notice the aperiodic large outbursts. (From Nussinovitch, A. et al., 2000, *J. Food Prot.*, 63, 1240–1247. With permission, courtesy of the International Association for Food Protection.)

Rating Raw Milk Sources

Examples of typical count records of bovine milk received at a dairy plant from commercial farms are shown in Figure 10.7 and their corresponding autocorrelation functions in Figure 10.8. These are two out of eight records examined by Nussinovitch et al. (2000) and Peleg (2002b). Again, the counts were not determined at equal intervals and they were not meant to serve as a database for a statistical study. They simply reflect the results of routine testing of incoming milk in the dairy plant.

The records show extremely irregular count fluctuations with occasional aperiodic, very large outbursts. The count series shown in the figure were examined by the methods described in Chapter 9 and the results of the statistical analysis are shown in Table 10.3 (Peleg, 2002b). Because a total count exceeding 50,000 per 100 ml requires lowering the milk quality classification, estimation of the frequency of such occurrences is of particular importance to the raw milk supplier and the receiving dairy plant

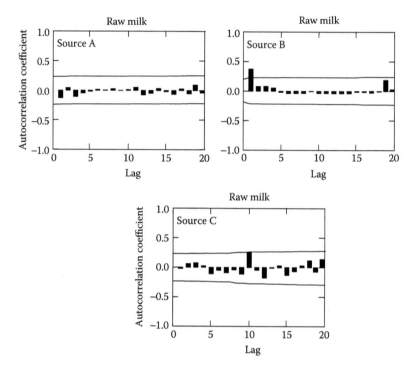

FIGURE 10.8
The autocorrelation functions of the count series of the three raw bovine milk samples shown in Figure 10.7. (From Nussinovitch, A. et al., 2000, *J. Food Prot.*, 63, 1240–1247. With permission, courtesy of the International Association for Food Protection.)

that processes the milk. In the three examples shown in Figure 10.7, the difference in the microbial quality of the milk coming from the farms is obvious even by casual visual inspection of the records. The statistical analysis in this case can be used to *quantify* the dissimilarities in the milk's microbial quality fluctuations.

Because of the relatively short records made available to us for the analysis and the need to further divide them in order to test the procedures' applicability, the estimates of the nonparametric and parametric models in these two cases did not differ much (Table 10.3). However, the examination of the counts' distribution by whatever method would still be beneficial because it would help to identify the risk of receiving an unacceptable or lower grade raw milk in terms of an expected probability or frequency. This information would have been lost if only the average performance were compared. A demonstration is given in Table 10.4, in which the counts in milk coming from eight different commercial farms are compared. Notice that, if one uses the 50,000 per 100 ml as a quality threshold, the milk coming from Farm G is more likely to have an unacceptable level

TABLE 10.3

Distribution Parameters of Total Microbial Counts in Commercial Raw Milk from Two Sources and Probabilities and Frequencies of Future High Counts Estimated by Different Methods

Source	Total no.	$\bar{N}_L$ First half	$\tilde{N}_L$ First half	s_L First half	N_c	$P(N_n \geq N_c)$				Estimated no. in second half				Actually observed in second half	
						Log normal MM and MLE	Log Laplace MM	MLE	Non-parametric	Log Normal MM and MLE	Log Laplace MM	MLE	Non-parametric	Freq.	N_o
A	75	0.847	0.895	0.319	15,000	0.152	0.100	0.151	0.132	5-6	3-4	5-6	5	0.19	7
					20,000	0.078	0.054	0.089	0.053	2-3	2	2-3	2	0.14	5
					25,000	0.042	0.034	0.059	0.026	1-2	1-2	2-3	1	0.05	2
					50,000	0.0038	0.008	0.016	0.026	1-2	0-1	0-1	1	0.03	1
B	75	1.247	1.120	0.393	30,000	0.279	0.237	0.200	0.289	10-11	9	7-8	11	0.16	6
					40,000	0.183	0.158	0.133	0.158	6-7	5-6	5	6	0.14	5
					50,000	0.125	0.115	0.097	0.105	4-5	4-5	3-4	4	0.05	2
					60,000	0.088	0.089	0.075	0.105	3-4	3-4	2-3	4	0.05	2

Sources: Data from Nussinovitch, A. et al., 2000, *J. Food Prot.,* 63, 1240–1247; Peleg, M., 2002, *Int. Dairy J.,* 12, 255–252. With permission, courtesy of Elsevier Ltd.

TABLE 10.4

Classification of Raw Milk Sources according to Probability that Total Microbial Count Will Exceed Two Selected Levels

Rank	Source	n	Max no. observed × 10³/ml	μ^a	$\mu_L{}^a$	$s_L{}^a$	$P(N \geq 50 \times 10^3)^b$	$P(N \geq 100 \times 10^3)$
1	A	75	79	10.2	0.845	0.369	0.010	0.00087
2	C	75	120	22.4	1.199	0.345	0.074	0.010
3	G	72	201	23.0	1.168	0.407	0.096	0.021
4	E	79	5800	22.4	1.122	0.503	0.122	0.042
5	F	76	178	28.8	1.235	0.459	0.156	0.048

[a] For counts expressed in multiples of 10^3/ml.
[b] A count of 50×10^3/ml is the upper limit of grade A milk.

Source: From Nussinovitch, A. et al., 2000, *J. Food Prot.*, 63, 1240–1247. With permission, courtesy of the International Association for Food Protection.

of microbial load than the milk that comes from Farm C, although, on average, the counts in the milk of source G are lower than those of the milk coming from source C.

Whether the potential frequency of unacceptable lots' arrivals can or should have administrative, legal, or financial consequences is obviously a managerial issue. Even without immediate repercussions, the analysis can alert the plant technical staff to a potential problem when the actual arrival of substandard milk from a given source has only rarely occurred or maybe even not at all. Once sanitary measures have been taken to improve the microbial quality of milk coming from a certain farm, their efficacy can be quantitatively assessed in terms of the new probability of unacceptable raw milk arrival, rather than in terms of mean performance only. As has been shown, the two need not rise and fall in unison because the mean does not reflect the counts' standard deviation and the skewness of their distribution.

Here, again, a mere superficial examination of the records' autocorrelation functions (Figure 10.8) can sometime reveal the existence of trends that might be desirable or undesirable, depending on their direction. The inspection of the autocorrelation function can also help to identify periodicities that might require special precautions at certain days of the week or particular times during the year. Needless to say, the described analyses should only supplement, not replace, current methods of statistical quality control. Implementation of the procedures to determine the autocorrelation function, construct the Q–Q plot, and calculate the probabilities of high counts will require neither extensive programming nor much time and effort. As has been mentioned in relation to the previous example (see "Microbial Counts in a Cheese-Based Snack"), count records with normal or log normal distribution can be analyzed and interpreted with a Microsoft Excel® program available as freeware on the Web.

Frozen Foods

The methods to analyze fluctuating microbial count records that are described in Chapter 9 can be applied to a variety of organisms found in frozen foods. For obvious reasons, long microbial quality control charts of industrial processed foods, meats included, are not accessible to researchers from outside the food industry. Thus, records with hundred of entries, of the kind described in "Microbial Counts in a Cheese-Based Snack," are very difficult to obtain. Industrial microbial records need not be kept beyond a certain time set by law. Therefore, historical records might be unavailable for analysis, even if they were in possession of a company willing to share them, because they have been destroyed. Still, we were very surprised by the uncooperative response of certain companies and, in one case, their association, who could only benefit from the (free) analysis and interpretation of their archived data.

Count records of various types of organisms found in frozen apple juice concentrate and carrots and their corresponding autocorrelation functions are shown in Figure 10.9 through Figure 10.12. Some of the outbursts seem to have had a noticeable duration; that is, the higher than usual levels were observed in several successive counts (see Chapter 11). Still, the counts in the records could be treated as practically independent because the series autocorrelation function with only a single exception showed no significant correlation coefficient for any lag (Figure 10.10, Figure 10.12, and Table 10.5). The count distribution could be considered as log normal, at least as first-order approximation (Nussinovitch and Peleg, 2000). Consequently, this distribution's parameters, the logarithmic mean (μ_L) and logarithmic standard deviation (σ_L), calculated from the first half of the data could be used to estimate the frequencies of high counts exceeding several chosen levels in the second half.

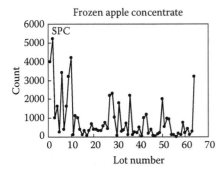 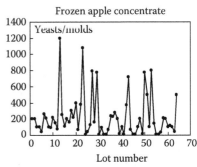

FIGURE 10.9

Examples of microbial count records of commercial frozen apple concentrate. (From Nussinovitch, A. and Peleg, M., 2000, *Food Res. Int.*, 33, 53–62. With permission, courtesy of Elsevier Ltd.)

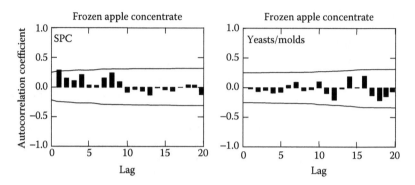

FIGURE 10.10
The autocorrelation functions of the count series of commercial frozen apple concentrate shown in Figure 10.9. (From Nussinovitch, A. and Peleg, M., 2000, *Food Res. Int.*, 33, 53–62. With permission, courtesy of Elsevier Ltd.)

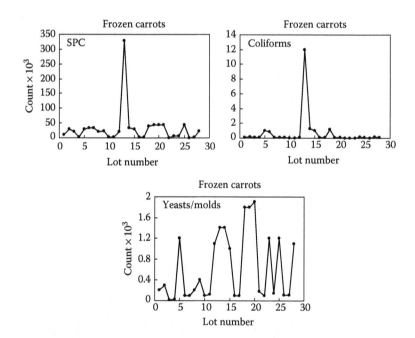

FIGURE 10.11
Examples of microbial count records of commercial frozen carrots. (From Nussinovitch, A. and Peleg, M., 2000, *Food Res. Int.*, 33, 53–62. With permission, courtesy of Elsevier Ltd.)

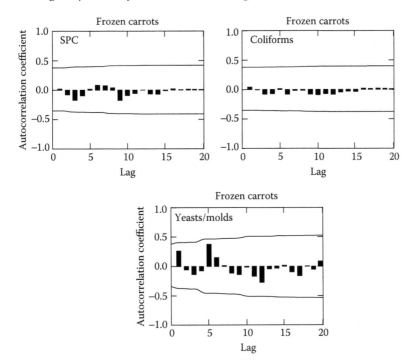

FIGURE 10.12
The autocorrelation functions of the count series of commercial frozen carrots shown in Figure 10.11. (From Nussinovitch, A. and Peleg, M., 2000, *Food Res. Int.*, 33, 53–62. With permission, courtesy of Elsevier Ltd.)

As shown in the table, general agreement was present between the estimates and actual values, albeit far from perfect in some cases. The relatively short records at hand and the suspicion that the entries were not truly independent despite passing the autocorrelation test suggest that the estimation method based on Equation 10.1 (and thus on the log normal distribution model) has been fairly robust. For these kinds of products, as has been shown in the analysis of other frozen foods, including ground meat, ice cream of various flavors, and frozen juice (Nussinovitch and Peleg, 2000), the shortest sequence that still allows one to estimate the frequency of future outbursts with any degree of acceptable accuracy is probably about 30 counts. However, one must remember that a long record is necessary not only to improve the estimate's quality, but also to establish whether the counts have a trend or underlying long-term oscillations, with a period or periods that might not be evident in a short record.

TABLE 10.5

Distribution Parameters of Microbial Counts in Commercial Frozen Foods with Estimated and Observed Numbers of Counts Exceeding Selected Levels

Food	Organism type	Total number of cases	Significant autocorrelation for lag(s)[a]	μ_L (First half)	s_L (First half)	N_c	$P(N \geq N_c)$	No. in second half[b] Estimated	Observed
Apple juice concentrate	SPC	64	1(0.29)	2.83	0.52	1,000	0.370	12	5
						1,500	0.252	8	3
						2,000	0.181	5	3
	Yeast/molds	64	None	2.15	0.52	200	0.385	12–13	13
						400	0.192	6	5
						600	0.113	3–4	3
						800	0.075	2–3	1
Baby carrots	SPC	28	None	4.26	0.57	35,000	0.309	4–5	5
						40,000	0.271	3–4	4
						45,000	0.242	3–4	1
	Coliforms	28	None	2.26	0.81	250	0.432	6	4
						500	0.295	4	2
						750	0.223	3	2
						1,000	0.181	2–3	2
	Yeast/molds	28	None	2.31	0.66	1,000	0.151	2–3	7
						1,250	0.116	1–2	3
						1,500	0.097	1–2	3
						1,750	0.078	1	2
Ground meat	SPC	57	None	6.17	0.65	2.0×10^6	0.419	12	13
						4.0×10^6	0.253	7	4
						6.0×10^6	0.174	4–5	3
						10.0×10^6	0.100	3–4	2
	Coliforms	57	None	3.04	0.55	2,500	0.261	7	6
						5,000	0.115	3–4	5
						10,000	0.041	1–2	2
						20,000	0.011	0–1	1

Ground meat	*Staphylococcus*	57	None	1.88	0.26	100	0.517	14–15	12
						125	0.202	5–6	6
						150	0.135	3–4	6
						175	0.081	2–3	6
						200	0.057	1–2	6
	Clostridium	57	None	1.40	0.41	50	0.232	6–7	13
						100	0.075	2	11
						150	0.029	0–1	4
						200	0.015	0.1	2
						250	0.008	0–1	2
	Enterococcus	57	None	2.38	0.77	625	0.294	8–9	14
						1,250	0.175	5	8
						2,500	0.094	2–3	2
						5,000	0.043	1–2	2
						10,000	0.018	0–1	1
Vanilla ice cream	SPC	46	None	3.96	0.63	12,000	0.426	10	4
						15,000	0.367	8	3
						20,000	0.296	6–7	2
						25,000	0.245	5–6	1
	Coliforms	46	1(0.36)	1.48	0.49	50	0.328	8	6
						75	0.212	5	4
						100	0.147	2–3	4
						150	0.080	1–2	1
						200	0.049	1	1
	Yeast/molds	46	None	1.37	0.46	50	0.239	5–6	4
						100	0.086	2	2
						150	0.040	1	2
						200	0.022	0–1	1
						250	0.029	0–1	1

[a] Numbers between parentheses are the corresponding correlation coefficients.

[b] Calculated with log normal distribution as the fluctuation model.

Source: From Nussinovitch, A. and Peleg, M., 2000, *Food Res. Int.*, 33, 53–62. With permission, courtesy of Elsevier Ltd.

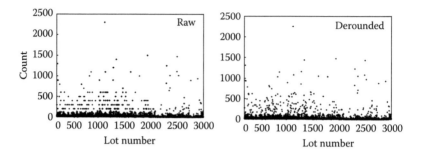

FIGURE 10.13

A record of *E. coli* counts in the wash water of a commercial poultry plant. Left: the original record; notice the occasional horizontal groupings of the counts, a result of uneven rounding. Right: the same record after derounding using the algorithm expressed by Equation 10.4. (From Corradini, M.G. et al., 2001, *Food Res. Int.*, 34, 565–572. With permission, courtesy of Elsevier Ltd.)

E. coli in Wash Water of a Poultry Plant

A record of 3000 successive *E. coli* counts in the wash water of a commercial poultry plant located in the Southeast U.S. was made available to us for analysis by courtesy of a company that preferred to remain anonymous. The "raw" record, which covers a period of several years, is shown in Figure 10.13 (right). The water was usually sampled two or three times daily when the plant was in operation. (The entries are in counts per 100 ml of water.) Again, the shown record reflects the results of a routine data collection procedure in the plant. The determinations were not part of any research project and were not originally intended to serve as a database for a statistical analysis of the kind discussed in this chapter. Even a brief visual examination of the record will reveal two salient features:

- Although outbursts can be observed all along the record, they seem to have been less frequent in the last third, where the background level of the counts was noticeably lower. This indicates that conditions in the plant (or the raw material) changed in a significant way around the corresponding time. Therefore, the whole record could not be considered as representing a situation in which the environment had remained unchanged — one of the key assumptions on which the discussed models are based. Consequently, the record, in which the counts are listed in a successive order, was divided into 10 equal segments of 300 counts each

(Corradini et al., 2001), and the predictive ability of the method was tested by comparing the number of outburst estimates derived from one segment with that actually observed in its successor (see below).

- The counts were unevenly rounded, i.e., they were sometimes rounded to the nearest 10, sometimes to the nearest 100, and sometimes not at all. This is evident in that many of the counts, but not all, appear to be aligned horizontally (see Figure 10.13). Consequently, the counts in the raw record examined as a whole, or after segmentation for that matter, could not pass the test for independence and could not be used to test the proposed method if left unchanged.

To overcome the uneven rounding problem, the record was derounded, as explained in Chapter 9, using the following algorithm (Corradini et al., 2001a):

$$\text{If } M_n < 10, \ N_n = M_n$$

$$\text{If } 10 \leq M_n \leq 140, \ N_n = M_n - 5 + 10R_n \tag{10.4}$$

$$\text{If } 140 > M_n, \ N_n = M_n - 50 + 100R_n$$

where
M_n is the original raw count
N_n is the derounded count to be used in the analysis
R_n is a random number generated between zero and one, $0 \leq R_n \leq 1$, with a uniform distribution

Application of this algorithm leaves any reported count below 10 as it is. However, it converts any count between 10 and 140, which had been frequently rounded to the nearest 10, into a new random number in the range of ±5 around the original entry. Similarly, counts above 140, which had been frequently rounded to the nearest 100, are replaced by a new random value in the range of ±50 around the original entry. The derounded record is shown in Figure 10.13 (right). Comparison of the record before and after the algorithm's application will immediately reveal that the order clearly evident as horizontal dotted lines in the original data (left) has totally disappeared after the derounding process.

Examples of the autocorrelation functions of the derounded record's segments are presented in Figure 10.14; they show that after derounding, the record's entries could be considered as independent for all practical purposes. Again, in the very few cases in which a significant autocorrelation

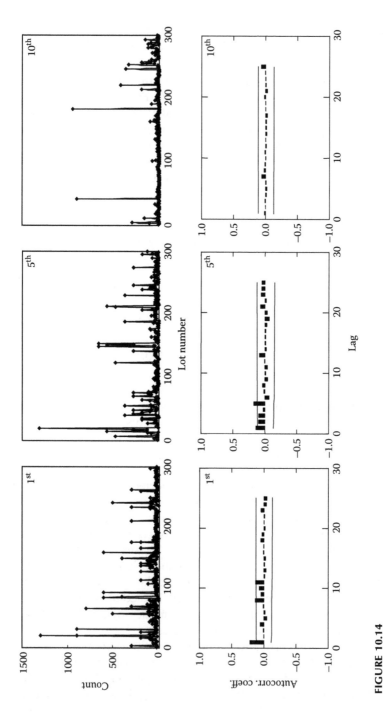

FIGURE 10.14
Examples of derounded segments of the record shown in Figure 10.13 (top) and corresponding autocorrelation functions (bottom). Notice that after derounding, the entries pass the test of independence. (From Corradini, M.G. et al., 2001, *Food Res. Int.*, 34, 565–572. With permission, courtesy of Elsevier Ltd.)

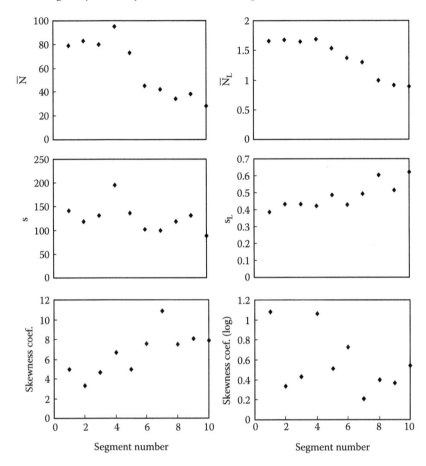

FIGURE 10.15
The mean, standard deviation, and skewness coefficient (linear and logarithmic) of the
derounded ten segments of the count record shown in Figure 10.13. Notice the clear trend.
(From Corradini, M.G. et al., 2001, *Food Res. Int.*, 34, 565–572. With permission, courtesy of
Elsevier Ltd.)

was found for certain lags, the correlation coefficient was very low. As
before, this slight correlation could be explained by that water samples
taken in succession are more likely to have a similar microbial load and
composition than samples taken several days or weeks apart.

Examination of the mean, standard deviation, and skewness of the
counts' distributions in the 10 segments or that of their logarithms' distributions reveals an unmistakable trend (Figure 10.15). Therefore, an
attempt to predict the frequencies of outbursts could only succeed in an
adjacent segment, if at all. Inspection of the counts' histogram and Q–Q

plots (Figure 10.16 and Figure 10.17) indicated that their distribution could be best described by the log Laplace distribution function:

$$f(N_L) = \frac{1}{2b}\exp\left(-\frac{|N_L - \mu_L|}{b}\right)$$ (10.5)

where
   N_L is the counts' logarithm
   b is the distribution's spread parameter
   μ_L is the counts' logarithmic mean

Notice that, unlike in the log normal distribution, the numerator of the exponential argument, $|N_L - \mu_L|$, *is the absolute difference* rather than the difference squared. Calculation of the distribution's parameters by the MM is based on:

$$\mu_L = \overline{N}_L$$ (10.6)

and

$$b = 2s_L^2$$ (10.7)

where N_L and s_L are, respectively, the logarithmic mean and standard deviation of the data — in our case, those of the counts in each segment.
   Calculation of the distribution parameters by the MLE is based on the fact that the estimated μ_L is:

$$\mu_L = \tilde{N}_L$$ (10.8)

where $\tilde{N}_L$ is the counts' logarithmic *median* and the spread parameter, b, is:

$$b = \frac{1}{n}\sum_{i=1}^{n}|N_{Li} - \tilde{N}_L|$$ (10.9)

where n is the number of counts in the segment at hand.
   A comparison of the predictions of the outbursts' frequencies in the various segments is given in Table 10.6. (Notice that there are only nine comparisons; the first segment was used to estimate the outbursts' frequencies in the second, the second to estimate the outbursts in the third, and so forth.) Also shown in the table are the estimates arrived at by the

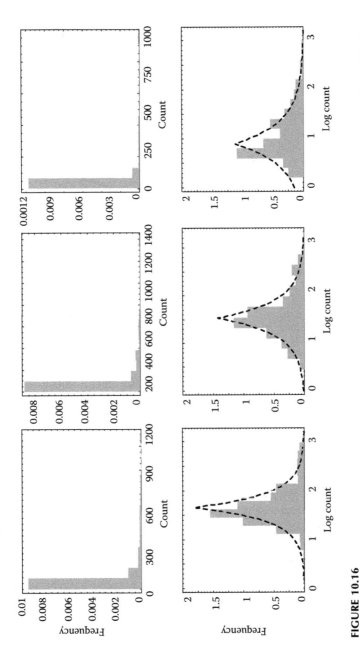

FIGURE 10.16
Examples of the histograms of segments of the count record shown in Figure 10.14, which indicate the candidacy of the log Laplace distribution function (Equation 10.15) as a model. (From Corradini, M.G. et al., 2001, *Food Res. Int.*, 34, 565–572. With permission, courtesy of Elsevier Ltd.)

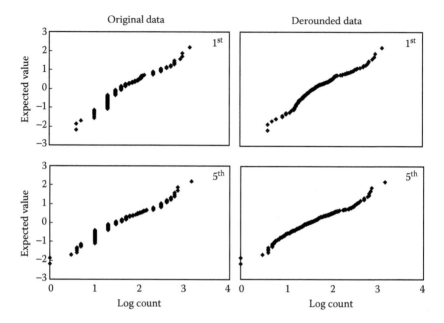

FIGURE 10.17
Examples of Q–Q plots of segments of the count record shown in Figure 10.14 supporting the choice of the log Laplace distribution function (Equation 10.15) as a model. (From Corradini, M.G. et al., 2001, *Food Res. Int.*, 34, 565–572. With permission, courtesy of Elsevier Ltd.)

nonparametric model. Recall that in the nonparametric approach, the assumption is that the frequency of counts exceeding any given level in a successive segment is expected to be the same as that observed in the previous segment and, thus, no formula is needed to determine it.

Although all the estimates given in Table 10.6 have been calculated from the derounded record, the comparisons shown are with the *actual* number of high counts observed, as they have been originally entered into the record, regardless of whether they were rounded or not. The table shows that the estimates calculated with the log Laplace distribution function as a model were in reasonable agreement with the values actually observed, at least as judged subjectively. The estimates calculated with the MLE method were generally more accurate than those obtained by the MM method, but not consistently. However, the differences, between the estimates calculated by these two procedures, whenever encountered, were not very large.

Despite the imperfections of the record and of the log Laplace distribution function as a model, the method based on this distribution function consistently yielded closer estimates of the outbursts' frequencies than those obtained by the nonparametric approach, irrespective of whether

TABLE 10.6

Comparison between Observed Frequencies of E. *coli* Counts in Industrial Poultry Plant Rinse Water and Those Calculated Using Nonparametric Method and Log Laplace Distribution (MM and MLE Methods)

Segment	μ	s	N_c	Nonparametric method P (%)	No.	MM P (%)	No.	MLE P (%)	No.	Observed P (%)	No.
2	79	140	600	2.7	8	0.8	2–3	0.6	1–2	1	3
			700	1.3	5–6	0.7	1–2	0.5	1–2	1	3
			800	1.3	4–5	0.5	1–2	0.4	1–2	0.7	2
			900	1.0	4–5	0.4	1–2	0.3	0–1	0	0
			1000	0.3	3–4	0.4	1–2	0.3	0–1	0	0
3	83	119	600	1.0	3	1.4	4–5	1.3	4	2	6
			700	1.0	3	1.2	3–4	1.1	3–4	1	3
			800	0.7	2	1	2–3	0.9	2–3	1	3
			900	0	0	0.8	2–3	0.8	2–3	1	3
			1000	0	0	0.7	2–3	0.7	2–3	0.7	2
4	78	129	600	2	6	1.3	3–4	1.5	4–5	3.3	10
			700	1	3	1	3–4	1.2	3–4	1.3	4
			800	1	3	0.8	2–3	1	3–4	1.3	4
			900	1	3	0.7	2–3	0.9	2–3	1.3	4
			1000	0.7	2	0.6	1–2	0.8	2–3	1	3
5	96	196	600	3.3	10	1.4	4–5	1	3–4	2	6
			700	1.3	4	1.1	3–4	0.8	2–3	1	3
			800	1.3	4	0.9	2–3	0.7	2–3	0.3	1
			900	1.3	4	0.8	2–3	0.6	1–2	0.3	1
			1000	1	3	0.7	2–3	0.5	1–2	0.3	1
6	73	138	600	2	6	1.4	4–5	1.5	4–5	1	3
			700	1	3	1.2	3–4	1.2	3–4	0.7	2
			800	0.3	1	1	2–3	1	3–4	0.7	2
			900	0.3	1	0.9	2–3	0.9	2–3	0.7	2
			1000	0.3	1	0.8	2–3	0.8	2–3	0.7	2
7	45	102	600	1	3	0.5	1–2	0.6	1–2	0.3	1
			700	0.7	2	0.4	1–2	0.5	1–2	0.3	1
			800	0.7	2	0.3	1–2	0.4	1–2	0.3	1
			900	0.7	2	0.3	0–1	0.4	1–2	0.3	1
			1000	0.7	2	0.2	0–1	0.3	0–1	0.3	1
8	42	101	600	0.3	1	0.8	2–3	1.1	3–4	1.7	5
			700	0.3	1	0.6	1–2	0.9	2–3	1.3	4
			800	0.3	1	0.5	1–2	0.8	2–3	1	3
			900	0.3	1	0.5	1–2	0.7	2–3	0.7	2
			1000	0.3	1	0.4	1–2	0.6	1–2	0.7	2
9	35	120	600	1.7	5	0.8	2–3	1.2	3–4	1	3
			700	1.3	4	0.7	2–3	1.1	3–4	1	3
			800	1	3	0.6	1–2	0.9	2–3	1	3
			900	0.7	2	0.5	1–2	0.8	2–3	1	3
			1000	0.7	2	0.5	1–2	0.8	2–3	1	3

TABLE 10.6 (continued)

Comparison between Observed Frequencies of *E. coli* Counts in Industrial Poultry Plant Rinse Water and Those Calculated Using Nonparametric Method and Log Laplace Distribution (MM and MLE Methods)

				Estimated with log Laplace distribution function						
				Nonparametric method		MM		MLE		Observed	
Segment	μ	s	N_c	P (%)	No.	P (%)	No.	P (%)	No.	P (%)	No.
10	38	130	600	1	3	1.3	3–4	2.1	6–7	0.7	2
			700	1	3	1.1	3–4	1.8	5–6	0.7	2
			800	1	3	1	2–3	1.7	5	0.7	2
			900	1	3	0.9	2–3	1.5	4–5	0.7	2
			1000	1	3	0.8	2–3	1.4	4–5	0	0

Source: From Corradini, M.G. et al., 2001, *Food Res. Int.*, 34, 565–572. With permission, courtesy of Elsevier Ltd.

the log Laplace distribution's parameters were calculated using the MM or the MLE. The superiority of the parametric approach, which is based on the use of a distribution function, over the nonparametric approach is not only that it gives better predictions, but also that it allows for the fact that counts of a magnitude not previously observed can appear in the future (Peleg and Horowitz, 2000). Even the highest experimentally recorded count rarely accounts for the maximum possible number of organisms in the medium examined; therefore, the possibility that still higher counts can be encountered at a later time cannot be discarded *a priori*.

Thus, using a parametric model of the kind presented here is probably a more prudent approach in risk assessment, provided that the chosen parametric model (the log Laplace distribution function in this particular case) indeed captures the statistical characteristics of the fluctuating counts' record. However, even if one does not adapt the parametric approach, the record presentation in the form of a histogram (with or without logarithmic transformation); examination of the autocorrelation function; and the plots of the moving mean, standard deviation, and skewness coefficient (see Figure 10.14) can reveal characteristics or patterns that might have safety implications, but which otherwise would have remained unnoticed. It would be appropriate to mention again that calculation of a distribution's parameters and generation of the autocorrelation function of a given record are integral parts of most if not all statistical packages and therefore do not require special programming by the user.

Also, notice that the method described in this section can be particularly useful for monitoring and predicting the level of fecal contamination in waters of industrial plants in which animals are processed. This is because

E. coli or other fecal organisms will always be present in such waters. In terms of the model, this entails that zero counts will be extremely rare or nonexistent; thus, a logarithmic transformation of the original count will almost always be a viable option. However, this might not be true for waters in which the presence of fecal organisms is not permanent. In such waters' records, zero entries might be quite frequent and thus alternative transformation might be required. The next section contains more on such scenarios.

Fecal Bacteria in Lake Kinneret

Lake Kinneret, known as the Sea of Galilee, is the main water reservoir of the State of Israel and its beaches are popular recreational sites. The Jordan River, which flows into and out of the lake, is also a major water source to the Kingdom of Jordan, which has a water-sharing agreement with Israel. Thus, the microbial quality of Lake Kinneret's water has direct public health impact and international ramifications in a very sensitive region of the world.

Fecal bacteria in the lake's water are monitored routinely by the Israeli Oceanographic and Limnological Research Institute. The water samples are collected at several locations around the lake at 2- to 4-week intervals, depending on the season, and analyzed at a laboratory located on the northwestern shore of the lake. The frequency of the sampling and subsequent analyses has been primarily determined by logistic considerations and has not been intended to follow any particular statistical design. The microbial count records determined by the institute, therefore, can be viewed as primarily a product of a routine monitoring program and not of a planned research project aimed at proving or disproving a statistical concept and mathematical models.

Because the fecal bacteria data in the lake's water were collected at unequal intervals, the abscissa of their plot in a successive order is only approximately a time coordinate. Consequently, any reference to the probability of encountering a count exceeding any given level will only pertain to the probability that a subsequent count will reach this level and not to the probability that such an event will occur within any specified time span.

The count records of fecal coliforms, *E. coli*, and *Enterococci* determined over a period of about 7 years in water samples taken at two different sites along the lake's shores (Hadas et al., 2004) are shown in Figure 10.18 and Figure 10.19. Irrespective of location and the organism counted, the 12 records had the following common features:

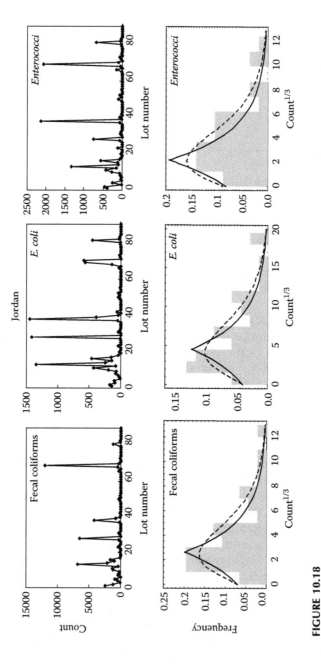

FIGURE 10.18

Count records of fecal organisms in the water of Lake Kinneret collected at the Jordan site. Top: the original data; bottom: the histogram of the cubic root of the counts with the truncated Laplace (solid line) and truncated EV distribution function (dashed line) superimposed. (From Hadas, O. et al., 2004, *Water Res.*, 38, 79–88. With permission, courtesy of Elsevier Ltd.)

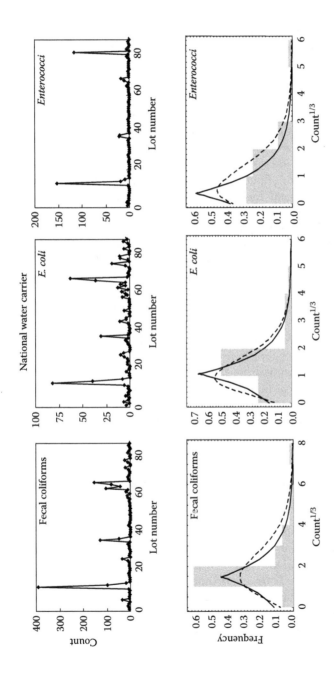

FIGURE 10.19

Count records of fecal organisms in the water of Lake Kinneret collected at the national water carrier site. Top: the original data; bottom: the histogram of the counts with the truncated Laplace (solid line) and truncated EV distribution function (dashed line) superimposed. (From Hadas, O. et al, 2004, *Water Res.*, 38, 79–88. With permission, courtesy of Elsevier Ltd.)

- They all had an irregular appearance with aperiodic outbursts of varying magnitude.
- Some of the outbursts had an apparent duration that was manifested in two and sometime even three or more successive high counts.
- Each record had a substantial number of zero entries.

Fecal contamination of an open water reservoir has certain haphazard characteristics, regardless of whether the discharge is of human or animal origin. Therefore, the observed pattern of long sequences of small fluctuating counts punctuated by sharp high peaks is not at all surprising. The small, or "background," fluctuations most probably represent the occasional contact with fecal material directly deposited into the lake or washed into it by rain, for example. The peaks, in contrast, most probably represent episodes of more massive discharge, by human failure to treat affluents effectively, the incidental presence of herds of cows or sheep in the lake's vicinity, etc. It is not surprising that some of the high counts lasted for 2 to 4 weeks, especially in a natural reservoir like a lake. It is possible that because of a lasting rain, for example, there had been a continuous flow of fecal material that otherwise would not have reached the lake.

Alternatively, once massive contamination of the water with fecal material has occurred, it takes time before the microbial population in the water returns to its normal undisturbed level. In fact, one can expect that *every* episode of massive fecal contamination (or contamination by any other type of organisms washed into a lake) will have a measurable duration. Because the outburst can be over before the next water sample is examined, its actual length and its true peak's magnitude need not be always manifested in the record (see Chapter 9). The situation in which an outburst lasts for a sufficient time so that the population's growth and decline are clearly evident in the record needs to be viewed as a different case and treated with different tools. How such records can be interpreted and described mathematically is the topic of the next chapter.

However, because none of the available data sets from Lake Kinneret had sufficient information to apply the methods described in Chapter 11, the statistical analysis of the records was done on the basis of the principles outlined in Chapter 9. Thus, even though at least some of the observed outbursts obviously had a significant duration and thus a clearly discerned structure (see below), the records were still treated as if the individual counts were independent, although in reality at least some of them were not. All the examined records of Lake Kinneret had a substantial number of zero entries; thus, it should be reiterated that this does not necessarily mean that the sought organism was totally absent in the water at the time at which the water samples were tested. The organism might have been present at an undetectable level or in a state that required a different and more sensitive recovery and counting procedure than the one used.

Even if the distinction between apparent and real absence might have some practical and safety implications, it will not concern us here. We will treat a zero count as a zero regardless of whether it means undetected or truly absent. Because the purpose of what follows is to present and evaluate a method to assess the frequency of outbursts, this approach is not unreasonable. Many zero counts eliminated the possibility of describing the records' fluctuations by the log normal, log Laplace, or any other standard parametric distribution function based on a logarithmic transformation of the independent variable. In other words, the most suitable distribution functions used to characterize the records discussed in "Microbial Counts in a Cheese-Based Snack" through "*E. coli* in the Wash Water of a Poultry Plant" cannot be used here, unless they are modified.

Characterization of Count Distributions

Nonlogarithmic Transformations of the Counts

When a set of numerical values covers several orders of magnitude, its range is frequently compressed to a manageable scale by the logarithmic transformation of its members. A most familiar example in microbiology is the presentation of counts or growth and survival ratios in terms of their base-10 logarithm. Obviously, the conversion would be impossible (or meaningless) whenever a zero is encountered, as in our case. However, there might be other reasons not to use a logarithmic transformation even when zeros are not involved. In certain systems, like the one at hand, the logarithmic transformation might simply result in *overcompression* of the data. The changes in the counts are simply not large enough to justify the use of a logarithmic scale.

A root transformation, $Z - N^{1/m}$, where Z is the transformed value of N and $m > 1$, is therefore an attractive alternative under such circumstances. It offers a way to tune the scale's compression to the characteristics of the actual record at hand and it applies to zero and nonzero values alike. Because negative counts are impossible, the complications that would have been encountered in such a case should not concern us here. Similarly, the inverse situation, where one wants to *expand* the scale rather than to compress it, is also not relevant to microbial records of the kind with which we are dealing here. Nevertheless, such a situation may arise in other microbial systems involving small numbers of organisms, at least in principle.

Finding a Truncated Distribution

Regardless of whether we attempt to characterize the distribution of the original or the transformed counts, the existence of a substantial number of zeros in the record indicates that there is a finite nonzero probability of encountering a zero count. This eliminates the possibility of using a distribution function of the Weibull or log normal type, which requires that $P(0) = 0$, i.e., that the probability of a zero entry is zero. For obvious reasons, the normal, logistic, Laplace, or any other symmetric distribution that allows for the existence of negative counts could not be used here. The solution was therefore to use a *truncated distribution function* as a model. As already mentioned in "Truncated Distributions" in Chapter 9, the frequency (density) form of a parametric distribution of a random variable Z, $f(Z)$, with a range $-\infty < Z < \infty$ truncated at zero is defined by:

$$f_{truncated}(Z) = \frac{f(Z)}{\int_0^\infty f(Z)dZ} \qquad (10.10)$$

The values of truncated distributions are not listed in standard statistical tables and calculations based on truncated distribution functions are not standard options or commands of most, if not all, available mathematical and statistical software. However, the values of truncated distribution functions can be easily calculated with a program like Mathematica®, using Equation 10.10 whenever the chosen "mother" distribution, $f(Z)$ (e.g., the normal or Laplace distribution function) is already defined in the program in its frequency (density) and cumulative forms. Once defined in this way, the truncated distribution function, $f_{truncated}(Z)$, can be used as such or be integrated like any other continuous function recognized by the program. In our case, the probability of encountering a count, N, equal to or larger than a chosen number, N_c, is:

$$P(N \geq N_c) = P(Z \geq Z_c) = \int_{Z(N_c)}^\infty f_{truncated} Z(N)dZ(N \qquad (10.11)$$

As before, the expected number of such counts in a record of length m will be $m \times P(Z \geq Z_c)$.

Distribution of Fecal Bacteria in the Lake's Water

Below the raw records of the fecal bacteria in the waters of Lake Kinneret (Figure 10.18 and Figure 10.19) are the histograms of the counts after their cubic root transformation, i.e., $Z = N^{1/3}$ (Hadas et al., 2004). This particular conversion was chosen after trying various alternatives, including $Z = N^{1/2}$. Although $Z = N^{1/3}$ might not be the best of all possible transformations, it could be used effectively for *all* the 12 records examined as judged by the corresponding Q–Q plots. Examples of Q–Q plots before and after the cubic root transformation are shown in Figure 10.20. The underlying mother distributions, before truncation, was the Laplace distribution:

$$f_{Laplace}(Z) = \frac{1}{2b} \exp\left(-\frac{|Z - \mu|}{2b}\right)$$ (10.12)

and the extreme value (EV) distributions:

$$f_{EV}(Z) = \frac{1}{b} \exp\left[-\exp\left(\frac{a - \mu}{b}\right) + \frac{a - Z}{b}\right]$$ (10.13)

where μ is the mean of the transformed values and a and b are constants.

The truncated versions of these two distributions functions (Equation 10.10) have been superimposed on the transformed counts' histograms given in Figure 10.18 and Figure 10.19. These show a reasonable fit as judged by subjective criteria. However, this observation is also strongly supported by the linearity of the corresponding Q–Q plots, examples of which are shown in Figure 10.20. This figure indicates that the two models, despite their mathematical dissimilarity, had a comparable fit, especially at the distributions' tail end — the region in which we are particularly interested.

It should be mentioned at this point that although defining a truncated parametric distribution function in Mathematica using any of the functions listed in the program (like the mentioned Laplace and extreme value distribution functions) is almost a trivial matter. Producing a Q–Q plot for a truncated distribution function, in contrast, requires special programming. The plots shown in Figure 10.20 were produced by such a program. To calculate the parameters of a truncated distribution function by the MM is fairly straightforward because Mathematica® has handy equation-solving procedures and one can write the standard deviation of a truncated distribution as an explicit expression using its basic definition. Thus, the spread parameter b of the truncated Laplace distribution can be easily extracted from the data.

The same procedure can be used to extract the b and a of the truncated extreme value distribution, except that another equation needs to be added, like that of the mode's or median's definition. It is by far more difficult to program a procedure to calculate the parameters of a truncated

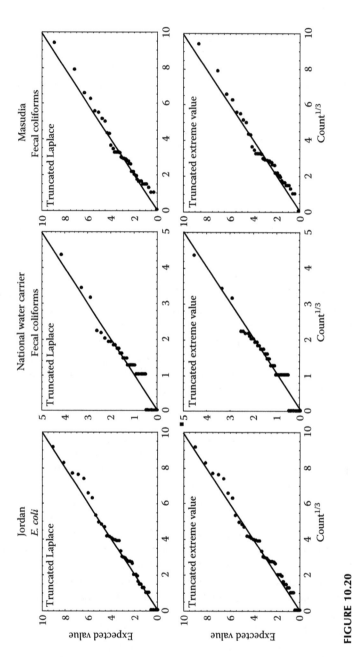

FIGURE 10.20

Examples of Q–Q plots of the coliform counts' cubic root (see Figure 10.18 and Figure 10.19), which supported the choice of the truncated Laplace and truncated EV distribution functions (Equation 10.13 and Equation 10.14), respectively, as models. (From Hadas, O. et al., 2004, *Water Res.*, 38, 79–88. With permission, courtesy of Elsevier Ltd.)

distribution function by the maximum likelihood estimation (MLE) method. Consequently, because the performance of the two procedures was not dramatically different when other records had been examined (see "*E. coli* in Wash Water of a Poultry Plant" above), the truncated Laplace and EV distribution parameters used to represent the counts shown superimposed on the histograms in (Figure 10.18 and Figure 10.19) were all calculated by the MM.

These MM parameters were subsequently used to estimate the future frequencies of outbursts listed in Table 10.7. If the results of the analysis described in "*E. coli* in Wash Water of a Poultry Plant" are any indication, it would be doubtful that the models' predictions would be much improved if the MLE had been used instead of the MM to determine their parameters. Consequently, the effort to develop an MLE program was deemed unworthy and the attempt abandoned.

Estimating the Frequency of Future Outbursts

The autocorrelation functions (ACFs) of the count records are shown in Figure 10.21. Surprisingly, only 5 out of the 9 records examined had a significant autocorrelation for any lag. Even in these five cases, the correlation coefficients were rather low. Some of the records show a certain degree of "waviness," which would be anticipated if seasonal changes played a major role in the lake's contamination pattern. Because the counts were not determined at fixed time intervals, however, the interpretation of the occasional significance of the autocorrelations may not be as straightforward as one would hope. More importantly, the autocorrelation functions did not reveal a consistent seasonal effect on the count records. This means that although climatic factors probably did influence the fecal contamination pattern, their effect was minor relative to that of the accidental discharge episodes.

Also, with probably very few possible exceptions, the outbursts in the different sites were not synchronized (see Figure 10.18 and Figure 10.19). This suggests that at least some of the observed outbursts were a result of local contamination rather than of major events that had affected the whole lake. The same conclusion can be reached by examining the synchronization of at least some of the outbursts of the three types of fecal bacteria recorded at the same site. These did not always coincide, suggesting that at least some of the higher than usual counts reflect contamination from different fecal sources. (If the origin of the discharge had been the same, it would be expected that the *ratios* between the different organisms' counts would be about the same, but this was not the case. The ratios should not have been expected to be exactly the same because the numbers of the three types of fecal bacteria could be differently affected by the local environmental conditions.)

TABLE 10.7

Estimated and Observed Fecal Counts in Various Sites around Lake Kinneret

Water source	Organism	No. observations in first and second halves	N_c	Estimated no. $N \geq N_c$ in second half		Observed in second half
				Truncated Laplace $Z(N) = N^{1/3}$	Truncated ext. value $Z(N) = N^{1/3}$	
Jordan	Fecal coliforms	46/45	500	2–3	3–4	3
			1000	1–2	1–2	2
			1500	1–2	1–2	1
	E. coli	46/45	600	1–2	1–2	2
			650	1–2	1–2	2
			700	0–1	0–1	1
	Enterococci	46/45	400	1–2	1–2	3
			500	1–2	1–2	2
			600	0–1	1–2	0
National water carrier	Fecal coliforms	46/46	90	1–2	2–3	2
			110	1–2	2–3	1
			130	0–1	1–2	1
	E. coli	46/46	30	2–3	3–4	2
			50	1–2	1–2	2
			70	0–1	1–2	1
	Enterococci	46/46	40	1–2	1–2	2
			60	0–1	1–2	1
			80	0–1	0–1	1
Masudia	Fecal coliforms	46/45	200	2–3	2–3	2
			300	1–2	1–2	1
			400	0–1	0–1	1
	E. coli	46/45	150	1–2	1–2	2
			200	0–1	0–1	1
			250	0–1	0–1	0
	Enterococci	46/45	60	3–4	4–5	1
			70	3–4	3–4	1
			80	2–3	3–4	0
Alumot	Fecal coliforms	44/43	130	2–3	3–4	3
			145	2–3	3–4	2
			160	2–3	2–3	0
	E. coli	44/43	70	1–2	1–2	2
			90	1–2	1–2	1
			110	0–1	0–1	1
	Enterococci	44/43	30	2–3	2–3	3
			40	1–2	1–2	1
			50	1–2	1–2	0

Source: From Hadas, O. et al., 2004, *Water Res.*, 38, 79–88. With permission, courtesy of Elsevier Ltd.

In light of the preceding, employing the described models to estimate the frequency of future episodes of high counts was not deemed an outrageous proposition. The results presented in Table 10.7 show that, despite

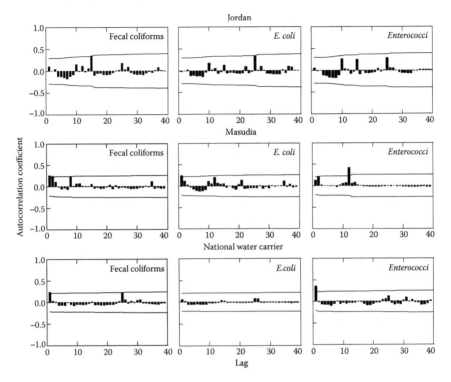

FIGURE 10.21
The autocorrelation functions of the fecal organisms in the water of Lake Kinneret at three different sites. (From Hadas, O. et al., 2004, *Water Res.*, 38, 79–88. With permission, courtesy of Elsevier Ltd.)

the records' obvious imperfections and the violation of at least some of the assumptions on which the calculation method had been based, the estimates and actually observed number of outbursts were in most cases in reasonable agreement. This was regardless of the specific distribution function chosen, i.e., neither the truncated Laplace nor the truncated extreme value (EV) distribution (Equation 10.12 and Equation 10.13, respectively) consistently gave better estimates. In the single case in which the predictions were substantially off the mark (the *Enterococci* record of Masudia), both models gave a similar overestimate of the frequencies. This suggests that exceptional conditions at the site were probably responsible for the discrepancy, rather than the mathematical properties of either model.

Issues of Concern

All the records whose interpretation is discussed in this chapter are of the kind that one would most likely encounter in real life: far from being ideal as far as experimental design is concerned and frequently in clear violation of the conditions required by the proposed calculation methods. Yet, it seems that as long as the infractions are not too serious, the described mathematical procedures to estimate the frequencies of future outbursts could still yield reasonable predictions. As has been amply shown, handy statistical tools, like the autocorrelation function and Q–Q plot, are at our disposal and they can be used to identify records for which the described procedures will be clearly inapplicable.

In many cases in which standard distribution functions, such as the log normal, log Laplace, or extreme value are used, much of the analysis can be performed with programs readily available as part of commercial mathematical and statistical software. There is little doubt that the list of distribution functions that can be tried as fluctuation models will continue to grow in the future and this might improve the described procedure's performance. Also, the statistical methods to test a record's suitability for the analysis are not limited to those discussed in this chapter. It can be argued, though, that the use of more sophisticated statistical methods might not be the solution if the available records are too short and/or in gross rather than in minor violation of the independence condition set for obtaining reliable predictions. At this point, let us highlight once more the most important points that need always to be considered:

- Reasonable predictions can be made only if no drastic change will occur in the system during the pertinent period. Obviously, this can never be guaranteed and dramatic changes, for better or for worse, can always occur whether by a major accident or as the result of deliberate action. However, whenever an outburst has an identifiable specific cause, any probabilistic model would become irrelevant anyway. This is because the model can only be used to predict the combined effect of many random factors, not of a single major event whose occurrence is not reflected in the previous record.

- An occasional failure of the described methods to provide satisfactory predictions of the frequency of future events does not necessarily invalidate the concept on which the methods are based. It is always possible (inevitable according to the laws of probability) that a finite segment taken from an infinite random series will have properties that differ considerably from those of the series as a whole. Of course, the probability of encountering such a segment in real life will dramatically decrease if the count record used to determine the distribution function and its parameters is sufficiently long.

The same can be said about the length of the record used to test the prediction's accuracy. If the model and assumptions on which it is based are correct, the probability of wrong prediction will decrease with the length of the inspected series of counts. (For more on this point, see Peleg and Horowitz, 2000.) At least in principle, a large discrepancy between the predicted and observed frequencies of high counts can be considered as evidence that conditions have improved or deteriorated, as the case might be, and not necessarily of the model's failure. In such a case, as already stated, the method can be used to characterize the change qualitatively and quantitatively — the former in terms of the distribution function kind and its characteristic parameters and the latter in terms of a new set of probabilities that outbreaks of any given magnitude will be encountered.

Obviously, if the length of the past and future records is determined by practical considerations, it is possible that suitable records for the analysis might not be available at all, and any attempt to use the methods will be futile. However, whenever one has control of the data acquisition, effort should be made to generate records that are as long and dense as practically possible so that the contemplated model's or models' applicability could be tested and confirmed.

- Especially when the available record is relatively short and/or has other imperfections, the counts' distribution can be frequently described, interchangeably, by more than one distribution function. If so, it is expected that several such distribution functions will provide very similar predictions of the frequencies of high counts. This will only be true for high counts that are lower or only slightly higher than the highest count already encountered. The same distribution functions may yield estimates that will differ significantly if they are used to predict extremely rare events. Thus, a correct prediction of the frequencies of outbursts of a magnitude far outside the range of counts in the record at hand might depend on the proper choice of the distribution function that will serve as a model.

The identification of the right model, however, may not be an easy task if rare events are not manifested in the record at hand. To eliminate this problem, one would need records long enough to establish which distribution function among the possible candidates is truly superior to all the others. However, such a long record might not be available or its determination, if intended, might not be a feasible option. A possible alternative in such a

Advanced Quantitative Microbiology for Foods and Biosystems

case, albeit not as sound on theoretical grounds, would be to repeat the calculation with a battery of several distribution functions that have a comparable fit. Then, one could select the most prudent estimate as a measure of the food's or water's microbial quality or of the risk level that it carries if left unchanged (Corradini et al., 2001b).

11

A Probabilistic Model of Historic Epidemics

Statistics: the only science that enables different experts using the same figures to draw different conclusions.

Evan Esarr

Historic endemic contagious diseases like smallpox or measles are frequently characterized by a cyclic mortality pattern. Thus, their fluctuating mortality records, retrieved from archives, have been traditionally described and characterized by population dynamics models and the results of spectral analyses (Mercer, 1990; Grenfel and Dobson, 1995; Scott and Duncan, 1998). Such analyses, especially when their results are supplemented by contemporaneous data on economic and demographic factors, can frequently provide an insight into the causes of the fluctuations and the disease spread pattern. Because the fluctuations are rarely perfectly regular, the autocorrelation functions (ACFs) of the records' time series have also been examined in an attempt to reveal whether the pattern is truly periodic or chaotic (Scott and Duncan, 1998; Ruelle, 1991). The oscillations' irregularity observed in epidemiological data can also be caused, at least partly, by unknown random factors that affect the disease's virulence and mode of transmission (Ruelle, 1992; Schaffer and Truty, 1989) and errors or uncertainties in the data themselves (Peleg et al., 1997).

Also, historic records of infection or death are usually presented in the form of total annual or other periodic counts. Therefore, they do not reveal the progress of the disease during any given year or the period entered. Such records (the same can be said about endemic contemporary diseases, especially in remote areas where there are no sufficient field data) cannot be used to construct a detailed mathematical model of the disease's spread and retreat. Also, complementary climatic, socioeconomic, and environmental information might be incomplete or missing altogether and thus cannot be incorporated in the model. In all such cases, one might need to invoke probabilistic considerations to account for the effects of unknown factors on the annual mortality pattern. Whenever the traditional epidemiological models have only limited applicability, a probabilistic model of fluctuations of the kind described in Chapter 9 can serve as an alternative, at least in principle. What follows in this chapter is an examination of this possibility.

The Model

Consider that, over the pertinent time period, the affected city's population in question is more or less stable, i.e., its size may have fluctuated, but had not experienced a dramatic growth or decline, and that a disease like smallpox or measles is endemic in the place. Individual adults or children succumb to the disease for reasons that may be well known medically, but are not fully documented epidemiologically. The number of people who die from the disease in a particular year (or any other given period) can be affected by numerous factors and their interactions. Under such circumstances, one can make the following assumptions:

- A general characteristic level of mortality is primarily determined by the virulence of the disease; the human population's susceptibility or immunity to it; and the socioeconomic, sanitary, and other conditions prevalent at the time.

- Many of the factors that affect the disease's lethality vary randomly. Thus, the probability that they will produce a coherent effect by coincidence diminishes as the magnitude of the effect increases. In other words, the probability of an increase or decrease in mortality in a particular year to a level well above or well below the characteristic level becomes progressively smaller as the deviation from this characteristic level increases in either direction.

- Because of the random nature of the factors that affect the mortality from the disease, which might include the weather, the fluctuations in the annual mortality records around the characteristic multiannual mean level are also random. Therefore, for the statistical analysis, the mortality in any particular year can be considered as independent of that of any previous year at least to a certain extent. (This assumption can be verified by examining the autocorrelation plot of the time series [see below]. This assumption of independence would almost certainly be judged inappropriate if the mortality data had been compiled daily, weekly, or perhaps even monthly; in which case, the periodicity that underlies the epidemic spread would probably have become clearly evident.)

- In the examined period, the mortality time series can be considered as stationary — that is, the statistical properties of the counts' record are time invariant.

- An asymmetry in the fluctuations' pattern is expected because the number of deaths can increase dramatically in a particular year, but can never be reduced to below zero.

One of the simplest models that incorporate these considerations is the one already presented and discussed in Chapter 9 and Chapter 10, except that here we are dealing with the fluctuating annual number of human victims of a disease instead of the counts of a living microorganism. Because in both cases the nature of the fluctuations is explained in a similar general manner, we can expect that the random log normal distribution would be an effective model, at least as a first-order approximation. Expressed mathematically and applied to the annual number of deaths in a given community, the model can be written as:

$$N(n) = 10^{\mu_L + \sigma_L z(n)} \qquad (11.1)$$

where
$N(n)$ is the number of deaths in a given year n
μ_L and σ_L are the logarithmic mean and standard deviation, respectively
$z(n)$ is a standard random number with a standard random number with a standard normal (Gaussian) probability (i.e., $\mu = 0$ and $\sigma = 1$)

This model is, of course, not unique. If the fluctuations are almost symmetric and within a narrow range around the characteristic level, a random normal (Gaussian) distribution would probably be a model just as appropriate. If the number of deaths has an upper limit determined by physical or demographic considerations, then, at least in principle, a random beta or log beta distribution could provide a similar or perhaps even a better account of the historical record. However, even in what would be expected to be the ubiquitous asymmetric scenario, an alternative distribution function, like the log Laplace or extreme value, might also be found applicable.

However, if Equation 11.1 is considered as a model, its validity can be easily established by testing the normality of the logarithmic annual dead counts distribution, which is a standard feature of almost all statistical packages. A visual comparison of simulated time series produced by Equation 11.1 as a model with the μ_L and σ_L values calculated from the data, with the actual mortality record, can then provide a subjective or informal indication of whether the random log normal distribution is indeed a realistic model.

Mortality from Smallpox and Measles in 18th Century England

Smallpox

Historic records of death from smallpox in London during four periods — 1661 to 1686, 1710 to 1740, 1750 to 1780, (before vaccination), and 1810 to 1880 (after vaccination) — taken from Scott and Duncan (1998) and Mercer (1990) are shown in the upper left side (a) of Figure 11.1 through Figure 11.4. They were all tested for log normality using the Lilliefors criterion (the analysis was performed with the Systat® software package) and the results are given in Table 11.1.

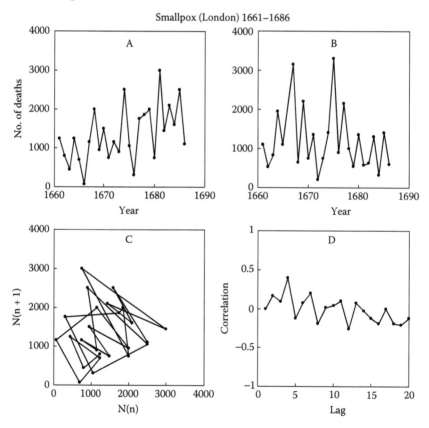

FIGURE 11.1

Records of mortality from smallpox in London 1661 to 1686 (before vaccination). A: the original data (From Scott, S. and Duncan, C.J., 1998, *Human Demography and Disease*. Cambridge University Press. Cambridge, U.K.); B: simulated data using Equation 11.1 as a model and the constants listed in Table 11.1; C: the Pachard–Taken plot for lag = 1 of the original data; D: the ACF plot of the original data.

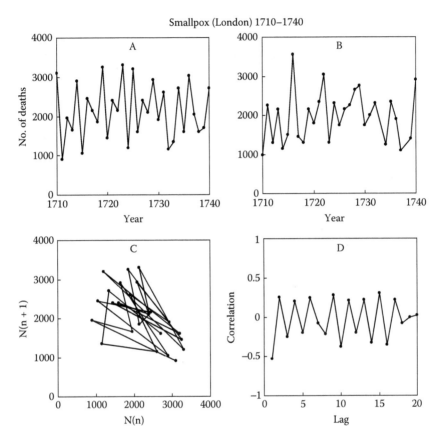

FIGURE 11.2

Records of mortality from smallpox in London 1710 to 1740 (before vaccination). A: the original data (From Scott, S. and Duncan, C.J., 1998, *Human Demography and Disease*. Cambridge University Press. Cambridge, U.K.); B: simulated data using Equation 11.1 as a model and the constants listed in Table 11.1; C: the Pachard–Taken plot for lag = 1 of the original data; D: the ACF plot of the original data.

The analysis confirmed that the log normal distribution model indeed captured the mortality pattern in all four periods. The records' Pachard–Taken plot with lag = 1 and the ACFs are shown at the bottom of the figures, marked as "C" and "D," respectively. Both indicate that the data had hardly any discernible structure and thus that the assumption that each annual record could be treated as independent is reasonable. There were some statistically significant correlations for lag one only, but they were very weak. The corresponding correlation coefficients were about 0.5, which indicates that only a small part of the variation could be explained by the mortality from the disease in a previous year.

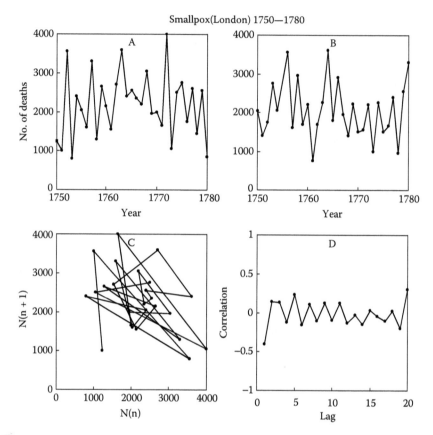

FIGURE 11.3
Records of mortality from smallpox in London 1750 to 1780 (before vaccination). A: the original data (From Scott, S. and Duncan, C.J., 1998, *Human Demography and Disease*. Cambridge University Press. Cambridge, U.K.); B: simulated data using Equation 11.1 as a model and the constants listed in Table 11.1; C: the Pachard–Taken plot for lag = 1 of the original data; D: the ACF plot of the original data.

The logarithmic mean and standard deviation (μ_L and σ_L) of each record, which are listed in the table, were used to create the simulated mortality records using Equation 11.1 as a model. They are shown next to the original record (marked "B") in Figure 11.1 and Figure 11.2. As can be seen in the figures, the simulated time series generated with the random log normal distribution as a model could be practically indistinguishable from actual historic records. They provide visual, albeit subjective, additional support to the notion that the random log normal distribution can be considered a viable model of this kind of historic epidemiological data.

Figure 11.1 through Figure 11.4 and Table 11.1 also indicate that the mortality patterns during the periods of 1710 to 1740 and 1750 to 1780

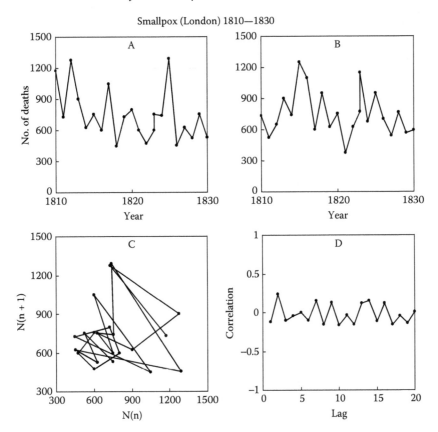

FIGURE 11.4
Records of mortality from smallpox in London 1810 to 1830 (after vaccination). A: the original data (From Mercer, A., 1990, *Disease, Mortality and Population in Transition*. Leichester University Press. Leichester, U.K.); B: simulated data using Equation 11.1 as a model and the constants listed in Table 11.1; C: the Pachard–Taken plot for lag = 1 of the original data; D: the ACF plot of the original data.

were fairly similar. It was therefore interesting to compare them with the pattern that emerged after the introduction of vaccination as a preventative measure.

The records of mortality from smallpox during the period of 1810 to 1830, obtained from Mercer (1990), are presented in Figure 11.4 (The data presented in Mercer's book cover the mortality from smallpox in London from 1700 to 1830, which includes previously discussed years.) As already mentioned, the analysis of these postvaccination mortality data showed that their distribution passed the test of log normality (Table 11.1). As judged by the autocorrelation functions, no lag has shown a significant correlation in this case and therefore the entries could be considered as

TABLE 11.1

Patterns of Mortality from Smallpox and Measles in two U.K. Cities during the late 17th, 18th, and early 19th Centuries[a]

Disease	City	Period	μ_L	σ_L	Lilliefors probability
Smallpox	London	(Before vaccination)			
		1661–1686	3.03	0.34	0.20
		1710–1740	3.31	0.152	0.56
		1750–1780	3.30	0.184	0.18
		(After vaccination)			
		1810–1880	2.85	0.134	0.29
	Glasgow	(Before vaccination)			
		1783–1800	2.50	0.15	0.09
		(After vaccination)			
		1805–1812	1.81	0.27	1.00
Measles	London	1750–1780	2.20	0.43	0.46
		1850–1900	2.25	0.082	0.12

[a] The original data are from Scott, S. and Duncan, C.J., 1998, *Human Demography and Disease.* Cambridge University Press. Cambridge, U.K., Mercer, A., 1990, *Disease, Mortality and Population in Transition.* Leichester University Press. Leichester, U.K.

random for all practical purposes (Figure 11.4). As could be expected, the postvaccination record reflects a considerable drop in the overall mortality level, expressed in a lower logarithmic mean (μ_L).

The vaccination also resulted in lowering the absolute fluctuations' amplitude, which was manifested in a lower magnitude of the logarithmic standard deviation (σ_L). Nevertheless, the *relative amplitude* of the fluctuations remained basically unchanged, i.e., $\sigma_L/\mu_L \approx 0.05$ in the records before and after vaccination. The decrease in the fluctuations' amplitude, however, was probably more dramatic than these numbers indicate because one would expect that the population of London in the years following the introduction of vaccination was probably larger than in the 18th century. Therefore, the *rate* of mortality from smallpox (e.g., the number per 100,000 inhabitants) was doubtlessly even smaller that the magnitudes of μ_L and σ_L indicate. Comparison of pre- and postvaccination mortality records from Glasgow (Table 11.1) indicated a dramatic drop of μ_L, which was expected. However, σ_L increased substantially, which is difficult to explain without supporting demographic data and information on whether the whole city population was vaccinated.

Measles

The annual mortality from measles in London during the periods of 1750 to 1780 and 1850 to 1900 is shown in Figure 11.5(A) and Figure 11.6(A). The statistical analysis of the data (Table 11.1) shows that they could be described adequately, at least as a first approximation, by the random log

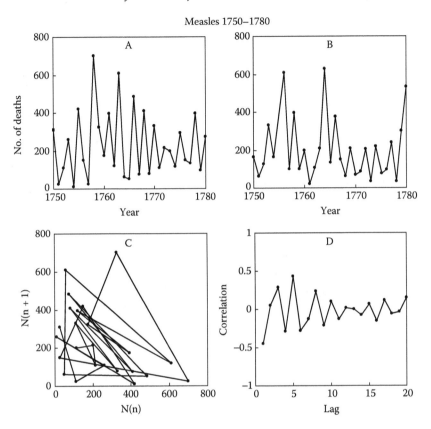

FIGURE 11.5
Records of mortality from measles in London 1750 to 1780. A: the original data (From Scott, S. and Duncan, C.J., 1998, *Human Demography and Disease*. Cambridge University Press. Cambridge, U.K.); B: simulated data using Equation 11.1 as a model and the constants listed in Table 11.1; C: the Pachard–Taken plot for lag = 1 of the original data; D: the ACF plot of the original data.

normal distribution (Equation 11.1). The Pachard–Taken and ACF plots of the records are shown at the bottom of the figures and marked "C" and "D." Examination of the autocorrelation function showed that none of the lags had a significant correlation, confirming that the annual mortality counts in these records could be considered as being independent. Again, simulation of the mortality patterns, using Equation 11.1 as a model, with the calculated μ_L and σ_L values listed in Table 11.1 could yield plots practically indistinguishable from the actual record, as demonstrated in the figures.

As could be expected, because of the lower virulence of the disease, the characteristic mortality level, expressed in terms of the logarithmic annual mean, μ_L, was much smaller than that of the smallpox (see table). Because vaccination against measles was not introduced in the intervening years

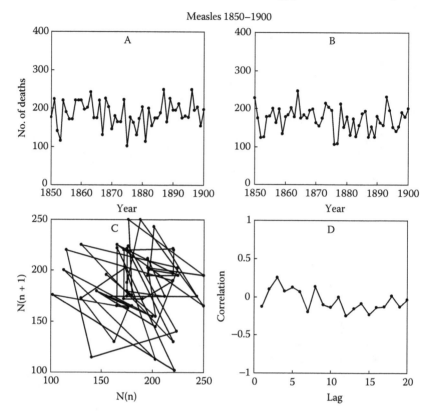

FIGURE 11.6

Records of mortality from measles in London 1850 to 1900. A: the original data (From Scott, S. and Duncan, C.J., 1998, *Human Demography and Disease*. Cambridge University Press. Cambridge, U.K.); B: simulated data using Equation 11.1 as a model and the constants listed in Table 11.1; C: the Pachard–Taken plot for lag = 1 of the original data; D: the ACF plot of the original data.

(the antimeasles vaccine only became largely available in the second half of the 20th century), the logarithmic mean values for the periods of 1750 to 1780 and 1850 to 1900 were about the same. Probably, they would have been somewhat different if adjusted for the changes in the city's population size. However, the data on which the analysis is based were reported in absolute terms — thus, the apparent similarity between the logarithmic means.

The outstanding difference between the 18th and 19th centuries' records was in the magnitude of the fluctuations' amplitudes of σ_L (see table). Because of the log normal distribution's asymmetry, a higher value of σ_L means a higher frequency or probability of outbreaks of a considerable magnitude, i.e., of much higher numbers of deaths in a particular year. This can be clearly seen in the original and simulated records, which are shown in Figure 11.5(A), Figure 11.5(B), Figure 11.6(A), and Figure 11.6(B).

Although the apparent "characteristic level" of mortality from measles remained roughly the same (about 120 deaths annually), years with a much higher than usual mortality were more frequent during the 18th century than during the 19th. Whether the changes in the mortality patterns could be attributed to human factors, such as improved sanitation and better nutrition and housing or to the diminishing virulence of the virus, is a question whose answer requires information not contained in the annual records. Yet, the change in the mortality pattern, whatever its cause or causes, could be quantified by the described model and procedure and could be compared to that of other diseases in the same or different location.

Potential Uses of the Model

It has been shown that models based on the random log normal distribution effectively captured the fluctuating mortality patterns of the two diseases as reported in historic records. It is quite conceivable that other endemic diseases or even these two diseases under different conditions may exhibit patterns that would require a different model. Consequently, establishing the model's validity by statistical criteria is an essential stage in the analysis. Fortunately, the statistical tests can be easily performed with commonly available statistical software.

Unlike rigorous epidemiological models based on population dynamics, the purely probabilistic model described here does not require that the model's mathematical structure be specified in advance. Also, once the distribution function has been chosen, the magnitude of its parameters can be determined from the available data directly, as in the two examples, by the method of moments (MM). If another distribution function is deemed more suitable, its parameters can be determined by the method of moments or the more elaborate maximum likelihood estimation method (MLE) discussed in Chapter 9 and Chapter 10. (Remember that, in the case of the normal and log normal distribution functions, the MM and MLE are identical and thus yield the same μ or μ_L and σ or σ_L and estimates. This would not be the case if an alternative distribution function is used.)

Because of its random element, the described annual mortality model (Equation 11.1) produces simulated data whose fluctuation pattern appears quite realistic. However, because it is merely probabilistic, the model does not provide any insight into the causes of the fluctuations and the mechanisms controlling the disease spread and virulence. Therefore, a model of the Equation 11.1 kind should be used if and only if the available database is unsuitable for the development of a more elaborate mechanistic model based on the principles of population dynamics. Historic and even contemporary epidemiological records (see below) can be unsuitable for more elaborate analysis. This might be due to insufficient resolution causing the pattern to be blurred or distorted. Also, the record might be too short for the true periodicity to be clearly discernible.

In principle at least, if the model is applied to contemporary epidemiologic data, it can be used to estimate the future probability of a disease's outbreak, provided that the assumptions on which the model is based remain valid. As shown in the previous chapters, this can be verified experimentally. Obviously, the level of infection or mortality that constitutes an outbreak is a matter for public health officials to decide.

The model as formulated in this chapter will lose its predictive ability in periods in which demographics change significantly — e.g., when the population grows or shrinks dramatically as a result of war or migration, or for transition periods, when a preventive measure like vaccination or water disinfection is introduced, or when a new medical treatment or drug becomes widely available to the population at risk. (In such cases, the condition of stationarity is strongly violated. However, the model can be used to quantify the efficacy of the preventive or remedial measures in terms of its parameters' magnitude if applied separately to the record before and after the event). The model's utility would be seriously curtailed in situations in which the periodic nature of the disease's spread is evident despite the record's imperfections. In principle at least (see Chapter 9 and Chapter 10), the model can be reformulated to account for aperiodic fluctuations superimposed on a known trend or regular long-term oscillations. However, this would require long records that, if they had been available in the first place, could probably also serve as a basis for a deterministic dynamic model development.

The described model's main advantage is that it enables one to account for the fluctuating patterns in terms of familiar statistical parameters and that these in turn can be used to quantify differences between epidemics in different locations and times. Such parameters — the μ_L and σ_L of the log normal distribution are just an example, as already mentioned — can also be used to assess the efficacy of prophylactic measures and medical treatments, which might be otherwise masked by the irregularity of the epidemiological records before and after their introduction.

Similarly, the magnitude of the distribution's parameters can be used to indicate a growing risk to a population, which also might escape attention because of the large fluctuations in what might be considered a "normal" pattern of the disease. Again, although the method's applicability has only been demonstrated with a limited number of historic epidemiologic records, the underlying concept would probably apply to contemporary diseases whenever the only available record is insufficient for the derivation of a more effective model.

12

Aperiodic Microbial Outbursts with Variable Duration

If we knew what it was we were doing, it would not be called research, would it?

Albert Einstein

The previous two chapters have dealt with microbial records in which each individual count was independent or could be considered as practically independent of its predecessor and any other previous counts. Theoretically, the models discussed in Chapter 10 and Chapter 11 should be applicable only if this condition of independence is satisfied. It so happens, though, that minor violations of this condition seem to have little influence on the models' ability to predict the frequencies of future outbursts. However, there is little doubt that such violations could affect the quality of the actual estimates.

The situation is totally different when the time intervals between successive counts are short enough to detect, consistently, the changes that occur in the population's size during the outbursts, but are too long to capture the details of the population's rise and fall. In other words, the population size, which probably grows and shrinks almost continuously (see Chapter 9 and Chapter 10), is sampled at a rate too slow to produce even an approximately smooth count record. Moreover, unlike in the laboratory when the microbial population's growth is followed at time intervals decided by the investigator so that the process's kinetics can be revealed, the sampling frequency of microorganisms in natural habitats, like water reservoirs and streams, is primarily determined by logistic considerations rather than by the desire to conform to any particular experimental design. This is especially the case when the counts are taken as part of a governmental or municipal monitoring system and are not intended to serve as a database for population dynamics models.

One of the most obvious consequences of such data-gathering routines is that the time intervals between successive counts quite frequently can vary considerably. The same can be said about other microbial quality monitoring systems, which might include that of sensitive foods. However,

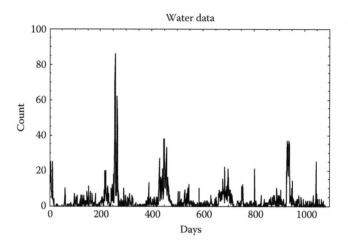

FIGURE 12.1
A record of the fecal coliform counts in a Massachusetts water reservoir. Notice that the original gaps due to weekends and holidays have been eliminated. (From Engel, R. et al., 2001, *Bull. Math. Biol.*, 63, 1005–1023. With permission, courtesy of Elsevier Ltd.)

whatever the reason for the sampling plan, in an intermediate class of microbial records, the counts are certainly not independent and also not dense enough to delineate continuity. What follows will describe how such records can be interpreted and modeled.

Microbial Fluctuations in a Water Reservoir

A record of fecal coliform counts in a Massachusetts water reservoir is shown in Figure 12.1. It was constructed from daily determinations performed by the Massachusetts Water Resources Authority (MWRA) and its analysis was reported by Engel et al. (2001a). The record shown in the figure is slightly distorted because the reservoir's water had not been monitored on weekends and holidays. Still, the record does retain the general character of the counts' fluctuating pattern: relatively small oscillations punctuated by large aperiodic outbursts of varying magnitude and *duration*. The latter could be on the order of days and sometimes weeks. During an outburst, the coliform numbers rarely if ever rose or fell smoothly. In most cases, the counts fluctuated widely and in what appears to be a random fashion. The coliform count is a measure of the water's level of contamination by fecal material. Thus, a dramatic rise in the coliform numbers is usually a manifestation of a massive discharge of fecal material into the reservoir.

The actual numbers can be viewed as rough indicators of the discharge magnitude and the reservoir's ability to disperse and eliminate it through natural biological processes. The structure of a peak is probably indicative of the discharge pattern. Its overall width is a reflection of the discharge duration and/or of the rate at which environmental conditions affect the spread and survival of the fecal organisms after they reached the water. Multiple peaks within the same outburst can be a manifestation of fecal loads discharged at close proximity, temporal or spatial, or a combination of local currents that had caused a temporary drop in the coliform number before it bounced back. Of course, other explanations are possible, but because our focus is on the mathematical description of records of this kind, the preceding list will suffice. Regardless of the actual causes of the observed pattern and its details, it has two discernible regimes:

- Irregular independent counts' oscillations having relatively small amplitude that can be dubbed 'background fluctuations.'
- A population explosion event or an outburst of several days' duration, in which the coliform number reaches levels that exceed the background fluctuation by a substantial margin.

To construct a single model that accounts for a pattern that consists of both regimes, consider the following assumptions:

- Under normal conditions — that is, when the water body is not experiencing massive contamination by fecal material — the coliform counts follow a fluctuation pattern obeying the previously described model based on the log normal distribution (Peleg et al., 1997; Peleg and Horowitz, 2000):

$$N(n) = 10^{\mu_L + \sigma_L Z(n)} \tag{12.1}$$

where $N(n)$ is the nth count, μ_L and σ_L are the count's distribution's logarithmic mean and standard deviation, respectively, and $Z(n)$ is a sequence of independent random numbers with standard normal distribution, i.e., with a zero mean ($\mu = 0$) and unit standard deviation ($\sigma = 1$). (As before, the model is formulated with the base-10 logarithm commonly used to present the size of microbial loads.) The model is formulated in this way to account for the asymmetry between growth and decay, that is, that a microbial population's size can increase dramatically during an outburst, but can never become negative after it is over (see Chapter 9). Embedded in Equation 12.1 is the notion that no detection threshold exists. The assumption, of course, might not always be correct, but this will hardly affect the discussion that will follow. We will address the detection threshold issue in relation to modeling the presence of pathogens in foods (see "A Model of Pathogen Outbursts in Foods" below). As to the coliforms in the water reservoir, we also assume that, on the pertinent time scale, their counts during a stationary stage between outbursts have no trend or long-term periodic oscillations.

- When the number of fecal organisms reaches or exceeds a certain high level, it might signal that an outburst is about to occur. At that point, the next count can further increase or, if it has been a false alarm, fall back to a level deemed normal; in this case, the counts will continue to fluctuate in the pattern described by Equation 12.1. Let us call this potential explosion level E. If any nth count reaches or exceeds this number, E, then there is a probability, p, that its successor will further increase, signaling the onset of an outburst. There is therefore a probability, $1 - p$, that the successor count will fall back to the normal level, in which case the fluctuations will return to the background pattern.

- Once an outburst has been set, there is still the probability, p, that the second successive count will continue to rise and a probability, $1 - p$, that it will decline, which *might* signal the end of the outburst and a return to the normal (background) fluctuation pattern.

- During an outburst, the fluctuating counts' *direction* is governed by the probability, p. However, their actual magnitude is determined by a log normal distribution, except that the random value rendered by Equation 12.1 is multiplied by a factor, k_g, when the counts continue to grow and by a factor, k_d, when they drop. (The subscripts g and d stand for growth and decline, respectively.)

- The same applies to the third, fourth, and any other successive count until the outburst ends and the counts return to their normal fluctuation pattern when they reach a level below E. In other words, when any declining count falls to below the explosion marker, E, the outburst is considered to be over and the record resumes its regular (i.e., random log normal) fluctuating mode governed by Equation 12.1, until the onset of the next outburst.

A schematic view of the preceding scenario is presented in Figure 12.2. Mathematically, it can be expressed as follows (Engel et al., 2001a):

- $L_1, L_2,...$ will denote a purely random sequence of log normally distributed random variables with logarithmic mean and standard deviation μ_L and σ_L. This means that L_n is governed by Equation 12.1.

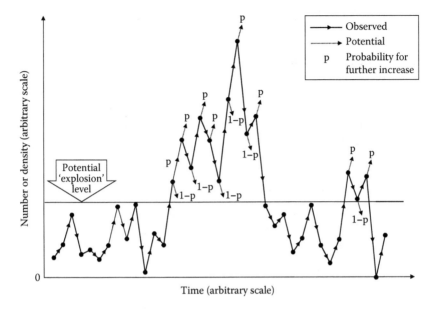

FIGURE 12.2
Schematic view of the construction of the algorithmic model that describes records of the kind shown in Figure 12.1. (From Engel, R. et al., 2001, *Bull. Math. Biol.*, 63, 1005–1023. With permission, courtesy of Elsevier Ltd.)

- In the explosion or outburst regime, triggered by $L_n > E$, the fluctuations are governed by p, k_g, and k_d, in addition to μ_L and σ_L.
- $N(n)$ denotes the actual recorded count at time n or that which corresponds to the nth member of the record.
- The constants k_g and k_d are the growth and decline factors that regulate, respectively, the magnitudes of the increases and decreases in the counts during an outburst.

For convenience, we will set the initial count $N(0)$ to an arbitrary value, e.g., $N(0) = 0$. For any $n \geq 1$:

$$\text{If } N_{n-1} \leq E, \text{ then } N_n = L_n \qquad (12.2)$$

$$\text{If } N_{n-1} > E, \text{ then } N_n = N_{n-1} + k_g L_n \text{ with probability } p \qquad (12.3)$$

and

$$N_n = (N_{n-1} - k_d L_n^+) \text{ with probability } 1 - p \qquad (12.4)$$

The sequence L_n is a population size that behaves as though no explosion threshold existed. The notation (x^+) indicates that $x = x$ if $x \geq 0$, and $x = 0$ if $x < 0$. Thus, Equation 12.4 accounts for the fact that the observed count can be zero but not negative.

The preceding is a formal manner to state that the sequence of counts is regulated by two sets of probabilities: one that that underlies the fluctuations below the explosive threshold, E, and the other governing the rises and falls during an outburst episode, considered here as any time when the counts exceed the threshold level, E.

Determination of Model Parameters

Conventional population models are usually presented in the form of a set of differential or difference equations (Murray 1989; Royama 1992; Brown and Rothery, 1993). In contrast, the model as expressed by Equation 12.2 through Equation 12.4 is in a form of an *algorithm*, and thus determination of its parameters from an actual record requires a special procedure of the kind originally proposed by Dr. Joseph Horowitz of the Mathematics and Statistics Department at the University of Massachusetts (see Engel et al., 2001a). A plot of the autocorrelation function of a counts record showing outbursts with a substantial duration of the kind shown in Figure 12.1 will immediately reveal that the counts are not independent (Figure 12.3).

Now, suppose that we remove from the record all counts exceeding a level, for example, E_1, that is visibly within the explosive regime (Figure 12.3) and plot the autocorrelation function again. Because this E_1 is well above the "true" explosive level, the autocorrelation will still fail any test for randomness. We can now set a new lower value, E_2, and remove from the record all the counts that are above it. Once these are removed, the autocorrelation function of the "culled again" record can be plotted and tested for randomness (see following). The process is then repeated for $E_3, E_4, \ldots E_n$ until the record passes the test of randomness; at this point, the chosen E_n is a close estimate of the threshold level, E. Once E is determined, the distribution of the remaining counts can be determined by the methods described in the previous chapter. If the counts' distribution is indeed log normal, the logarithmic mean, μ_L, and standard deviation, σ_L, can be readily calculated.

The magnitude of the probability of a further increase in the explosive regime, p (see Figure 12.2), can be estimated by counting the number of increments during all the recorded outbursts, u, and the number of decrements, d. The probability estimate of $\hat{p}$ will then be:

$$\hat{p} = \frac{u}{d+u} \qquad (12.5)$$

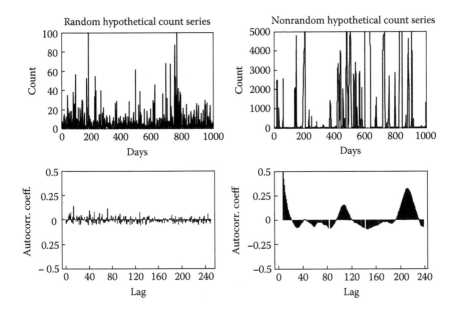

FIGURE 12.3
Simulated random and nonrandom fluctuating count series and their corresponding auto-correlation function. Notice that records of the kind shown in Figure 12.1 and the one at the right of this figure can be trimmed from the top and the autocorrelation of the remaining entries examined. The trimming can then be repeated until the remainders' autocorrelation function will pass the test for randomness, in which case the upper limit of the untrimmed residuals will mark the explosion threshold (*E*). (From Engel, R. et al., 2001, *J. Sci. Food Agric.*, 81, 1250–1262. With permission, courtesy of Wiley Publishing.)

The magnitudes of the increment and decrement factors k_g and k_d, respectively, can be estimated by comparing the mean observed increment or decrement in the explosive regime with the corresponding values in the regular or normal background regime (that is, when calculated for the counts when they are fluctuating below the threshold level *E*). Said differently, the estimate of k_g is the ratio between the mean rise at the explosive regime and that at the regular background fluctuations. Similarly, the estimate of k_d is the ratio between the mean drop at the explosive regime and that at the regular background fluctuations. Thus, even though the fluctuation model is in the form of an algorithm expressed by Equation 12.2 through Equation 12.4, its six parameters — namely, E, μ_L, σ_L, p, k_g, and k_d — can be estimated from a given record by a well defined mathematical method. The estimates will be meaningful, of course, only if the analyzed record has long enough segments of regular fluctuations to calculate μ_L and σ_L and a sufficient number of outbursts for calculating p, k_g, and k_d.

Autocorrelations — water data

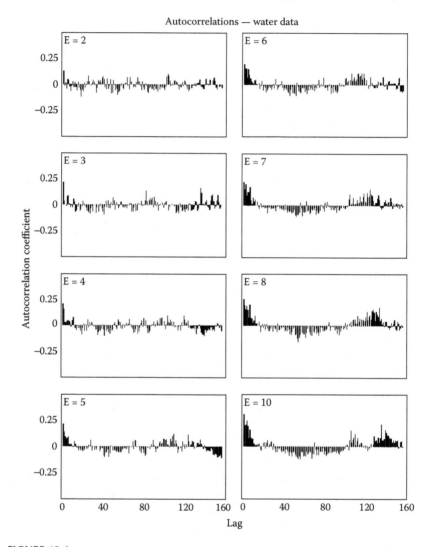

FIGURE 12.4
The autocorrelation functions of the water reservoir data shown in Figure 12.1 after trimming at various explosion threshold levels (E). Notice the progressive loss of waviness as E drops from 10 (lower right) to 2 (upper left). (From Engel, R. et al., 2001, *Bull. Math. Biol.*, 63, 1005–1023. With permission, courtesy of Elsevier Ltd.)

Fluctuation Parameters of the Massachusetts Water Reservoir

The autocorrelation functions of the coliform record in the reservoir with different assumed levels of thresholds in the range of 2 to 10 are shown in Figure 12.4. Even casual visual inspection would reveal that when the

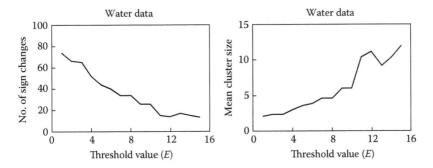

FIGURE 12.5
The mean number of sign changes and mean cluster size in the autocorrelation functions of the original water reservoir record, trimmed at various cutoff levels of E. (From Engel, R. et al., 2001, *Bull. Math. Biol.*, 63, 1005–1023. With permission, courtesy of Elsevier Ltd.)

selected cutoffs were four and above ($E \geq 4$), the autocorrelation function had a clear "waviness," which indicated that the counts of the corresponding trimmed record could not be considered as independent.

However, when the cutoff was a count of two or three, the remaining record needed a closer look. When the number of sign changes and mean cluster size were plotted against the contemplated threshold value, the results (Figure 12.5) showed the expected rise and decline. However, there was no sharp rise or a drop in these parameters that would identify the threshold level unambiguously. The culled data with the various levels of presumed E used as the cutoff value were therefore subjected to the "portmanteau test" (Brockwell and Davis, 1987), whose results are summarized in Table 12.1. They identified the threshold level as being two, i.e., the level at which the remaining counts series' autocorrelation passed the randomness test. The low value means that a massive outburst can occur without warning, even when all preceding counts are fairly low. This is consistent with the nature of fecal contamination, which is largely caused by accidental discharge of fecal material into the reservoir. With the threshold level for explosion determined, i.e., $E = 2$, all other parameters could be estimated as explained earlier: $\mu_L = 0.335$, $\sigma_L = 0.330$, $\hat{p} = 0.256$, $k_g = 4.5$, and $k_d = 3$.

In this case, the available database for the analysis was sufficiently large so that the described procedure would be applied. Also, although the gaps in the data did not render the method inoperable, they probably did affect the quality of the parameter estimates. Some statistical methods can estimate missing data, but their practicality in water analysis has yet to be demonstrated. Nevertheless, even with all the record's imperfections, the described method still allowed quantification of the reservoir microbial

TABLE 12.1

Results of Portmanteau Tests of Massachusetts Reservoir Water Data
after Being Trimmed at Various Contemplated Explosive Thresholds, E

E	No. counts left	Degr. freedom	Q	chi^2 ($p = 0.05$)	Random
1	167	12	7.3	21.03	+
2	296	17	10.9	27.59	+
3	382	19	30.9	30.14	−
4	452	21	46.1	32.67	−
5	496	22	55.3	33.92	−
6	529	23	79.9	35.17	−
7	562	23	115.0	35.17	−
8	581	24	154.1	36.42	−
9	597	24	162.8	36.42	−
10	612	24	212.5	36.42	−
11	627	25	273.8	37.65	−
12	637	25	283.1	37.65	−
13	643	25	287.3	37.65	−
14	650	25	313.8	37.65	−
15	652	25	339.2	37.65	−

Source: From Engel, R. et al., 2001, *Bull. Math. Biol.*, 63, 1005–1023. With permission, courtesy of Elsevier Ltd.

contamination pattern, which cannot be described by any conventional population dynamics model unless its parameters are adjusted.

The algorithmic model and, of course, its simpler predecessors (discussed in the two previous chapters) can be used to simulate long-range patterns of contamination. These in turn can be used to estimate whether the frequency of massive outbursts and their duration poses a significant health hazard that will require preventive action. If and when remedial measures are taken, their efficacy could be assessed quantitatively in terms of the same model's parameters. Once these are determined, they can be used to generate future scenarios that, upon examination, will reveal whether the risk of a massive presence of coliforms in the water has been reduced significantly (see below).

Validation of the Threshold Estimation Method

Experimental confirmation of the model is difficult for technical and logistic reasons; it will take a few years to produce a similar record from the same reservoir, for example. However, one can use simulated data to test the mathematical method, at least. This can be done by generating random records with known parameters and attempting to retrieve their magnitude by the described procedure. Examples of generated parameters using the algorithmic model (Equation 12.2 through Equation 12.4) are shown in Figure 12.6 and Figure 12.7. With the exception of E, they had the same

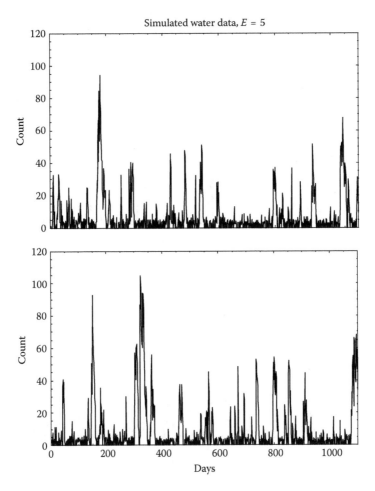

FIGURE 12.6
Examples of simulated random count records using the algorithmic model (Equation 12.1 through Equation 12.4) with $E = 5$, $\mu_L = 0.335$, $\sigma_L = 0.330$, $p = 0.256$, $k_g = 3.0$, and $k_d = 3.4$. Notice the large differences in the details of the two records even though the generation parameters were identical. (From Engel, R. et al., 2001, *Bull. Math. Biol.*, 63, 1005–1023. With permission, courtesy of Elsevier Ltd.)

parameters — namely, μ_L, σ_L, p, k_g, and k_d — as those of the reservoir's water. They do not closely resemble the real record visually partly because they were created without any gap between successive entries. Still, they share the common feature that the outbursts are randomly spaced and have variable magnitude and duration.

The autocorrelation functions after trimming the records at various levels of E are shown in Figure 12.8 and Figure 12.9. A plot of the number of sign changes and mean cluster sizes as a function of the presumed

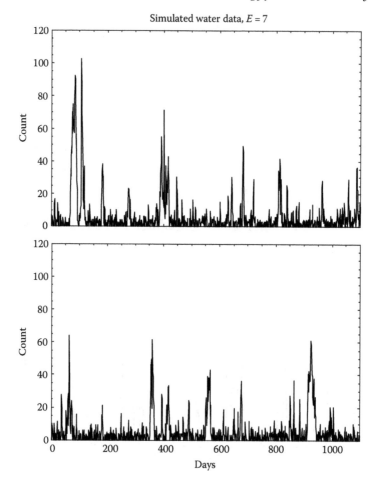

FIGURE 12.7
Examples of simulated random count records using the algorithmic model (Equation 12.1 through Equation 12.4) with $E = 7$, $\mu_L = 0.335$, $\sigma_L = 0.330$, $p = 0.256$, $k_g = 3.0$, and $k_d = 3.4$. Notice the large differences in the details of the two records even though the generation parameters were identical. (From Engel, R. et al., 2001, *Bull. Math. Biol.*, 63, 1005–1023. With permission, courtesy of Elsevier Ltd.)

threshold value, E (see Figure 12.10), again showed the expected rise or decline but no clear sharp rise or drop that would identify the correct value of E. Results of the portmanteau test are given in Table 12.2 and Table 12.3. They show that, in three cases, the identification of E was accurate and that, in one case, the estimate was only slightly higher than the correct value. The same was observed in the analysis of additional simulations generated with different values of E not shown here. In all the cases in which there was a discrepancy, it was small, in absolute terms, and always an overestimate (see Engel et al., 2001a).

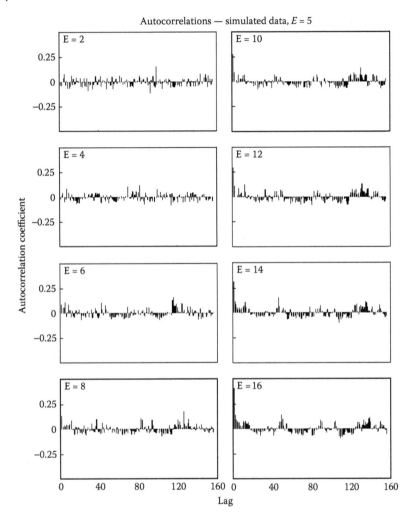

FIGURE 12.8

The autocorrelation functions of the simulated records shown in Figure 12.6 ($E = 5$) after various levels of trimming. Notice that the clear waviness, which is evident when E is above 8, disappears when the cutoff is below 6. (From Engel, R. et al., 2001, *Bull. Math. Biol.*, 63, 1005–1023. With permission, courtesy of Elsevier Ltd.)

Notice that the shape of random time series can vary dramatically (see Figure 12.6 and Figure 12.7) even when generated with a model having exactly the same parameters. Thus, an occasional discrepancy between the calculated and expected values is not at all surprising. The problem can be solved by running the model's program repeatedly until enough parameter estimates can be examined statistically. Examples of such analyses are given in the next paragraphs.

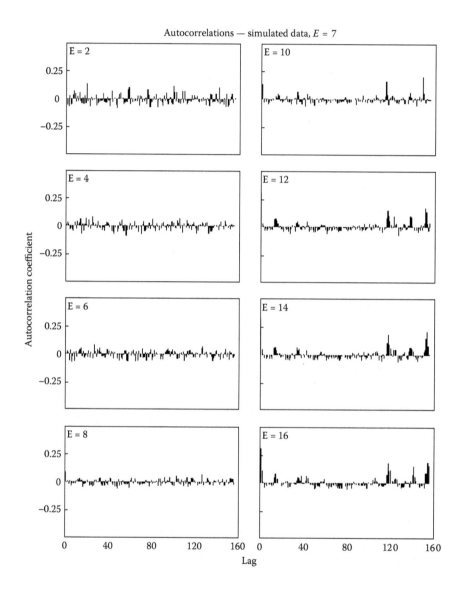

FIGURE 12.9
The autocorrelation functions of the simulated records shown in Figure 12.7 ($E = 7$) after various levels of trimming. Notice that the clear waviness, which is evident when E is above 10, disappears when the cutoff is below 8. (From Engel, R. et al., 2001, *Bull. Math. Biol.*, 63, 1005–1023. With permission, courtesy of Elsevier Ltd.)

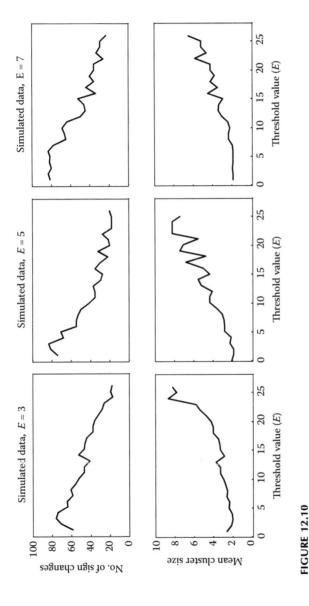

FIGURE 12.10

The number of sign changes and mean cluster size in the autocorrelation function of simulated data generated with the algorithmic model (Equation 12.1 through Equation 12.4), with three levels of trimmings. (From Engel, R. et al., 2001, *Bull. Math. Biol.*, 63, 1005–1023. With permission, courtesy of Elsevier Ltd.)

TABLE 12.2

Results of Portmanteau Tests of Simulated Data with $E = 5$ after Being Trimmed at Various Guessed Values of Explosive Threshold, E

Simul. no.	E	No. counts left	Degr. freedom	Q	chi² ($p = 0.05$)	Random
1	1	178	13	10.33	22.36	+
	2	428	20	13.14	31.41	+
	3	592	24	25.51	36.42	+
	4	714	26	23.08	38.89	+
	5	795	28	33.35	41.33	+
	6	822	28	42.06	41.33	−
	7	839	28	37.83	41.33	+
	8	852	29	41.94	42.56	+
	9	866	29	76.43	42.56	−
	10	875	29	101.60	42.56	−
	11	880	29	119.69	42.56	−
	12	887	29	127.23	42.56	−
	13	890	29	153.43	42.56	−
	14	897	29	140.64	42.56	−
	15	902	30	190.77	43.77	−
2	1	181	13	12.04	22.36	+
	2	408	20	21.10	31.41	+
	3	566	23	14.73	35.17	+
	4	671	25	17.70	37.65	+
	5	727	26	16.51	38.89	+
	6	750	27	45.13	40.11	−
	7	764	27	50.58	40.11	−
	8	779	27	58.40	40.11	−
	9	788	28	60.09	41.33	−
	10	796	28	59.60	41.33	−
	11	807	28	68.45	41.33	−
	12	819	28	73.77	41.33	−
	13	826	28	74.06	41.33	−
	14	830	28	81.55	41.33	−
	15	834	28	93.48	41.33	−

Source: From Engel, R. et al., 2001, *Bull. Math. Biol.*, 63, 1005–1023. With permission, courtesy of Elsevier Ltd.

A Model of Pathogen Outbursts in Foods

Avoiding food poisoning outbreaks has been the prime goal of all food manufacturers, distributors, and handlers, on the one hand, and of public health authorities, on the other. When an outbreak does happen, its cause is usually traced to a specific pathogen. In many cases, the investigation that follows an outbreak also reveals a chain of events that has allowed the pathogen's presence in the food and enabled its proliferation to a level

TABLE 12.3

Results of Portmanteau Tests of Simulated Data with $E = 7$ after Being Trimmed at Various Guessed Values of Explosive Threshold, E

Simul. no.	E	No. counts left	Degr. freedom	Q	Chi² ($p = 0.05$)	Random
1	1	144	12	10.53	21.03	+
	2	420	20	22.68	31.41	+
	3	618	24	30.07	36.42	+
	4	753	27	30.01	40.11	+
	5	827	28	21.23	41.33	+
	6	880	29	32.39	42.56	+
	7	908	30	16.17	43.77	+
	8	922	30	19.34	43.77	+
	9	931	30	25.21	43.77	+
	10	941	30	30.55	43.77	+
	11	948	30	92.09	43.77	−
	12	957	30	120.80	43.77	−
	13	959	30	138.77	43.77	−
	14	960	30	130.33	43.77	−
	15	967	31	126.80	44.99	−
2	1	162	12	9.74	21.03	+
	2	430	20	11.45	31.41	+
	3	617	24	30.80	36.42	+
	4	716	26	34.07	38.89	+
	5	795	28	46.38	41.33	−
	6	832	28	27.51	41.33	+
	7	867	29	31.52	42.56	+
	8	882	29	57.55	42.56	−
	9	889	29	61.40	42.56	−
	10	899	29	77.90	42.56	−
	11	905	30	89.60	43.77	−
	12	909	30	95.96	43.77	−
	13	912	30	131.51	43.77	−
	14	915	30	134.25	43.77	−
	15	923	30	287.33	43.77	−

Source: From Engel, R. et al., 2001, *Bull. Math. Biol.*, 63, 1005–1023. With permission, courtesy of Elsevier Ltd.

at which its ingestion would result in acute poisoning. Yet, some pathogens seem to be almost always present in certain foods — *Salmonella* in poultry and eggs is a classic example — but this pathogen's ubiquitous presence causes only occasional episodes of actual poisoning and these, when they occur, have varying severity.

The conditions that promote a food-poisoning outbreak and how it can be avoided are usually well known. The same can be said about the interactions of the organism with the human body after it is ingested. These have been studied extensively and should not concern us here. The issue that we would like to address is whether a food-borne pathogen's history at the *population level* could be analyzed in a way that would explain the sporadic nature of food-poisoning outbreaks and could also help in the quantification of the associated risk.

Risk assessment is a topic covered by a large body of research in epidemiology and public health. What follows is only a proposal of an alternative approach that might be useful in elucidating the unpredictable character of pathogen outbursts in foods. We will refer to an outburst as a situation in which the number of pathogens is sufficiently high to be of safety concern. This does not mean that once the contaminated food is consumed, it will necessarily poison the person. The development of clinical symptoms may depend on the amount of the infected food ingested and the state of the person's immune system, for example. The latter is a factor clearly outside the scope of the discussion here and has been addressed in a totally different way.

As far as the pathogen's population size in a given food is concerned (ground meat, raw chicken, and homemade potato salad are a few foods known to have caused food poisoning), consider the scenario shown schematically in Figure 12.10. If sampled periodically under conditions nonconducive to an explosion, the pathogen's population size would most likely fluctuate according to the model described by Equation 12.1 or in a manner governed by another asymmetric distribution function. We can also add the assumption that there is a minimum detection level, D, below which the presence of the pathogen cannot be revealed by a routine assay. (As always in microbiology, a zero reading can mean that the organism has been absent or that its presence could not be detected.) We also assume, that for an outburst to occur, the population must exceed a threshold level that, again, we will call E. Once this level is passed, there is a probability, p, that the population will continue to grow and a probability of $1 - p$ that it will start to decline. (The reasons for the decline should not concern us here. They can be resource exhaustion, crowding, release of inhibiting metabolites, etc.)

During an outburst, the growth and decline can be governed by k_g and k_d, respectively, which are factors that multiply the amplitude of the usual random fluctuations in the pathogen's population size. After the growth has reached its peak and once the declining population has fallen to below the explosive threshold level E (Figure 12.11), the regular random fluctuations will resume. In theory, the past outburst fluctuations can continue until the onset of a new outburst. In reality, of course (see below), once a serious food poisoning outbreak has occurred, it is unlikely that the situation will be allowed to continue.

Thus, as far as commercial operations such as in a restaurant, hotel, hospital, school, or a military base are concerned, the described scenario will usually come to an end after or perhaps even during the first explosive episode. The same can be said about other kinds of outbreaks not caused by food-borne pathogens, but these are outside the scope of this book.

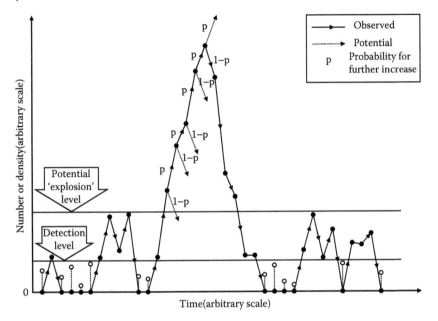

FIGURE 12.11

Schematic view of the algorithmic pathogen outburst model's construction (Equation 12.6 through Equation 12.12). It differs from that shown in Figure 12.2 in that it allows for a threshold detection level or the pathogen's true absence. (From Engel, R. et al., 2001, *J. Sci. Food Agric.*, 81, 1250–1262. With permission, courtesy of Wiley Publishing.)

Nevertheless, a possibility that cannot be ruled out is that records of the kind qualitatively described in Figure 12.10 (or Figure 12.2, which is a special case in which $D = 0$) would be formed in the course of certain epidemics.

Mathematically, hypothetical scenarios of the kind depicted previously can be expressed by the following algorithmic model:

- $L_1, L_2,...$ will be a sequence of log normally distributed random variables with logarithmic mean and standard deviation μ_L and σ_L, i.e., each of its members, L_n, is governed by Equation 12.1.

- L'_n and L''_n are two similar sequences, except that their members are governed by Equation 12.1 with μ'_L and σ'_L and μ''_L and σ''_L, respectively.

- D and E are the detection and explosion threshold values, respectively, as explained earlier.

- N_n is the recorded nth count (or the count after time n).

- k_g and k_d are the growth and decline factors that control the steepness of the increase and decrease in the pathogen's population size during an outburst, as explained earlier.

The sequence starts with an arbitrary number, N_0 (e.g., $N_0 = 0$), after which:

$$\text{If } L_1 \leq D, \text{ then } N_1 = 0 \tag{12.6}$$

$$\text{If } L_1 > D, \text{ then } N_1 = L_1 \tag{12.7}$$

The following statements pertain to $n > 1$:

$$\text{If } N_{n-1} \leq E \text{ and } L_n \leq D, \text{ then } N_n = 0 \tag{12.8}$$

$$\text{If } N_{n-1} \leq E \text{ and } L_n > D, \text{ then } N_n = L_n \tag{12.9}$$

$$\text{If } N_{n-1} > E \text{ and } N_{n-2} > N_{n-1}, \text{ then } N_n = (N_{n-1} - k_d L''_n)^+ \tag{12.10}$$

$$\text{If } N_{n-1} > E \text{ and } N_{n-2} \leq N_{n-1}, \text{ then } N_n = N_{n-1} + k_g L'_n$$
$$\text{with a probability } p \tag{12.11}$$

and

$$N_n = (N_{n-1} - k_d L''_n)^+ \text{ with probability } 1 - p \tag{12.12}$$

According to this model, L_n is the actual population size that would be observed if no detection level and outburst threshold had existed. The notation $(x)^+$, as before, means that $x = x$ if $x \geq 0$ and $x = 0$ if $x < 0$. It is introduced so that the population's size would never be negative. During an outburst, the count can increase by the amount $k_g L'_n$ with a probability of p and decline by a factor of $k_d L''_n$ with a probability of $1 - p$. In reality, the three pairs of μ_L and σ_L (the parameters of Equation 12.1 that govern the fluctuations during the nonexplosive and explosive regimes) need not be the same and most likely they are not. This is to say that the fluctuations are governed by different sets of random processes when the pathogen's population is in a normal or explosive state. Yet, for simplicity, we will assume that $\mu_L = \mu'_L = \mu_L''$ and $\sigma_L = \sigma'_L = \sigma''_L$ — that is, there is only one underlying distribution and the rise and fall of the population during an outburst is solely regulated by the probability, p, and the magnitude of the parameters k_g and k_d.

Examples of simulated count records using the preceding model are shown in Figure 12.12 and Figure 12.13. Such simulations can be used to investigate how the magnitude of the different model parameters affects the number of outbursts in a given sequence length, their duration, and the peak number reached (Engel et al., 2001b). Examples are shown in Figure 12.14 and Figure 12.15. They can also be used to check the reliability

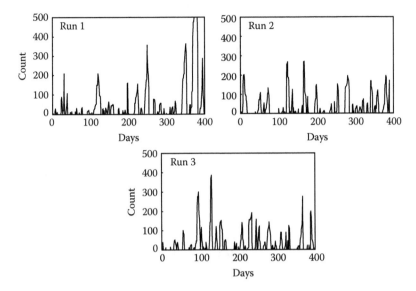

FIGURE 12.12
Simulated records of pathogen outburst generated with the algorithmic model (Equation 12.6 through Equation 12.12) with $D = 7$, $E = 0$, $\mu_L = 0.3$, $\sigma_L = 0.6$, and $p = 0.7$. (From Engel, R. et al., 2001, *J. Sci. Food Agric.*, 81, 1250–1262. With permission, courtesy of Wiley Publishing.)

of the method to estimate the model's parameters and the range in which they are expected to be found. This is done by generating a large number of simulated records, applying the method, and creating a histogram of the estimated values. Examples of such histograms are shown in Figure 12.16.

The estimates' frequencies are the black bars and the "correct" values (the ones used for generating the simulated records) are the shaded bars. The histograms demonstrate that the calculated values indeed form a bell-shaped distribution with a center quite close to the correct value. The ease at which sequences can be generated and studied once the program is in place make the method a potential tool in microbial risk assessment. If reasonable estimates of a pathogen's parameters are available or can be derived from an actual record or records, the program can be run for as many times as needed to account for a year or any other period of interest. Alternatively, the program can be run to account for many locations — the number of homes, restaurants, or school cafeterias, for example. The resulting simulations will produce a breakdown of the number of potential outbursts of different magnitudes and duration. These, in turn, can be used to estimate quantitatively the risk of encountering an outburst, which might lead to a food poisoning outbreak.

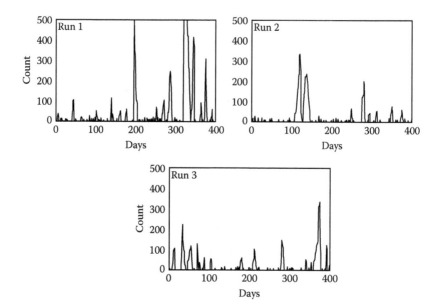

FIGURE 12.13
Simulated records of pathogen outburst generated with the algorithmic model (Equation 12.6 through Equation 12.12) with $D = 7$, $E = 20$, $\mu_L = 0.3$, $\sigma_L = 0.6$, and $p = 0.8$ (From Engel, R. et al., 2001, *J. Sci. Food Agric.*, 81, 1250–1262. With permission, courtesy of Wiley Publishing.)

The simulated examples given in Figure 12.14 and Figure 12.15 show, as expected, that the distributions of the outbursts' magnitude and duration are asymmetric with a very strong skewness to the right. In other words, one can expect, even on a theoretical basis, that there will always be many more small and short outbursts than explosive situations, which are likely to result in an actual food-poisoning outbreak. Interestingly, the general shape of the histograms shown in these figures is almost identical to the one actually observed in a study of the presence of *Samonella enteritidis* in eggs and egg products (Baker et al., 1998). This is an unlikely coincidence. If the assumptions on which the model is based are correct (i.e., the concept that at least two random fluctuation regimes governed by two probabilistic mechanisms operate between and during the outbursts is valid), then the patterns shown in Figure 12.14 and Figure 12.15 are an inevitable outcome.

Notice, though, that in the case in which an outburst of a food-borne pathogen has resulted in an actual episode of food poisoning, it is very unlikely that it will be allowed to run its course uninterrupted. In most cases, public health authorities will intervene and preventive measures will be taken, including temporarily or permanently shutting down the operation. This may not be the case in developing countries, especially in rural areas or low-income neighborhoods in large cities, where patterns

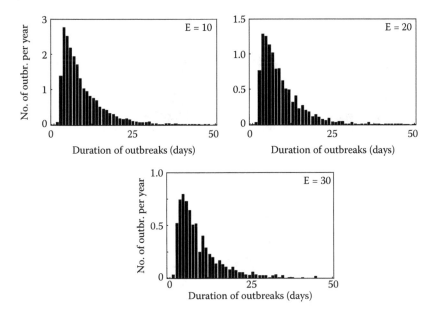

FIGURE 12.14
Effect of the explosion threshold (*E*) on the outbursts' duration distribution extracted from simulated data using the algorithmic model (Equation 12.6 through Equation 12.12) with *D* = 0; *E* = 10, 20, and 30; μ_L = 0.3; σ_L = 0.6; and *p* − 0.8. Notice the distributions' skewness to the right. (From Engel, R. et al., 2001, *J. Sci. Food Agric.*, 81, 1250–1262. With permission, courtesy of Wiley Publishing.)

of the kinds depicted by the model can be realistic scenarios. Either way, analyses of the kind demonstrated earlier can be a handy tool to assess the seriousness of an existing situation and the efficacy of remedial measures like disinfection of a water source or improving sanitary conditions in a school cafeteria, etc. Comparison of the histograms before and after the measures' introduction and their statistical evaluation will immediately reveal whether the probability of outbursts exceeding any given magnitude and lasting for more than a given time has been really and significantly reduced. However, the significance of an improvement should be judged primarily in practical terms rather than by statistical criteria alone.

Other Potential Applications of the Model

Most epidemiological models of endemic disease outbreaks are based on population dynamics (see Chapter 11). When periodicity can be detected

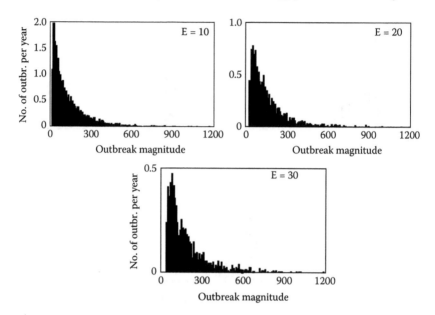

FIGURE 12.15

Effect of the explosion threshold (E) on the outbursts' magnitude distribution extracted from simulated data using the algorithmic model (Equation 12.6 through Equation 12.12 with D = 0; E = 10, 20, and 30; μ_L = 0.3; σ_L = 0.6; and p = 0.8. Notice the distributions' skewness to the right. (From Engel, R. et al., 2001, *J. Sci. Food Agric.*, 81, 1250–1262. With permission, courtesy of Wiley Publishing.)

and there are enough data to follow the disease's detailed natural history at a given location, the appropriate dynamic model possibly can be constructed and the magnitude of its parameters estimated from the epidemiological data. However, if the outbreak pattern shows no discernable periodicity and it is known that it is strongly affected by external factors (e.g., intermittent application of counter measures, erratic entry and exit of biological vectors, etc.), then a probabilistic model of the kind described in this chapter might be required.

This would particularly be the case if the collected data are not dense enough to follow the outbreak's evolution and if data concerning fluctuations in the pathogen's numbers between outbreaks are missing. Again, models of the kind shown and discussed in this chapter can be used to create, or recreate, scenarios that resemble the situations that actually exist or might have existed. At least theoretically, the generated scenarios upon examination can reveal the magnitude of the danger and help in its reduction or elimination.

The models also provide a conceptual framework to simulate and study patterns that do not conform to classic theories of microbial evolution dynamics. The construction of the models need not be based on the log

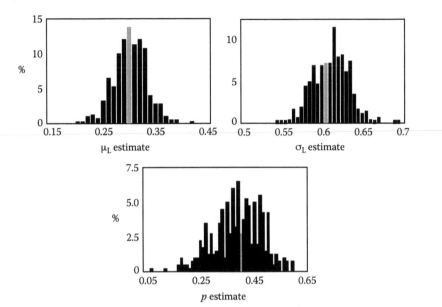

FIGURE 12.16

The distributions of the algorithmic model's μ_L, σ_L, and p estimates in repeated simulations. The "correct" values (the ones used for the records' generation) are marked in gray. They coincide with the histograms' centers, as would be expected when the method to estimate the model's parameter is effective. Notice that histograms of this kind can help to estimate the model's parameter spread in changing fluctuation patterns. (From Engel, R. et al., 2001, *J. Sci. Food Agric.*, 81, 1250–1262. With permission, courtesy of Wiley Publishing.)

normal distribution; alternative distribution functions will probably be just as effective. Also, the clear distinction between the explosive and nonexplosive regimes, primarily introduced for the sake of simplicity, is to a certain extent arbitrary. One can easily come up with models that allow for intermediate explosive regimes to exist. Whether such models would be practical is another matter and the same can be said about the more elaborate ways to account for the statistical properties of the fluctuations during the explosive and nonexplosive regimes. At this time, with hardly any experimental evidence to support the model, it should be considered as merely a potential mathematical tool whose utility in resolving actual outstanding issues concerning food poisoning has yet to be demonstrated.

13

Outstanding Issues and Concluding Remarks

There are two possible outcomes: if the result confirms the hypothesis, then you have made a measurement. If the result is contrary to the hypothesis, then you have made a discovery.

Enrico Fermi

The proposed models that are described and evaluated in this volume's chapters address only a narrow range of the spectrum of mathematical problems that quantitative microbiology poses. Many of these, notably the analysis of bacterial and other microbial genomes, the biochemical/biophysical control of intracellular events, and the kinetics of industrial fermentation processes, have already become fertile fields of research in their own right. What follows is a list of the particular issues that this book raises in relation to microbial growth and inactivation at the population rather than cellular or molecular level. They can be divided into two kinds:

- Problems whose solution will require expansion of concepts that already exist and extension of mathematical models already outlined.
- Problems whose solution will require a new approach and methodology.

Inactivation Models

Determination of Survival Parameters from Inactivation Curves Determined under Nonisothermal Conditions

It has been demonstrated in Chapter 4 that, at least theoretically and in some instances in practice, survival parameters can be determined from nonisothermal data. Yet, the database for testing this concept and resulting models is extremely limited at the present time. Surprising as it may sound, it is difficult to find suitable sets of nonisothermal survival data and corresponding accurate descriptions of the temperature profiles for proper analysis. This is primarily because the vast majority of the publications reporting thermal inactivation data were not intended to prove or disprove a modeling concept. Many were written with the first-order kinetics and log linearity of the D values' temperature dependence taken for granted. Thus, they only contain lists of D and z values, the validity of which cannot be affirmed. Similar difficulties are encountered when one tries to interpret published results of nonthermal inactivation experiments and when the Arrhenius equation's energy of activation (of inactivation) is reported — no pun intended.

The issue has become particularly acute with the growing applications of ultrahigh temperatures in the pasteurization and sterilization of foods and perhaps in other industrial products of safety concern. To calculate survival curves at such temperatures by the currently accepted methods, one needs to obtain isothermal data at the pertinent temperature range. However, this might be physically impossible. For example, when one pasteurizes a food to rid it of *Salmonella* at 70 to 80°C or sterilizes a soup to rid it of *C. botulinum* spores at 130 to 140°C, even a few seconds of exposure at these temperatures or a deviation of a only single degree from the target temperature might have a significant effect on the measured survival ratio. Thus, the isothermal data's reliability can be seriously compromised by the come-up and cooling times, unless special techniques are employed. It would not be at all surprising that in at least some experimental isothermal treatments, the nominal temperature's effect might be small or even negligible relative to that of the temperatures to which the product has actually been exposed during its heating and possibly cooling stages.

The same kind of concern arises, as already mentioned, in the interpretation of experimental chemical treatments with a dissipating chemical agent. The prevalent underlying assumption that the survival ratio is uniquely defined by the product of the agent's momentary concentration multiplied by the exposure time makes many published data unsuitable for a kinetic model derivation. Thus, to predict the outcome of water disinfection processes effectively, for example, a new database in which the momentary survival ratio and simultaneously determined agent concentration are reported together will be required. Only such data can be used to validate an inactivation model and determine its parameters. The alternative — to design and carry out experiments in which the chemical agent's concentration is maintained constant or almost constant throughout the experiment's duration — is much less attractive for obvious technical reasons.

However, the method to extract survival parameters form nonisothermal experimental data discussed in Chapter 4 is only applicable to organisms known *a priori* to have a Weibullian spectrum of resistances with a shape factor (power) that can be reasonably estimated and fixed for the calculation. It should also be known in advance or be legitimately assumed that the temperature dependence of these organisms' Weibullian rate parameter follows the log logistic pattern. The preceding also applies to the estimation of survival parameters from experimental data obtained with actually dissipating chemical agents, as discussed in Chapter 5.

Most of the dynamic inactivation data currently available in the literature are insufficient to determine the shape factor (power) and the other survival parameters (namely, k and T_c or C_c of the log logistic equation) simultaneously. This is because the reported survival curves do not have enough data points or because the reported counts or survival ratios have an excessive scatter. Since all the three survival parameters of the Weibull log logistic model can be determined simultaneously from simulated data, the problem is not theoretical, but rather, practical. Future research aided by computer simulations, will probably establish the kinds of experimental records needed to determine all three survival parameters reliably.

Terra incognita is how to find a survival model and calculate its parameters from dynamic inactivation data when the survival pattern of the organism in question is known to have or is suspected of having a non-Weibullian survival pattern. Obviously, one can use trial and error to screen for and eventually find a working model. The test of a candidate model, though, will remain the same — that is, whether it could predict survival patterns under varying conditions different from those under which its parameters have been determined. A mathematical challenge to future researchers will be to automate the identification and testing procedure so that an appropriate model can be found within reasonable time and with minimal effort.

An even more challenging problem is how to determine the survival parameters of an organism from experimental data gathered when several factors changed simultaneously. The simultaneous temperature and pressure rise during ultrahigh pressure treatment or disinfection with a volatile chemical agent while the temperature changes are two illustrative examples (see below).

The question of what constitutes a satisfactory prediction still has some unresolved aspects. There are, of course, standard statistical measures, and in microbiology, the *accuracy* and *bias factors* originally proposed by Ross (1996) are often used for the purpose. Although these factors can be easily calculated and expressed numerically, their translation into practical criteria may still require a certain degree of subjective judgment. Statistical measures of the kind mentioned can be used to compare the overall degree of agreement between a curve predicted by a model and the actually observed experimental values. However, in the case of a crossover of the theoretical and experimental curves, such indices might not be sensitive to *where* the discrepancy between the prediction and actual data is more notable.

Needless to say, a model that errs when the inactivation sets on but accurately predicts the final survival ratios is superior, from a practical viewpoint, to a model that accurately predicts the onset of inactivation, but fails when it comes to the residual survival. Such problems can be solved by altering the relative weight of different regions in the survival curves, for example. However, whether such methods would be useful in quantitative microbiology, in which the experimental data are notoriously scattered and in many cases too few and widely spaced, remains an unresolved issue.

Modeling and Predicting Survival Patterns when Several Influential Factors Vary Simultaneously

In some processes, the lethality of an agent strongly depends on the intensity of other lethal agents or on other factors. A notable example is the frequently mentioned microbial inactivation by ultrahigh pressure. Because of adiabatic heating, it is practically impossible to assess the effect of pressure in isolation — a well-known fact whose quantitative implications have been largely overlooked in much of literature on the subject. Because temperature affects the lethality of high pressure (and vice versa when the temperature is high enough), the rate model, a differential equation, must account for the synergism. As has already been shown, if temperature and pressure can be expressed as functions of time, models of the kind described in Chapter 6 have a good chance of predicting correctly the outcome of the combined effect. (Regrettably, at the time at which this book is being written, evidence that the approach might work is scant and informal.)

Another example is the chemical treatment of water and food with a dissipating agent under conditions in which the temperature fluctuates, especially in a partially random fashion. The same can be said about a similar situation in water treatment, when oxygen is depleted by reduced solubility or added through aeration. In processes like drying or smoking, the water activity as well as the solute concentration is continuously changing; the temperature might change too. To account for how microorganisms respond to the changing combination of conditions will again require a new modeling approach. In all probability, chemical disinfection of spaces, as well as fumigation of grains, legumes, and other commodities is performed under conditions in which the agent's concentration is continuously changing and temperature and relative humidity of the surroundings might vary also.

These changes may affect the agents' lethality and thus ought to be taken into account. In all the preceding operations, one cannot assume *a priori* that the effects of various factors are independent as the structure of many currently used model implies. It is more likely that they are not. To account correctly for the interactive roles of several lethal factors will require that they be included in the inactivation rate model in the form of mathematical expressions incorporated into *the differential equation's coefficients*. The same applies to any synergistic or antagonistic factor and will be true for growth as well (see below).

Currently, the dominant concept is that the survival ratio or rate can be written in a form of a product or sum, such as:

$$\text{Rate constant or a function of the rate constant} =$$
$$f_1[T(t)] \cdot f_2[\text{pH}(t)] \cdot f_3[a_w(t)]... \tag{13.1}$$

or

$$\text{Rate constant or a function of the rate constant} =$$
$$f_1[T(t)] + f_2[\text{pH}(t)] + f_3[a_w(t)]... \tag{13.2}$$

where the f's are different functions of temperature, pH, water activity, etc.

In some models, linear interactions, such as $k[T(t)\text{pH}(t)]$, $k[T(t)^2]$, etc., are also included in the model. Even if the inactivation rate could be unambiguously defined, expressions of the kind in Equation 13.1 or Equation 13.2 impose mathematical restrictions on the manner in which the effects of the various factors can be manifested in the survival pattern. These, obviously, cannot be known *a priori* and the targeted organism or spores need not abide by them.

As has been stressed repeatedly in this book, a proclaimed kinetic model (in contrast with an empirical model) and the meaning of its parameters ought to be established by independent tests. The rationale that the effects of pH and temperature, for example, are multiplicative or additive as in the preceding model equations needs explanation. There is certainly no compelling reason to assume that they are either. The fit of any such model, whose number of terms can be modified at will, does not validate the assumption regarding the nature of the interactions between the various factors. Without an independent proof, one cannot assume that the found model has general applicability or that it is unique. (Models like Equation 13.1 or Equation 13.2 can be used for interpolation, but this is not what most scientists and engineers expect from a predictive kinetic model.)

In contrast, consider the Weibullian–log logistic model with a fixed power (see Chapter 2) as an example. Under nonisothermal conditions, where the pH, for example, remains constant, the survival parameters, n, k, and T_c, are also constants. However, if the pH also changes during the process — that is, pH = pH(t) — the model should be written in a form in which these survival parameters become functions of time because of their pH dependence. In other words, because $n = n(\text{pH})$, $k = k(\text{pH})$, and $T_c = T_c(\text{pH})$, whenever the pH also changes they become $n(t) = n[\text{pH}(t)]$, $k(t) = k[\text{pH}(t)]$, and $T_c = T_c[\text{pH}(t)]$. The result is a rate model's equation whose simplest form is:

$$\frac{d\log_{10} S(t)}{dt} = -\log_e[1 + \exp\{k[pH(t)] \cdot [T(t) - T_c][pH(t)]\}]$$

$$\cdot n[pH(t)] \cdot \left[\frac{-\log_{10} S(t)}{\log_e[1 + \exp\{k[pH(t)] \cdot [T(t) - T_c[pH(t)]]\}]} \right]^{\frac{n[pH(t)]-1}{n[pH(t)]}} \quad (13.3)$$

Such a model requires that the effect of pH on the survival parameters n, k, and T_c should be found *experimentally* to produce the terms $k(\text{pH})$, $T_c(\text{pH})$, and $n(\text{pH})$. Once these forms are determined independently and the temperature profile, $T(t)$, and the pH profile, $\text{pH}(t)$, are known, the differential equation can be solved to produce the nonisothermal "noniso pH" survival curve. Notice that the degree of curvature of the isothermal semilogarithmic survival curves need not be the same at different pH levels and thus, unless proven otherwise, $n[\text{pH}(t)]$ cannot be assumed to be constant.

The manner in which the pH affects the magnitude of n and that of the other survival parameters, k and T_c, is determined by the organism, its growth stage, and the medium. It is not expected to follow any preconceived pattern or model, although certain trends might be guessed. Again, any rate equation constructed would remain a tentative model until it is demonstrated that it can predict correctly the outcome of processes in which the temperature and pH vary simultaneously. The preceding applies to any primary and secondary model combination constructed on the basis of similar principles.

Despite the enormous complexity that the resulting model might have, especially if the pH and temperature link is not unique, there is no reason why it could not be solved numerically with software like Mathematica®. The same can be said of similar models that might include osmotic pressure and/or other factors that may affect the organism's survival pattern. Again, with the power of current algorithms and the speed of calculation of even the least expensive personal computers, the model's mathematical complexity should not be expected to be a hindrance to its solution. Thus, it is necessary to redirect the effort in order to determine how factors like pH, water activity, or salt concentration affect the actual survival parameters of the organism at hand without taking for granted the applicability of any primary or secondary model.

Similar models should be developed for water and space disinfection or commodity fumigation, for example, in which the agent dissipates while the temperature (and/or humidity) changes monotonically or fluctuates. Because the (hypothetical) semilogarithmic isoconcentration curves cannot be assumed to exhibit the same degree of curvature at different temperatures, the power, n, would need to be expressed as $n[C(t),T(t)]$, $n[C(t), T(t), RH(t)]$, etc., just like the other two survival parameters, namely, $k(t)$ and $T_c(t)$.

A special challenge to future research would be the calculation of the basic survival parameters directly from experimental inactivation data in which two or more influential factors vary simultaneously. Although not an impossible task in principle, the practicality of such a procedure is not at all guaranteed, unless extremely dense and reproducible survival data could be obtained.

Non-Weibullian Inactivation Patterns

Although the Weibullian–power law was found, empirically, to be applicable to a large number of organisms and under a variety of conditions (see Chapter 1 through Chapter 5), there is no theoretical reason to believe that it must apply to all survival curves, especially when recorded under conditions in which the effects of various influential factors change simultaneously. The same can be said about the log logistic equation as a secondary model. How and when one should try alternative models is still an open question. Similarly, how the model's choice will affect the prediction's quality is yet to be demonstrated.

Because there is no reason to assume that there must be a unique superior model for any particular inactivation pattern, the issue of how to choose an *optimal* model will naturally arise. 'Optimal' in this context refers to the model's simplicity and its performance. To reach an optimal model, one would need to ask the question, "What is the minimum number of parameters that can be used without compromising the model's predictive ability and at what point will the complexity of the model's mathematical structure render it totally impractical except for qualitative demonstrations?"

The place of certain traditional survival parameters in modeling would need to be reassessed. As has been shown, parameters such as "lag time" and "specific maximum inactivation (or growth) rate" are so entrenched that they will be difficult to dislodge despite their obvious theoretical shortcomings. This is not because their weaknesses have not been exposed, but rather because of their attraction as "intuitive parameters." How to make their replacements intuitively meaningful, yet free of their predecessors' faults, will be a challenge to future model development.

Systems in which the Inoculum Size May Affect Inactivation

All the models discussed in Chapter 1 through Chapter 6 are expressed in terms of the *survival ratio*, $S(t)$, change with time. Thus, regardless of whether the lethal agent's intensity is constant or changing, the use of $S(t)$ as the dependent variable implies that the actual number of cells or spores is immaterial. In other words, the survival pattern would be the same if the initial inoculum were large or small. This assumption, as previously explained, clearly breaks down when a very small number of organisms or spores (e.g., five) are involved. In such a case, they can be totally wiped out or partially (or even completely) survive the treatment in a manner that cannot be accounted for by models based on a continuous distribution function.

The statistics of small numbers coupled with the potential role of microenvironment in relation to microbial survival under adverse conditions is not yet fully developed. However, it might be essential to the understanding of sporadic spoilage that cannot be explained by other means. Remember that discrete distribution functions like the Poisson distribution become practically indistinguishable from continuous distribution functions when the numbers reach the several 10s level. Thus, the use of continuous distribution functions like Weibull's or the logistic is fully justified, even though microorganisms and spores are countable individuals.

However, this will not be the case when the number is in the range of 1 to 20, for example, *especially* when the microorganisms' environment might vary (see below). In such a case, they cannot be even assumed to belong to the same population, a prerequisite for the application of all the previously discussed models. The issue may arise when one scrutinizes the meaning of very low survival ratios, especially when variations in the local microenvironment may produce very different growth or inactivation conditions. A more detailed discussion of this topic can be found in the works of Hills et al. (1997ab).

A different issue may emerge when the initial size of a *large inoculum* plays a role in the inactivation kinetics. An illustrative example is the response of certain bacteria to cold. In certain cases, after being subjected to freezing temperatures, the dead cells can contain metabolites that act as cyroprotectants, thus increasing the survival probability of those that are still alive. In such cases, the *survival ratio* will depend not only on the usual factors (temperature, pH, etc.) but also on the organism's initial number or density. At least theoretically, such scenarios can exist when an agent other than cold is applied. Fortunately, testing this possibility is straightforward; all that one must do is to attempt to reproduce the survival curve with inocula of different sizes. If the attempt fails, the mathematical survival models as described in the first chapters of this book will not apply. They should have to be modified so that the survival parameters are functions of temperature or chemical agent concentration as well as of the population's initial density.

Similar considerations (see below) will also apply to the characterization of microbial growth, although the role of the initial inoculum size in that case might be regulated by very different mechanisms. The shape of dose–response curves (Chapter 7), especially when the lethal agent is an antimicrobial chemical compound, can also be affected by the targeted organism's density. However, here too the problem will not be the mathematical formulation of the model, but rather how to determine experimentally its coefficients' dependence on the population's density.

Robustness and Sensitivity

The first stage in all the inactivation models discussed in this book is the derivation of a "primary model" from experimental data, which requires linear or more frequently nonlinear regression. Therefore, the resulting inactivation parameters have an uncertainty range determined by the original data scatter and the mathematical model tried. This range will most likely grow if the experiments are repeated with tested organisms of a different origin and at different growth stages and if the treatments are performed in freshly prepared media with ingredients from different sources.

The practical significance of the differences and how they should be translated into process parameters and safety factors are issues that will require a more systematic database than that currently available in the standard literature and from Websites. Computer simulations can come to our aid by allowing us to reproduce the inactivation and process with as many variations in the survival parameters. The theoretical range of the resulting survival ratios at any given process can then be easily calculated using the methods described in the book. Yet, only microbiologists can answer the question of what constitutes a reasonable range of parameters. Thus, cooperation between modelers and those who have more intimate familiarity with the targeted organism or spores will be required in order to implement the proposed methodologies fully.

The potential variability of the secondary models' parameters is another case in point. Again, its role can be easily assessed mathematically by running simulations with models having coefficients of different magnitudes. However, the *safety implications* of the range could only be assessed by microbiologists. They alone can answer the question of what constitutes a safety margin and recommend an appropriate safety factor. Put differently, the survival ratio that can be considered indicative that a given organism or spore has been practically destroyed is yet to be agreed upon.

Relationship between Survival Parameters and Inactivation Mechanism

All the models presented in Chapter 1 through Chapter 7 are of the kind known as 'phenomenological,' that is, they *describe* the inactivation process (the phenomenon) but do not explain it. As already mentioned, one can expect that the shape of the survival curve is a manifestation of events that occur at the molecular and cellular levels, as is the temperature (or concentration or pressure) dependence of the survival parameters. An interesting topic for investigation would be how what happens to individual cells or spores translates into the survival curve characteristics, such as its concavity direction (as expressed by the power, n, of the Weibullian model), the inactivation's marker (T_c, C_c, or P_c) in the log logistic secondary model, and/or the inactivation acceleration rate beyond it (the k of the log-logistic model). The results *might* help future researchers to predict inactivation patterns from the fundamentals of cell biology and the biophysics of sporal inactivation, at least qualitatively.

The previously mentioned models are just an example; an equally unresolved issue is whether the emergence of a Weibullian vis-a-vis a log linear or a logistic inactivation pattern can be traced to cellular and molecular mechanisms. The emphasis on 'might' stems from the fact that there need not be a unique relationship between events at the molecular and cellular levels and the response of a population to a lethal agent in a given environment, which can be nonhomogenous on the cellular length scale. Yet, models of the kinds discussed in Chapter 1 through Chapter 8 would be by far more adequate *tools* to reveal such a relationship, if it exists, than the first-order kinetic model, for example.

Alternative Inactivation Technologies

Data on how treatments with unconventional lethal agents affect microorganisms are fast accumulating. They cover treatments with electron beams, ultraviolet light, pulsed electric and even high-intensity magnetic fields. In comparison with heat and chemical treatments, the inactivation kinetics of these processes is scantly covered in the literature — especially in respect to the role of secondary factors such as pH, dissolved oxygen, and the like. Development of predictive inactivation models for such treatments with all relevant factors taken into account (including temperature, of course) will be a challenge to future researchers, especially if combined with traditional or other unconventional methods. Again, a distinction must be made between descriptive models that are only applicable within a narrow range of conditions, such as the response surface methodology, and truly kinetic models that can be used to predict the outcome of treatments not yet tried.

Growth Models

Terminology

Much of the modeling of microbial growth, especially in food, has been focused on sigmoid growth patterns. Thus, the vocabulary of the field contains terms such as 'lag time', 'specific maximum growth rate', and maximum growth level. As has been shown in Chapter 8, these have serious theoretical shortcomings and limited applicability at best. If the 'lag time' is redefined as the time needed to double, triple, or multiply the original inoculum size by any other chosen factor, the lag time will become model independent. Whether such a definition, which is not new, will be acceptable to the majority of microbiologists remains to be seen. The "maximum" or "average" (linear or logarithmic) growth rate has been and will remain a loose growth parameter, but probably useful in comparing orders of magnitude, e.g., tenfold increase in 6 h vs. in 2 days. However, it might be less useful to distinguish between similar growth curves in which the mathematical model chosen might affect its calculated value or to account for the fact that the exponential growth stage duration might be very different.

Whether there is a predictable true maximum growth level (rather than what is actually an asymptotic level but dubbed erroneously as maximal) is another unresolved issue. How to predict the onset of a population's decline and at what rate it will proceed once started are two more such problems. Their solution might not be relevant to foods' shelf life, but it certainly is to fermentation processes. Because of the adherence to the logistic growth model and its various modifications, many experiments described in the literature were stopped well before the expected decline in the population's size could be observed. Here, again, the practical implications of the distinction between an asymptotic level and a true peak growth would probably depend on whether the issue is spoilage or harvest.

Growth under Changing Conditions

Publications that include complete sets of isothermal and nonisothermal growth data are rather scarce in the literature. Consequently, establishing the predictive ability of a kinetic growth model is rather difficult unless one can produce the growth data experimentally. The same can be said about growth in the presence of protective agents with diminishing potency. Temperature alone can generate situations whose mathematical modeling is a challenge. A most notable one is the common scenario in which the temperature increase or decrease causes the microbial population to grow and then die out or vice versa (Corradini and Peleg, 2006).

Two typical examples involve microbial growth in food treated with an unstable chemical preservative like certain cured meats or in treated water during the stage in which the agent's concentration approaches an ineffective level. In both cases, inactivation is followed by growth; these are qualitatively different responses. Even more scarce are data sets that can be used to model microbial growth when several influential factors in the habitat change simultaneously, e.g., temperature, dissolved oxygen, and the concentration of an inhibitor. In principle, the same approach that has been proposed to modeling inactivation under changing conditions can also be used to model growth. Nevertheless, as already stated, growth models are expected to be more complicated than those used for predicting inactivation. This is because the description of even the simplest asymmetric sigmoid growth curves requires at least three adjustable parameters while that of the majority of inactivation curves require only two.

With a true maximum growth, the minimum number of adjustable parameters can rise to four or more, in which case the very possibility to determine them in an unambiguous manner might come into question (Corradini and Peleg, 2005). This is because determination of several adjustable parameters simultaneously from noisy data frequently yields totally unrealistic values and/or values that cannot be used to determine meaningful secondary models. Some mathematical and statistical methods can find parameters that make sense, but their success will largely depend on the density and scatter of the experimental growth data. For logistic reasons, generating suitable sets of growth and inactivation data under a variety of constant and changing conditions might not be feasible. Thus, another unresolved issue in predictive microbiology concerns the point at which one would need to replace experimental data with computer simulations and how reliable these would be.

All the preceding is equally applicable to the converse — that is, to scenarios in which growth precedes the inactivation, such as in very slow heating of foods or a slow dispersion of a disinfectant in water. A pathogenic bacterium exposed to an antibiotic drug or a parasite to an antimicrobial agent *in vivo* may also exhibit regimes of alternating growth and inactivation depending on the drug or agent's administration regime, at least in principle.

Growth under Arbitrary Conditions

Like an inactivation model, once a growth model is constructed, it can be used to generate growth patterns under randomly changing conditions. Examples are random temperature fluctuations that one might expect in unrefrigerated food transportation or random temperature fluctuations superimposed on the diurnal oscillations, as in an open water reservoir or stored foods. Again, the issue here is not the mathematical models, but how to obtain reliable data regarding the amplitude and frequencies of the random fluctuations so that the computer simulations will yield realistic results. Comparison of the models' predictions with actual growth patterns will obviously be needed; this in turn will require the creation of new databases where the record will include both the microbial counts and the temperature detailed history.

Simultaneous Growth and Inactivation or Inactivation and Growth

The need of models for situations in which the same microbial population experiences growth followed by inactivation or vice versa is mentioned in "Growth under Changing Conditions." However, in some situations, growth and inactivation occur simultaneously at different locations. For example, the microbial population is not completely destroyed in a partially cooked food and the remnants can grow during the cooling period or during storage. In certain sliced or ground meats that are slowly cooked, the organisms at the center may grow while those at the exterior are destroyed by the heat. (The "showarma," a Mediterranean food gaining much popularity in Europe as well as in big cities in the U.S., is made of meat and fat layers exposed to radiant heat from the outside through rotation. The cooked edge is sliced and served when ready, but the interior might be left for hours at a lower than lethal temperature.) In both cases, a single temperature profile induces growth and inactivation *at different times and locations within the same food.*

Similar situations can arise in fermentation when one tries to increase the temperature in a large vessel to improve efficiency. The kinetics of microbial growth and inactivation combined, or what happens at temperatures where the transition between the two occurs, has been studied but to a far lesser extent than that of each alone (see McKellar and Lu, 2004; Corradini and Peleg, 2006c). In most of the works reported in the literature, what followed was the growth and inactivation rate, expressed in terms of a single rate constant. Such a rate constant must change sign at the transition from growth to lethality or vice versa. One must expect, though, that an accurate description and prediction of such scenarios will require kinetic models that take into account the inevitable changes in the entire growth and inactivation *patterns.* These, of course, cannot be accounted for by a single rate parameter and thus modification of the modeling approach will be required.

It seems that the solution will come from a *set* of secondary models that will account for the transition from a growth to inactivation mode or the other way around. To avoid 'if statements' that might make the model less attractive, the growth and inactivation stages or modes would need to be described by the same phenomenological (primary) model, i.e., in terms of a single mathematical expression. This is not impossible, however, because a sigmoid growth pattern can be easily transformed into a clearly monotonic inactivation within any given region, for example, by adjusting the model's parameters.

A more challenging task for future research will be to take the location into consideration, i.e., to follow the population evolution (or fate) everywhere in the bulk, rather than only at the coldest or any other single point. This will enable one to estimate or predict the total number of organisms *within a given volume of a food*, for example, and the safety implications of its consumption or storage. Again, the mathematical and computational tools to combine kinetic models with heat transfer theories are available (see Amézquita et al., 2005, for example). One might expect that it will not be long before three-dimensional mapping of the temporal microbial population distribution within the bulk could be generated and studied.

Fluctuating Records in Water and Foods

Censored Data

Chapter 9 through Chapter 12 showed how probabilistic models can be used to estimate the frequency of future outbreaks of various magnitudes. The reliability of the estimates, as has been stated, primarily depends on the database quality and the correctness of the assumptions on which the models are based, especially with regard to the future stability of the external conditions. Obviously, the latter is beyond our control. There is no way to guarantee that the conditions existing when the data were collected will remain unchanged in the future. In fact, one can expect continuous attempts to improve the sanitary conditions, for example, in an orderly food operation. Conversely, one can expect deteriorating conditions when a sanitary system collapses in a time of economic crisis. Thus, any prediction reached using the described models will only be valid if no drastic change in the environment will occur during the coming days, weeks, or months, whichever the pertinent time scale might be.

However, data quality has a lot of room for improvement. Preferably, the sampling and testing should always be done at fixed time intervals so that the characteristics of the fluctuation patterns will be clearly manifested in the experimental record. This, however, might not be feasible because of logistic considerations, e.g., sampling the water of municipal open reservoirs during weekends and holidays might not be an option because of a labor contract. The remedy might be the application of statistical methods to estimate missing data. Mathematical procedures can be used to process what are known as 'censored data.' To the best of my knowledge, though, application of such methods to the analysis of fluctuating microbial records and their interpretation is still a topic for future research. Yet, such methods can and perhaps should be of interest to biostatisticians and public health officials so that more useful information could be extracted from microbial counts records already collected.

Sampling at Different Locations

A basic underlying assumption of the methods described in Chapter 9, Chapter 10, and Chapter 12 is that the counts fluctuations have a distribution in time but not in space — a rather idealized scenario. For example, in food production, ground meat samples are frequently mixed and the count represents an average of the microbial load at different locations. In a large water reservoir like a lake, there must be significant differences in various geographical locations (see Chapter 10). However, significant differences might also exist at the same place at which current strength, water depth, and other topographical features produce local conditions that can affect the microbial presence and growth pattern differentially.

All this is, of course, well known and thus has brought about the great effort to establish a collection procedure that will produce a representative sample. How to combine the probabilistic approach with the spatial distribution and/or how to include known factors such as weather data in the model are still unresolved issues at the present time. At the other end, if the sampling frequency can be increased, then at what point should the probabilistic approach be replaced by a proper population dynamics analysis? This topic will need to be addressed in future studies. It seems that the statistical tools to resolve the preceding issues already exist and therefore it will be interesting to compare the predictions of both kinds of models when applied to the same data.

Risk Assessment

At last, some of the methods described in this book can be used to estimate the probability of future outbreaks from past records in two ways:

- Directly from the counts distribution as described in Chapter 9 and Chapter 10.
- By generating numerous hypothetical scenarios using models whose parameters have been derived from past records and counting the number of outbursts produced (see Chapter 12).

Either way, the information retrieved or generated would be in the form of the estimated frequency of outbursts exceeding any given magnitude. If the models described in Chapter 12 are used, the estimates might include the frequencies of outburst exceeding any given duration as well. However, what constitutes an outburst of public health risk is yet to be determined. Obviously, it will be different for different organisms and for different foods or waters (drinking vs. irrigation or bathing, for example).

Also, in most cases, what is actually monitored is the number of coliforms or *E. coli*, which are considered a measure of the fecal contamination level. The relation between the coliform or *E. coli* counts and the actual health risk is a continuously debatable issue. The associated risk might be different in different geographical regions and implemented by local conditions. The issue might be resolved in the future when typing the organisms will be more commonplace than it is now, at least with regard to routine analysis of fresh and partially processed foods (the presence of pathogenic *E. coli* 0157:H7 is a different case, of course) and drinking water. Until then, even when microbial standards do exist, what constitutes a "tolerated frequency" before corrective measures are implemented will need to be clarified and codified.

A Few Last Remarks

This book presents a personal view on selected topics of a wide field that should be properly called quantitative microbiology. The chapters present methods to describe an assortment of microbial growth and inactivation patterns developed at our laboratory. They also provide demonstrations that, in certain cases, the proposed methods can be used to predict growth and inactivation patterns with reasonable accuracy. (Remember that a 'prediction' in the scientific sense is a correct estimate of the outcome of experiments not used to determine the model and its parameters.)

Because microbial growth and inactivation are processes affected by numerous factors (not all of them known in detail or tightly controlled), an occasional discrepancy between prediction and observation should not come as a surprise. Yet, in *all* the systems examined, there was an agreement between the calculated estimates and actual experimental results even if only rarely perfect. This refers to the continuous kinetic models of growth and inactivation and the probabilistic models used in the interpretation of the discrete irregular counts' fluctuation in foods and water. Considering the unavoidable scatter even in the most reproducible microbial counts, this repeatedly observed agreement suggests that the proposed models, sometimes despite their obvious crudeness, can still capture the essence of the phenomena involved.

The word 'advanced' in this book's title was suggested by the publisher on the basis of the reviewers' comments on this book's proposal. (My original suggestion for the title was "Alternative Models in Quantitative Microbiology.") The use of a term like 'advanced' in a title of any publication can be problematic in more than one way. Except for a very few specific cases in which currently held concepts have been openly criticized, the term's inclusion in the title does not mean that others' ideas and models are necessarily less advanced. Certainly, this has not been the intended message.

Whenever the word 'advanced' is uttered, the question "relative to what?" naturally arises. In the case of the inactivation models, there is little doubt in my mind that those promoted in Chapter 1 through Chapter 5 are more advanced than those based on the grossly oversimplified first-order kinetics, which are still taught in microbiology courses and used in the food and pharmaceutical industries. The proposed new models are based on what I believe is the correct observation that survival curves are the cumulative form of the temporal distribution of lethal or inactivation events and that the concavity direction of their semilogarithmic form can be interpreted in terms such as damage accumulation or weeding out the sensitive members of the population.

The new modeling approach eliminates the unnecessary implicit analogy, still held in the field, between a microbial cell's death and a spore's inactivation, both complex biophysical processes with more than a single path, and radioactive decay or a simple chemical reaction. The same can be said about the recommended abandonment of the z value and the Arrhenius or WLF equation for expressing the temperature (or pressure, or concentration of water activity) dependence of the "rate constant." This is because the momentary logarithmic inactivation rate of a real microbial population, with few exceptions, *must* depend on its momentary survival ratio and not on temperature (or any other equivalent factor) alone.

Thus, in my opinion, the expression of the rate equation in terms of state-dependent coefficients is a step forward. It is to be hoped that the general concept, at least, will replace that which is still commonplace in all standard textbooks and industrial applications. (As mentioned earlier, it is very peculiar that the idea that microbial inactivation needs to be described by a rate equation that accounts for the population's state has drawn so much resistance and even hostile reaction, while the very same principle is widely accepted in modeling bacterial growth!)

In contrast with the proposed inactivation models, the ones used in this book to predict microbial growth would be almost certainly considered by some readers, with a certain degree of justification, as a *step backward*. This is because they have been deliberately constructed as purely empirical models, eschewing any mechanistic considerations in their derivation.

This approach has been dictated by the principle of parsimony, known as Ockham's razor, which requires that when several models can describe the same data, one should choose the one with the smallest number of adjustable parameters. The choice of the empirical primary models has also been aided by the fact that, contrary to common belief, the introduction of mechanistic parameters into a model's mathematical formula does not mean that they can be determined by curve fitting of growth data. At the present time, none of the existing growth models, including the ones described in Chapter 8 can be considered unique. This fact should be highlighted, not hidden, and serve as a warning against unwarranted interpretation of a growth model's equation.

It is to be hoped that future research will lead to the construction of microbial growth models based on fundamental principles or demonstrate convincingly that the parameters of current models can be related to underlying mechanisms in the manner claimed. However, as has been repeatedly stated in this volume, any truly mechanistic model will need to be validated by independent experimental tests. These will validate the model and also confirm the interpretation of its parameters' magnitude.

All the critical comments on the traditional secondary models of inactivation are just as relevant to the microbial growth models. Here, too, the implied analogy between the temperature's effect on microbial growth rate and on simple reactions between gases must be abandoned for theoretical, as well as practical, reasons. The same can be said about complex biological processes and biochemical reactions in general (Peleg et al., 2004).

Despite the empirical and thus less glamorous character of the alternatives, releasing growth models from the constraints of the Arrhenius equation and the unnecessary transformations of the rate and temperature scales that it requires might be a step forward after all. Whatever the theoretical shortcomings of the described empirical approach might be, the resulting models are derived from the actually recorded growth patterns. Thus, the primary and secondary models need not conform to any preconceived kinetics that may or may not be applicable to the particular organism in the particular habitat.

The probabilistic models of fluctuating microbial counts (Chapter 9 through Chapter 11) can be viewed in a similar light. They have been proposed as a replacement of the more elaborate population dynamic models simply because, in all the records that we have examined, the data were insufficient to construct such models. The problem of imperfect data is real and commonplace. The data that we have presented are all actual records of industrial food plants and water-quality monitoring systems in two countries. Probably, they are typical rather than exceptional.

Thus, the demonstration that the fluctuations of microbial counts can be translated into a set of probabilities of future problems should certainly be considered as a step in the right direction. The resulting method to estimate these probabilities will enable those interested to extract useful information from data now discarded because of their irregularity and randomness. The concept behind these fluctuation models can also help in the interpretation of quality control charts, which need not be restricted to microbial counts records (Gonzales–Martinez et al., 2003). In any case, the proposed analysis can reveal the existence of a potential danger even when all the recorded entries are within the margins allowed by the specification.

Chapter 12 deals with "structured" microbial outbursts and their modeling. The advance and departure from the traditional approach here is that the models are expressed as an *algorithm* and not as an algebraic equation, a set of a differential or difference equations or a distribution function. To the best of my knowledge, algorithmic models are a novelty in quantitative water and food microbiology and their potential is yet to be assessed. Perhaps what is more important is the demonstration that the parameters of such algorithmic models can be determined from experimental data by statistical and algebraic methods that bear no resemblance to regression. The algorithmic models may open new ways to interpret microbial outbreak data and could, in principle at least, be used in risk assessment.

As far as the mathematical level or depth is concerned, very few if any of the models presented in this book are really advanced. They have all been derived from a qualitative assessment of how microbial populations rise and fall under favorable and adverse conditions and are all expressed mathematically in what we believe is the simplest possible form. The resolution of the outstanding issues listed in this chapter probably will require a more advanced level of mathematics. The prediction of simultaneous growth and inactivation within an arbitrary *volume* of food or water subjected to varying temperature and changes in other relevant conditions may serve as a typical example. The prediction of microbial (bacterial and fungal) growth patterns on nonuniform solid surfaces is just another example. It will probably require the development and application of modified cellular automata models, which have not even been mentioned in this book but have already been used by several researchers in the field.

Much of what is discussed in this book might serve as a bridge between the traditional concepts of quantitative microbiology and those of the future. These will almost certainly exploit the combined powers of mathematics and the sciences and will produce models that explain as well as predict the dynamics of microbial populations under conditions relevant to the safety of food, water, and pharmaceuticals.

References

Abernethy, R.B. 1996. *The New Weibull Handbook: Reliability and Statistical Analysis for Predicting Life, Safety, Survivability, Risk, Cost and Warranty Claims*. Gulf Publications, North Palm Beach, FL.

Albert, I. and Mafart, P. 2005. A modified Weibull model for bacterial inactivation. *Int. J. Food Microbiol.*, 100, 197–211.

Albiol, J., Robust, J., Carus, C., and Poch, M. 1993. Biomass estimation in plant cell cultures using an extended Kalman filter. *Biotechnol. Prog.*, 9, 174–178.

Amézquita, A. 2004. Development of an integrated model for heat transfer and dynamic growth of *Clostridium perfringens* during the cooling of cooked boneless ham. Ph.D. dissertation. University of Nebraska–Lincoln.

Amézquita, A., Weller, C.L., Wang, L., Thippareddi, H., and Burson, D.E., 2005. Development of an integrated model for heat transfer and dynamic growth of *Clostridium perfringens* during the cooling of cooked boneless ham. *Int. J. Food Microbiol.*, 101, 123–144.

Anderson, W.A., Hedges, N.D., Jones, M.V., and Cole, M.B. 1991. Thermal inactivation of *Listeria monocytogenes* studied by differential scanning clorimetry. *J. Gen. Microbiol.*, 137, 1419–1424.

Anderson, W.A., McClure, P.J., Baird–Parker, A.C., and Cole, M.B. 1996. The application of a log-logistic model to describe the thermal inactivation of *Clostridium botulinum* 213B at temperatures below 121.1°C. *J. Appl. Bacteriol.*, 80, 283–290.

Anonymous. 2000. Kinetics of microbial inactivation for alternative food processing technologies. U.S. Food and Drug Administration Center for Food Safety and Applied Nutrition. Washington, D.C. (Web URL: http://vm.cfsan.fda.gov/~comm/ift-over.html).

Arnoldi, A. 2001. Thermal processing and food quality: analysis and control, in, Richardson, P. (Ed.). *Thermal Technologies in Food Processing*. Woodhead CRC, Cambridge. U.K.

Arroyo, F.N., Dur'n Quintana, M.C., and Fernandez, A.G. 2005. Evaluation of primary models to describe the growth of *Pichia anomala* and study of temperature, NaCl and pH effects on its biological parameters by response surface methodology. *J. Food Prot.*, 68, 562–570.

Astro, A.J., Barbosa–Canovas, G.V., and Swanson, B.G. 1993. Microbial inactivation of foods by pulsed electric fields. *J. Food Process Preserv.*, 17, 47–73.

Augustin, J.C., Carlier, V., and Rozier, J. 1998. Mathematical modeling of the heat resistance of *L. monocytogenes*, *J. Appl. Microbiol.*, 84, 185–191.

Baker, A.R., Ebel, E.D., Hogue, A.T., McDowel, R.M., Morales, R.A., Schlosser, W.D., and Whiting, R. 1998. *Salmonella enteritidis* risk assessment. Shell eggs and egg product. A final report for the USDA Food Safety and Inspection Service. Washington, D.C.

Balaban, N.Q., Merrin, J., Chait, R., Kowalik, L., and Leibler, S. 2004. Bacterial persistence as a phenotypic switch. *Science*, 305, 1622–1625.

Ballestra, P., Abrue da Silva, A., and Cuq, J.L. 1996. Inactivation of *Escherichia coli* by carbon dioxide under pressure. *J. Food Sci.*, 61, 829–831, 836.

Baranyi, J. and Roberts, T.A. 1994. A dynamic approach to predicting bacterial growth in food. *Int. J. Food Microbiol.*, 23, 277–294.

Bartlett, M.S. 1960. *Stochastic Population Models in Ecology and Epidemiolgy*, Chapman & Hall, London.

Bazin, A. 1983. *Mathematics in Microbiology.* Academic Press, London.

Bilbia, T.A., Ranucci, C., Glazinitsky, K., Buckland B.C., and Aunins, J.G. 1994. Monoclonal antibody process development using medium concentrates. *Biotechnol. Prog.*, 10, 87–96.

Brock, T.D., Madigan, M.T., Martinko, J.M., and Parker, J. 1994. *Biology of Microorganisms* (7th ed). Prentice Hall, Englewood Cliffs, NJ.

Brockwell, P.J. and Davis, R.A. 1987. *Time Series: Theory and Methods*, Springer–Verlag, New York.

Brown, D. and Rothery, P. 1993. *Models in Biology — Mathematics, Statistics and Computing.* John Wiley & Sons, Chichester, U.K.

Buchanan, R.L., Whiting, R.C., and Damert, W.C. 1997. When is simple good enough: a comparison of the Gompertz, Baranyi, and three-phase linear models for fitting bacterial growth curves. *Food Microbiol.*, 14, 313–326.

Campanella, O.H. and Peleg, M. 2001. Theoretical comparison of a new and the traditional method to calculate *C. botulinum* survival during thermal inactivation. *J. Sci. Food Agric.*, 81, 1069–1076.

Casolari, A. 1988. Microbial death, in Bazin, M.J. and Prosser, J.I. (Eds.). *Physiological Models in Microbiology*, CRC Press, Boca Raton, FL, 1–44.

Chorianopoulos, N.G., Boziaris, I.S., Stamatiou, A., and Nychas, G.J.E. 2005. Microbial association and acidity development of unheated and pasteurized green-table olives fermented using glucose or sucrose supplements at various levels. *Food Microbiol.*, 22, 117–124.

Clark, D.S. and Takacs, J. 1980. Ozone, in *Microbial Ecology of Foods*, vol. 1. *Factors Affecting Life and Death of Microorganisms*. Publication of the International Commission on Microbial Specifications for Foods. Academic Press, New York, p. 191.

Coote, P.J., Holyoak, C.D., and Cole, M.B. 1991. Thermal inactivation of *Listeria monocytogenes* during a process simulating temperatures achieved during microwave heating. *Appl. Microbiol.*, 70, 489–494.

Corradini, M.G. and Peleg, M. 2003a. A model of microbial survival curves in water treated with a volatile disinfectant. *J. Appl. Microbiol.*, 95, 1268–1276.

Corradini, M.G. and Peleg, M. 2003b. A theoretical note on the estimation of the number of recoverable spores from survival curves having an activation shoulder. *Food Res. Int.*, 36, 1007–1013.

Corradini, M.G. and Peleg, M. 2004a. A model of nonisothermal degradation of nutrients, pigments and enzymes. *J. Sci. Food Agric.*, 84, 217–226.

Corradini, M.G. and Peleg, M. 2004b. Demonstration of the Weibull-Log logistic survival model's applicability to nonisothermal inactivation of *E. coli* K12 MG1655. *J. Food Prot.*, 67, 2617–2621.

Corradini, M.G. and Peleg, M. 2005. Estimating nonisothermal bacterial growth in foods from isothermal experimental data. *J. Appl. Microbiol.*, 99, 187–200.

Corradini, M.G. and Peleg, M. 2006a. The nonlinear kinetics of microbial inactivation and growth, in Brul, S., Zwietering, M., and van Grewen, S. (Eds.). *Modelling Microorganisms in Food*. Woodhead Publishing, Cambridge, U.K. (in press).

Corradini, M.G. and Peleg, M. 2006b. Prediction of vitamins loss during non-isothermal heat processes with non-linear kinetic models. *Trends. Food Sci. Technol.*, 17, 24–34.

Corradini, M.G. and Peleg, M. 2006c. On modeling and simulating transitions between microbial growth and inactivation or vice versa, *Int. J. Food Microbiol.*, (in press).

Corradini, M.G., Amézquita, A., Normand M.D., and Peleg, M. 2006. Modeling and predicting nonisothermal microbial growth using general purpose software. *Int. J. Food Microbiol.* (in press).

Corradini, M.G., Engel, R., Normand, M.D., and Peleg, M. 2002. Estimating the frequency of high microbial counts from records having a true or suspected trend or periodicity. *J. Food Sci.*, 67, 1278–1285.

Corradini, M.G., Horowitz, J., Normand, M.D., and Peleg, M. 2001a. Analysis of the fluctuating pattern of *E. coli* counts in the rinse water of an industrial poultry plant. *Food Res. Int.*, 34, 566–572.

Corradini, M.G., Normand, M.D., Nussinovitch, A., Horowitz, J., and Peleg, M. 2001b. Estimating the frequency of high microbial counts in commercial food products using various distribution functions. *J. Food Prot.*, 64, 674–681.

Corradini, M.G., Normand, M.D., and Peleg, M. 2005. Calculating the efficacy of heat sterilization processes. *J. Food Eng.*, 67, 59–69.

Corradini, M.G., Normand, M.D., and Peleg, M. 2006. Expressing the equivalence of non-isothermal and isothermal heat sterilization processes. *J. Sci. Food Agric.* (in press).

Datta, A.K. 1993. Error estimates for approximate kinetic parameters used in the food literature. *J. Food Eng.*, 18, 181–199.

Davidson, P.M. 2000. Antimicrobial compounds, in Francis, F.J. (Ed.). *Encyclopedia of Food Science and Technology*. (2nd ed.). Vol. 1, pp. 63–75. John Wiley & Sons, New York.

Davis, B.D., Dulveco, R., Eisen, H.M., and Ginsberg, H.S. 1990. *Microbiology* (4th ed.). J.B. Lippincott, Philadelphia, PA.

de Heij, W.B.C., van Schepadel, L.J.M.M., Moezelaar, R., Hoogland, H., Master, A.M., and van den Berg, R.W. 2003. High-pressure sterilization: maximizing the benefits of adiabatic heating. *Food Technol.*, 57, 37–41.

Denys, S., Ludikhuyze, L.R., van Loey, A.M., and Hendrickx, M.E. 2000. Modeling conductive heat transfer and process uniformity during batch high-pressure processing of foods. *Biothechnol. Prog.*, 16, 92–101.

Devleghere, F., Francois, K., Vereecken, K.M., Geeraerd, A.H., van Impe, J.F., and Debevere, J. 2004. Effect of chemicals on the microbial evolution in foods. *J. Food Prot.*, 67, 1977–1990.

Dougherty, D.P., Breidt, F., McFeeters, R.F., and Lubkin, S.R. 2002. Energy-based dynamic model for variable temperature batch fermentation by *Lactococcus lactic*. *Appl. Environ. Microbiol.*, 68, 2468–2478.

El-Kest, S.E. and Marth, E.H. 1988. Inactivation of *Listeria monocytogenes* by chlorine. *J. Food Prot.*, 51, 520–524.

Elliot, D.K. 1986. (Ed.). *Dynamics of Extinction*. John Wiley & Sons, New York.

El-Shenawy, M.A. and Marth, E.H. 1988. Inhibition of *Listeria monocytogenes* by sorbic acid. *J. Food Prot.*, 51, 842–846.

Engel, R., Normand, M.D., Horowitz, J., and Peleg, M. 2001a. A model of microbial contamination of a water reservoir. *Bull. Math. Biol.*, 63, 1025–1040.

Engel, R., Normand, M.D., Horowitz, J., and Peleg, M. 2001b. A qualitative probabilistic model of microbial outbursts in foods. *J. Sci. Food Agric.*, 81, 1250–1262.

Evans, D.A. 2000. Disinfectants, in Francis, F.J. (Ed.). *Encyclopedia of Food Science and Technology*. (2nd. ed.). Vol. 1. John Wiley & Sons, New York, pp. 501–509.

Everitt, B.S. and Hand, D.J. 1981. *Finite Mixture Distributions*. Chapman & Hall, London.

Farkas, D.F. and Hoover, D.G. 2000. High-pressure processing, in Kinetics of microbial inactivation for alternative food processing technologies. *J. Food Sci.* Special Suppl., Chicago, Il: Institute of Food Technologists. pp. 46–64.

Fernandez, A., Salmeron, C., Fernandez, P.S., and Martinez, A., 1999. Application of the frequency distribution model to describe the thermal inactivation of two strains of *Bacillus cereus*. *Trends Food Sci. Tech.*, 10, 158–162.

Fujikawa, H., Kai, A., and Morozumi, S. 2004. A new logistic model for *Escherichia coli* growth at constant and dynamic temperatures. *Food Microbiol.*, 21, 501–509.

Gaze, J.E., Brown, G.D., and Holdworth, S.D. 1990. Microbiological assessment of process lethality using food alginate particles. Campden Food and Drink Research Association, technical memorandum 580, 1–66.

Geeraerd, A.H., Herremans, C.H., and Van Impe, J.F. 2000. Structural model requirements to describe microbial inactivation during a mild heat treatment. *Int. J. Food Microbiol.* 59, 185–209.

Geeraerd, A.H., Valdramidis, V.P., Bernaerts, K., Debevere, J., and Van Impe, J.F. 2004. Evaluating microbial inactivation models for thermal processing, in Richardson, P. (Ed.). *Improving the Thermal Processing of Foods*. Woodhead Publishing, Cambridge, U.K., pp. 427–453.

Goel, N.S. and Richter–Dyn, N. 1979. *Stochastic Models in Biology*. Academic Press, New York.

Gonzalez–Martinez, C., Corradini, M.G., and Peleg, M. 2003. Probabalistic models of foods microbial safety and nutritional quality. *J. Food Eng.*, 56, 135–142.

Grenfell, B.T. and Dobson, A.P. (Eds.). 1995. *Ecology of Infectious Diseases in Natural Populations*. Cambridge University Press. Cambridge, U.K.

Gulik, W.M., Nuutila, A.M., Vinke, Ko L., tenHoopen, H.J.G., and Heijnen, J.J. 1994. Effects of carbon dioxide, air flow and inoculation density on batch growth of *Catharantis roseus* cell suspensions in stirred fermentors. *Biotechnol. Prog.*, 10, 335–339.

Hadas, O., Corradini, M.G., and Peleg, M. 2004. Statistical analysis of the fluctuating counts of fecal bacteria in the water of Lake Kinneret. *Water Res.*, 38, 79–88.

Hassani, M., Mañas, P., Raso, J., Condèn, S., and Pagan, R. 2005. Predicting heat inactivation of *Listeria monucytogenes* under nonisothermal treatments. *J. Food Prot.*, 68, 736–743.

Hills, B.P., Arnauld, L., Bossu, C., and Ridge, Y.P. 2000. Microstructural factors controlling the survival of food-borne pathogens in porous media. *Int. J. Food Microbiol.*, 66, 163–173.

Hills, B.P., Manning, C.E., Ridge, Y., and Brocklehurst, T. 1997a. NMR water relaxation, activity and bacterial survival in porous media. *J. Sci. Food Agric.*, 71, 185–194.

Hills, B.P., Manning, C.E., Ridge, Y., and Brocklehurst, T. 1997b. Water availability and the survival of *Salmonella typhimurium* in porous systems. *Int. J. Food Microbiol.*, 36, 187–198.

Ho, K–L. G. 2004/2005. Dense phase carbon dioxide processing for fruit juice. *Int. Rev. Food Sci. Technol.*, IUFoST (winter) pp. 90–95.

Holdsworth, S.D. 1997. *Thermal Processing of Packaged Foods*. Blackie Academic Professional. London.

Horowitz, J., Normand, M.D., and Peleg, M. 1999. On modeling the irregular fluctuations in microbial counts. *Crit. Rev. Food Sci. Nutr.*, 39, 503–517.

Huemer, I.A., Llijn, N., Vogelsang, W.J., and Langeveld, L.P.M. 1998. Thermal death kinetics of spores of *Bacillus sporothermodurans* isolated from UHT milk. *Int. Dairy J.*, 8, 851–855.

Hunt, N.H. and Mariñas, B.J. 1997. Kinetics of *Escherichia coli* inactivation with ozone. *Water Res.*, 31, 1355–1362.

Ingham, S.C. and Uljas, H.E. 1998. Storage conditions influence on the destruction of *Escherichia coli* 0157:H7 during heating of apple cider and juice. *J. Food Prot.*, 61, 390–394.

Ingraham, J.L., Maaløe, O., and Neidhart, F.C. 1983. *Growth of the Bacterial Cell*. Sinauer Associates, Sunderland, MA.

Ingram, M. and Roberts, T.A. 1980. Effect of ionizing radiation on microorganisms, in *Microbial Ecology of Foods*. vol. 1. *Factors Affecting Life and Death of Microorganisms*. Publication of the International Commission on Microbial Specification for Foods. Academic Press, New York, p. 191.

Jay, J.M. 1996. *Modern Food Microbiology*. Chapman & Hall, New York.

Juneja, V.K., Snyder, O.P., Jr., and Mariner, B.S. 1997. Thermal destruction of *Escherichia coli* 0157:H7 in beef and chicken: determination of D- and Z-values. *J. Food. Prot.*, 35, 231–237.

Kauffman, S.A. 1992. *The Origin of Order — Self-Organization and Selection in Evolution*. Oxford University Press, Oxford, U.K.

Kauffman, S.A. 1995. *At Home in the Universe – the Search for the Laws of Self Organization and Complexity*. Oxford University Press, Oxford, U.K.

Knorr, D., 1995. Hydrostatic pressure treatment of foods — microbiology, in Gould, G.W. (Ed.). *New Methods of Food Preservation*. Blackie Scientific, London, pp. 159–175.

Knorr, D. 2000. Process aspects of high-pressure treatment of food systems, in Barbosa–Canovas, G.V. and Gould, G.W. (Eds.). *Innovations in Food Processing*. Technomic, Lancaster, PA, pp. 13–30.

Körmedy, I. and Körmedy, L. 1997. Considerations for calculating heat inactivation processes when semilogarithmic thermal inactivation models are nonlinear. *J. Food Eng.*, 34, 33–40.

Koutchma, T. and Murakami, E. 2004/2005. Effect of carbon dioxide and pressure processing on microbial and enzyme inactivation in food and beverages. *Int. Rev Food Sci. Technol.*, IUFoST (winter), 78–86.

Koutsoumanis, K. 2001. Predictive modeling of the shelf life of fish under nonisothermal conditions. *Appl. Environ. Microbiol.*, 67, 1821–1829.

Lambert, R.J.W., Johnston, M.D., and Simons, E.A., 1999. A kinetic study of the effect of hydrogen peroxide and peracetic acid against *Staphylococcus aureus* and *Pseudomonas aeruginosa* using the bioscreen disinfection method. *J. Appl. Microbiol.*, 87, 782–786.

LeClair, K., Heggart, H., Oggerl, M., Barlett, F.M., and McKellar, R.C. 1994. Modeling the inactivation of *Listeria monocytogenes* and *Salmonella typhimorium* in simulated egg wash water. *Food Microbiol.*, 11, 345–349.

LeDantec, C., Dauet, J.P., Montiel, A., Damoutier, N., Dubrou, S., and Vincent, V. 2002. Chlorine disinfection of atypical mycobacteria isolated from a water distribution system. *Appl. Environ. Microbiol.*, 68, 1025–1032.

LeJean, G., Abraham, G., Debray, E., Candau, Y., and Piar, G. 1994. Kinetics of thermal destruction of *B. stearothermophilus* spores using a two-reaction model. *Food Microbiol.*, 11, 229–237.

Lewis, M. and Heppel, N. 2000, *Continuous Thermal Processing of Foods: Pasteurization and UHT Sterilization.* Aspen Publishers, Gaithersburg, MD.

Linton, R.H., Carter, W.H., Pierson, M.D., Hackney, C.R., and Eifert, J.D. 1996. Use of a modified Gompertz equation to model nonlinear survival curves of *Listeria monocytogenes* Scott. A. *J. Food Prot.*, 59, 16–23.

López, S., Prieto, M., Dijkstra, J., and Dhanoa, M.S. 2004 Statistical evaluation of mathematical models for microbial growth. *Int. J. Food Microbiol.*, 96, 289–300.

Mafart, P., Couvert, O., Gaillard, S., and Leguerinel, I. 2002. On calculating sterility in thermal preservation methods: application of the Weibull frequency distribution model. *Int. J. Food Microbiol.*, 72, 107–113.

Mattick, K.L., Legan, J.D., Humphrey, T.J., and Peleg, M. 2001. Calculating *Salmonella* inactivation in nonisothermal heat treatments from non-linear isothermal survival curves. *J. Food Prot.*, 64, 606–613.

McKellar, R. and Lu, X. (Eds.). 2003. Modeling *Microbial Responses on Foods.* CRC Press, Boca Raton, FL.

McMeekin, T.A., Olley, J.N., Ross, T., and Ratkowsky, D.A. 1993. *Predictive Microbiology Theory and Applications.* John Wiley & Sons, New York.

Mercer, A. 1990. *Disease, Mortality and Population in Transition.* Leichester University Press. Leichester, U.K.

Meyer, R.S., Cooper, K.L., Knorr, D., and Lelieveld, H.L.M. 2000. High-pressure sterilization of foods. *Food Technol.*, 11, 67–72.

Mizrahi, S. and Karel, M. 1978. Evaluation of a kinetic model for reactions in moisture-sensitive products using dynamic storage conditions. *J. Food Sci.*, 43, 750–753.

Murphy, R.Y., Marks, B.P., Johnson, E.R., and Johnson, M.G. 1999. Inactivation of *Salmonella* and *Listeria* in ground chicken breast meat during thermal processing. *J. Food Prot.*, 62, 980–985.

Murray, J.D. 1989. *Mathematical Biology,* Springer–Verlag, New York.

Nelder, J.A. and Meade, R. 1965. A simplex method for function minimization. *Comp. J.*, 7, 308–312.

Neyens, P., Messens, W., and De Vuyst, L. 2003. Effect of sodium chloride on growth and bacteriocin production by *Lactobacillus amylovorus* DCE 471. *Int. J. Food Microbiol.*, 88, 29–39.

Nisbet, R.M. and Gurney, W.S.C. 1982. *Modeling Fluctuating Populations*, John Wiley & Sons, New York.

Normand, M.D. and Peleg, M. 1998. Kauffman's abstract model of phase transitions. *J. Texture Stud.*, 29, 375–386.

Nussinovitch, A. and Peleg, M. 2000. Analysis of the fluctuating patterns of microbial counts in frozen industrial food products. *Food Res. Int.*, 33, 53–62.

Nussinovitch, A., Curasso, Y., and Peleg, M. 2000. Analysis of the fluctuating microbial counts in commercial raw milk — a case study. *J. Food Prot.*, 63, 1240–1247.

Ocio, M.J., Fernández, P.S., Alvarruiz, A., and Martínez, A., 1994. Comparison of TDT and Arrhenius models for rate constant inactivation predictions of *Bacillus stearothermophilus* heated in mushroom alginate substrate. *Lett. Appl. Microbiol.*, 19, 114–117.

Patel, J.K., Kapadia, C.H., and Owen, D.B., 1976. *Handbook of Statistical Distributions*. Marcel Dekker, New York.

Peleg, M. 1995. A model of microbial survival after exposure to pulsed electric fields. *J. Sci. Food Agric.*, 67, 93–99.

Peleg, M. 1996a. A model of microbial growth and decay in a closed habitat based on combined Fermi and the logistic equations. *J. Sci. Food Agric.*, 71, 225–230.

Peleg, M. 1996b. Evaluation of the Fermi equation as a model of dose–response curves. *Appl. Microbiol. Biotechnol.*, 46, 303–306.

Peleg, M. 1997. Modeling microbial populations with the original and modified versions of the continuous and discrete logistic equations. *Crit. Rev. Food Sci. Nutr.*, 37, 471–490.

Peleg, M. 2000a. Microbial survival curves — the reality of flat shoulders and absolute thermal death times. *Food Res. Int.* 33, 531–538.

Peleg, M. 2000b. Modeling and simulating microbial survival in foods subjected to a combination of preservation methods, in *Innovations in Food Processing*. Barbosa–Canovas, G.V. and Gould, G.W. (Eds.). Technomic, Lancaster, PA, pp. 163–181.

Peleg, M. 2002a. A model of survival curves having an "activation shoulder." *J. Food Sci.*, 67, 2438–2443.

Peleg, M. 2002b. Interpretation of the irregularly fluctuating microbial counts in commercial dairy products. *Int. Dairy J.*, 12, 255–262.

Peleg, M. 2002c Modeling and simulation of microbial survival during treatments with a dissipating lethal chemical agent. *Food Res. Int.*, 35, 327–336.

Peleg, M. 2002d. Simulation of *E. coli* inactivation by carbon dioxide under pressure. *J. Food Sci.*, 67, 896–901.

Peleg, M. 2003a. Calculation of the nonisothermal inactivation patterns of microbes having sigmoidal isothermal semilogarithmic survival curves. *Crit. Rev. Food Sci. Nutr.*, 43, 645–658.

Peleg, M. 2003b. Microbial survival curves: interpretation, mathematical modeling and utilization. *Comments Theor. Biol.*, 8, 357–387.

Peleg, M. 2004. Analyzing the effectiveness of thermal preservation processes, in, Richardson, P. (Ed.). *Improving Thermal Processing*. Woodhead Publishing, Cambridge, U.K., pp. 411–426.

Peleg, M. and Cole, M.B. 1998. Reinterpretation of microbial survival curves. *Crit. Rev. Food Sci. Nutr.*, 38, 353–380.

Peleg, M. and Horowitz, J. 2000. On estimating the probability of aperiodic outbursts of microbial populations from their fluctuating counts. *Bull. Math. Biol.*, 62, 17–35.

Peleg, M. and Penchina, C.M. 2000. Modeling microbial survival during exposure to a lethal agent with varying intensity. *Crit. Rev. Food Sci. Nutr.*, 40, 159–172.

Peleg, M. and Normand, M.D. 2004. Calculating microbial survival parameters and predicting survival curves from non-isothermal inactivation data. *Crit. Rev. Food Sci. Nutr.*, 44, 409–418.

Peleg, M., Corradini, M.G., and Normand, M.D. 2004 Kinetic models of complex biochemical reactions and biological processes. *Chemie Ingen. Technik*, 76, 413–423.

Peleg, M., Engel, R., Gonzalez–Martinez, C., and Corradini, M.G. 2002. Non Arrhenius and non WLF kinetics in food systems. *J. Sci. Food Agric.* 82, 1346–1355.

Peleg, M., Normand, M.D., and Campanella, O.H. 2003. Estimating microbial inactivation parameters from survival curves obtained under varying conditions — the linear case. *Bull. Math. Biol.*, 65, 219–234.

Peleg, M., Normand, M.D., and Corradini, M.G. 2005. Generating microbial survival curves during thermal processing in real time. *J. Appl. Microbiol.*, 98, 406–417.

Peleg, M., Normand, M.D., and Damrau, E. 1997. Mathematical interpretation of dose–response curves. *Bull. Math. Biol.*, 59, 747–761.

Peleg, M., Normand, M.D., and Tesch, R. 1996. Simulation of fluctuating populations of micro- and macroorganisms with models having a normal random variate term. *J. Sci. Food Agric.*, 73, 17–20.

Peleg, M., Nussinovitch, A., and Horowitz, J. 2000. Interpretation and extraction of useful information from irregular fluctuating industrial microbial counts. *J. Food Sci.*, 65, 740–747.

Peleg, M., Penchina, C.M., and Cole, M.B. 2001. Estimation of the survival curve of *Listeria monocytogenes* during nonisothermal heat treatments. *Food Res. Int.* 34, 383–388.

Periago, P.M., van Zuijlen, A., Fernandez, P.S., Klapwijk, P.M., ter Steeg, P.F., Corradini, M.G., and Peleg, M. 2004. Estimation of the nonisothermal inactivation patterns of *Bacillus sporothermodurans* IC4 spores in soups from their isothermal survival data. *Int. J. Food Microbiol.*, 95, 205–218.

Prescot, L.M., Harley, J.P., and Klein, D.A. 1996. *Microbiology* (3rd ed.) WCB, Duboque, IA.

Raup, D.M. 1991. *Extinction: Bad Genes or Bad Luck?* W.W. Norton, New York.

Rice, J.A. 1995. *Mathematical Statistics and Data Analysis*. Brookslide Publishing, Duxbury, MA.

Robinson, R.K., Batt, C.A., and Patel, P.D. (Eds.). 2000. *Encyclopedia of Food Microbiology*. Academic Press, New York. (See entries for "Preservatives," pp. 1710–1783.)

Rodriguez, A.C., Smerage, G.H., Texeira, A.A., and Busta, F.F. 1992. Population model of bacterial spores for validation of dynamic thermal processes. *J. Food Proc. Eng.*, 15, 1–30.

Ross, T. 1996. Indices for performance evaluation of protective models in food microbiology. *J. Appl. Microbiol.*, 81, 501–508.

Ross, T. and Dalgaard, P. 2004. Secondary models, in McKellar, R.C. and Lu, X. (Eds.). *Modeling Microbial Responses in Food*. CRC Press, Boca Raton, FL, pp. 63–150.

Royama, T. 1992. *Analytical Population Dynamics*. Chapman & Hall, London.

Ruelle, D. 1989. *Chaotic Evolution and Strange Attractors*. Cambridge University Press, Cambridge, U.K.

Ruelle, D. 1991. *Chance and Chaos*. Princeton University Press, Princeton, NJ.

Ruelle, D. 1992. Deterministic chaos: the science and the fiction. *Proc. R. Soc. Lond.* A, 427, 241–247.

Sapru, V., Texeira, A.A., Smerage, G.H., and Lindsay, J.A. 1992. Predicting thermophilic spore population dynamics for UHT sterilization processes. *J. Food Sci.*, 57, 1248.

Sapru, V., Texeira, A.A., Smerage, G.H., and Lindsay, J.A. 1993. Comparison of predictive models for bacterial spore population resources to sterilization temperature. *J. Food Sci.*, 58, 223–228.

Schaffer, W.M. and Truty, G.L. 1989. Chaos vs. noise-driven dynamics, in *Models in Population Biology*, Vol. 20, American Mathematical Society, Providence, RI, pp. 77–96.

Schubert, H., Wachler, E., and Krug, H. 1984. Erich Rammler — a pioneer of particulate technology. *Particulate Sci. Technol.*, 2, 2–17.

Scott, S. and Duncan, C.J. 1998. *Human Demography and Disease*. Cambridge University Press. Cambridge, U.K.

Seaborg, G.T. and Loveland, W.D. 1990. *The Elements beyond Uranium*. John Wiley & Sons, New York.

Shull, J.J., Cargo, G.T., and Ernst, R.R. 1963. Kinetics of heat activation and thermal death of bacterial spores. *Appl. Microbiol.*, 11, 485–487.

Smith, D. and Stoddard, W. 2000. Ozone and food processing, in Francis, F.J. (Ed.). *Encyclopedia of Food Science and Technology*. (2nd ed.). Vol. 3. John Wiley & Sons, New York, pp. 1801–1805.

Smith, S. and Schaffner, D.W. 2004. Evaluation of the predictive model for *Clostridium perfringens* growth during cooling. *J. Food Prot.*, 67, 1133–1137.

Stephens, P.J., Cole, M.B., and Jones, M.V., 1994. Effect of heating on the thermal inactivation of *Listeria*. *J. Appl. Microbiol.*, 77, 702–710.

Stumbo, C.R. 1973. *Thermobacteriology in Food Processing* (2nd ed.), Academic Press. New York.

Sugihara, G., Grenfell, B.T., and May, R.M. 1990. Distinguishing error from chaos. *Phil. Trans. R. Soc.*, 330B, 235–251.

Tailor, P.D. (Ed.). 2004. *Extinctions in the History of Life*. Cambridge University Press, Cambridge, U.K.

Taub, L.A., Feeherry, F.E., Ross, E.W., Kustin, K., and Doona, C.J. 2003. A quasi-chemical kinetic model for the growth and death of *Staphylococcus aureus* in intermediate moisture bread. *J. Food Sci.*, 68, 2530–2537.

Texeira, A. 1992. Thermal processing calculations, in *Handbook of Food Engineering*, Heldman, D.R. and Lund, L.B. (Eds.), Marcel Dekker, New York, pp. 563–619.

Texeira A.A. and Busta, F.F. 1992. Population model of bacterial spores for validation of dynamic thermal processes. *J. Food Proc. Eng.*, 15, 1–30.

Thompson, W.S., Busta, F.E., Thompson, D.R., and Allen, C.E. 1979. Inactivation of *Salmonella* in autoclaved ground beef exposed to constant rising temperature. *J. Food Prot.*, 42, 410–415.

Toledo, R.T. 1999. *Fundamentals of Food Process Engineering*. (2nd ed.) Aspen, Gaithersburg, MD, pp. 315–397.

Valdramidis, V.P., Geeraerd, A.H., Bernaerts, K., and Van Impe, J.K. 2004. Dynamic vs. static thermal inactivation: the necessity of validation some modeling and microbial hypotheses. Paper no. 434. *Proceedings of the 9th International Conference of Engineering and Food ICEF 9*, Montpellier, France.

van Boekel, M.A.J.S. 2002. On the use of the Weibull model to describe thermal inactivation of microbial vegetative cells. *Int. J. Food Microbiol.*, 72, 159–172.

van Boekel, M.A.J.S. 2003. Alternate kinetics models for microbial survivor curves, and statistical interpretations (summary of a presentation at IFT Research Summit on kinetic models for microbial survival). *Food Technol.*, 57, 41–42.

van Grewen, S.J.C. and Zwietering, H.N. 1998. Growth and inactivation models to be used in quantitative risk assessment. *J. Food Prot.*, 61, 1541–1549.

Vessoni Penna, T.C. and Moraes, D.A. 2002. The influence of nisin on the thermal resistance of *Bacillus cerius*. *J. Food Prot.*, 65, 415–418.

Wagner, M., Brumelis, D., and Gerh, R. 2002. Disinfection of wastewater by hydrogen peroxide or peracetic acid: development of procedures for measurement of residual disinfectant and application to a physicochemically treated municipal effluent. *Water Environ. Res.*, 74, 33–50.

Whiting, R.C. 1995. Microbial modeling in foods. *Crit. Rev. Food Sci. Nutr.*, 35, 467–494.

Yang, X. and Tsao, G.T. 1994. Mathematical modeling of inhibition kinetics in acetone–butanol fermentation by *Clostridium acetobutylicum*. *Biotechnol. Prog.*, 10, 532–538.

Freeware

This section lists freeware for quantitative microbiology written in Microsoft Excel® and posted on the Web on behalf of the Food Science Department at the University of Massachusetts at Amherst.

1. Estimating the probability of high microbial counts from fluctuating counts' records (Program written by M.D. Normand and M. Peleg in 2000)

   http://www-unix.oit.umas.edu/~aew2000/microbecounts.html

   The user can paste a list of counts. The program tests the normality and log normality of their distribution and estimates the probabilities of encountering counts exceeding five different levels specified by the user.

2. Generation of nonisothermal microbial survival curves of an organism whose isothermal inactivation curves follow the Weibullian–log logistic model (Program written by M.D. Normand and M. Peleg in 2004)

   The organism's or spore's survival parameters can be entered by the user. The program comes in two versions: (1) the chosen temperature profile is generated by a formula, and (2) the user pastes his or her experimental time–temperature data.

   Real-time generation of microbial survival or microbial inactivation curves during heat pasteurization:

   http://www-unix.oit.umass.edu/~aew2000/SalmSurvival.html

   Real-time generation of bacterial spores' survival or bacterial spores' inactivation curves during heat sterilization:

   http://www-unix.oit.umass.edu/~aew2000/CBotSurvival.html

3. Generation of microbial survival curves of an organism whose inactivation follows the Weibullian–log logistic model when exposed to a dissipating chemical disinfectant (Program written by M.D. Normand and M. Peleg in 2004)

The organism's survival parameters can be entered by the user. The program comes in two versions: (1) the chosen concentration profile is generated by a formula, and (2) the user pastes his or her experimental time–concentration data.

Generation of microbial survival or microbial inactivation curves during disinfection with a volatile (dissipating) chemical agent:

http://www-unix.oit.umass.edu/~aew2000/ConcSurvival.html

4. Generation of nonisothermal microbial growth curves of an organism whose isothermal growth curves follow a sigmoid pattern (Program written by M.G. Corradini, M.D. Normand, and M. Peleg in 2005)

The organism's growth parameters can be entered by the user. The program comes in two versions: (1) a chosen oscillating temperature profile is generated by a formula, and (2) the user pastes his or her experimental time–temperature data.

Generation of microbial growth curves (long and short lag time — model A):

http://www-unix.oit.umass.edu/~aew2000/
MicrobeGrowthModelA.html

Generation of microbial growth curves (short lag time — model B):

http://www-unix.oit.umass.edu/~aew2000/
  MicrobeGrowthModelB.html

Index